INTERNATIONAL SERIES OF MONOGRAPHS IN

NATURAL PHILOSOPHY

GENERAL EDITOR: D. TER HAAR

VOLUME 26

AN INTRODUCTION TO THE PHENOMENOLOGICAL THEORY OF FERROELECTRICITY

AN INTRODUCTION TO THE PHENOMENOLOGICAL THEORY OF FERROELECTRICITY

BY

J. GRINDLAY

Physics Department, University of Waterloo, Ontario

THE QUEEN'S AWARD TO INDUSTRY 1966

PERGAMON PRESS

OXFORD · LONDON · EDINBURGH · NEW YORK
TORONTO · SYDNEY · PARIS · BRAUNSCHWEIG

Pergamon Press Ltd., Headington Hill Hall, Oxford
4 & 5 Fitzroy Square, London W.1
Pergamon Press (Scotland) Ltd., 2 & 3 Teviot Place, Edinburgh 1
Pergamon Press Inc., Maxwell House, Fairview Park, Elmsford, New York 10523
Pergamon of Canada Ltd., 207 Queen's Quay West, Toronto 1
Pergamon Press (Aust.) Pty. Ltd., 19a Boundary Street, Rushcutters Bay, N.S.W. 2011, Australia
Pergamon Press S.A.R.L., 24 rue des Écoles, Paris 5e
Vieweg & Sohn GmbH, Burgplatz 1, Braunschweig

First Edition 1970

Library of Congress Catalog Card No. 72–90455

Printed in Hungary

08 006362 4

To GLORIA

GRINDLAY: An Introduction to the Phenomenological Theory of Ferroelectricity

ERRATA

Page 91. Remove commas occurring in middle of equations 3.3.3, 3.3.4 and 3.3.5.

Page 155. First footnote, line 2 -- remove symbol "E_i".

Page 196. Line 3, "$\kappa = \partial E/\partial D$", not "$\kappa = \partial E/\partial D_{E=0}$".

Contents

Contents

Contents

Preface

IN THIS monograph we have attempted to describe the basis and derivation of the macroscopic or phenomenological theory of the elastic, dielectric and thermal properties of crystals as applied in the field of ferroelectricity. Most of the results and ideas we describe are scattered through the literature of this subject and it was thought worth while to present them together with their physical background in one reference. The discussion is restricted to the theory required to describe the properties of homogeneous specimens subject to low frequency fields—inhomogeneous effects such as domains and space charges receive only passing mention.

The plan of the monograph is as follows: Chapter 1 contains a brief introduction to the field of ferroelectricity. Many of the terms and results used in this chapter are introduced in later chapters. In Chapter 2 the strain, stress, electric and displacement fields are defined and the linear field equations and equations of state are derived. The thermodynamics of the elastic dielectric is described in Chapter 3 with particular reference to thermodynamic identities and the idea of thermodynamic stability. Chapter 4 contains a discussion of a rigorous non-linear theory of the static elastic–dielectric. The results in this chapter give some criteria for the range of validity of the quasi-linear theory which is based on the linear theory of Chapter 2. In the final chapter, Chapter 5, we introduce phenomenological models which display ferroelectric behaviour. These models provide examples of some problems and results peculiar to ferroelectrics. At a number of points in the text various types of identities are derived, particularly in Chapters 3 and 5. We have made no attempt to provide a complete set of these identities.

It is hoped that sufficient detail has been given that the reader can immediately derive for himself whatever other identity he wishes. At the end of each chapter there is a bibliography containing what we regard as the major references for the material in that chapter. These should provide at least an entrée to the area.

It is appropriate at this point to comment on the major omissions in this monograph. With a phenomenological theory of crystals we seek to develop a mathematical apparatus which describes the macroscopic response of crystals to external stimuli. Out of such a theory emerge, among other things, certain identities between the various properties of the crystal. Additional identities occur if the equations of state are given polynomial representations. Both types of identities are open to experimental verification and, indeed, play a major role in testing the validity of the phenomenological theory. In our opinion what is badly needed in the literature at this time (and what is not given here) is an application of the present body of experimental data on the ferroelectrics to a critical test of the appropriateness of the standard theory described in Chapters 2, 3 and 5. Such a test would underscore discrepancies and gaps in the data and at the same time would possibly indicate where additional theoretical work is required. At various places in the text we have introduced experimental data, but here the intention is only to give the reader some idea of the magnitudes involved. Domain theory, anti-ferroelectricity and high frequency dielectric properties are topics of considerable interest in the field of ferroelectricity, and these are virtually ignored in this text. However, it is hoped that what has been discussed will prove useful both to the beginner and the expert in this field.

Since our aim is an introduction to the phenomenological theory we have tried to keep the mathematical notation as simple and straightforward as possible. Nevertheless, some mathematical complexity is the price one pays for the description of the interaction of the elastic, dielectric and thermal properties of crystals. All definitions are italicized. Footnotes contain extraneous but interesting points, cross-references or minor proofs. The rationalized MKS system is used throughout.

Part of the text was written while the author was at the National Research Council in Ottawa. I would like to express my gratitude to

Dr. J. Morrison and his group for their kind hospitality during my visit. In particular Dr. V. Janovec provided a number of stimulating discussions. It is also a pleasure to acknowledge the constructive criticism and proof-reading assistance provided by Dr. V. Dvorak while he was a post-doctoral fellow at the University of Waterloo.

Physics Department J. G.

University of Waterloo

CHAPTER 1

Ferroelectricity

THE material in this chapter is intended to serve as a brief introduction to the elastic, dielectric and thermal properties of ferroelectric crystals. It is assumed that the reader is familiar with the contents of Chapters 2 and 3.

In § 1.1 ferroelectricity is defined and the dielectric hysteresis loop and its significance are described. Sections 1.2, 1.3 and 1.4 contain summaries of the elastic, dielectric and thermal properties of the ferroelectric crystals KH_2PO_4, Rochelle salt and $BaTiO_3$. In the final section, § 1.5, we discuss the work of Mueller on Rochelle salt. His treatment clarified the thermodynamic analysis of the ferroelectric phase transitions of Rochelle salt and opened the way for the more detailed analysis of Devonshire (see Chapter 5).

1.1 Ferroelectricity

1.1.1 FERROELECTRICITY

The characteristic feature of *ferroelectric crystals* is the appearance of a spontaneous electric dipole moment which can be reversed, with no net change in magnitude, by an applied electric field. Because of this reversibility the ferroelectrics form a distinct subclass of the pyroelectrics. Many ferroelectrics possess a spontaneous dipole moment only in certain temperature ranges. At the boundaries of these ranges the crystal undergoes a phase transition to a non-pyroelectric phase.

The phases with and without a spontaneous dipole moment are referred to as ferroelectric and *paraelectric* respectively.† In addition to the dielectric properties, the ferroelectrics also possess anomalous‡ elastic and calorimetric properties, but it should be emphasized that in the present understanding of ferroelectric behaviour the dielectric properties are the distinguishing characteristics of the ferroelectrics and the other anomalous properties stem from or can be accounted for in terms of the dielectric anomalies (see § 1.5). At the present time the list of known ferroelectrics contains over 100 materials.

1.1.2 Dielectric Hysteresis

The measurement of the properties of the ferroelectric phase presents problems because one cannot always ensure that the specimen of interest remains homogeneous—a necessary condition to obtain accurate single crystal data. We shall illustrate this point with reference to the spontaneous dipole moment.

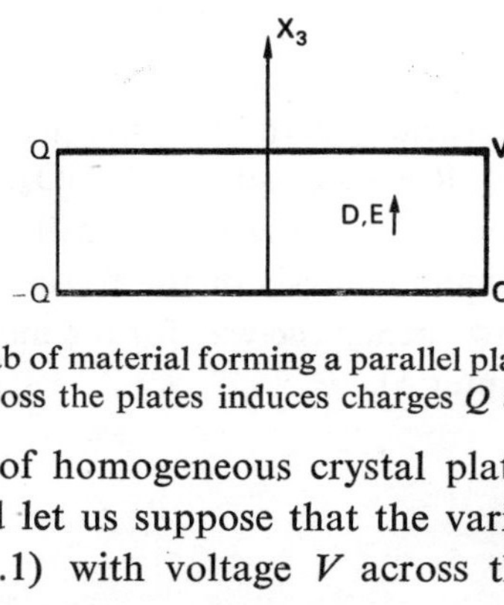

FIG. 1.1.1. Plated slab of material forming a parallel plate condenser. A potential difference V across the plates induces charges Q and $-Q$ on the plates.

Consider a slab of homogeneous crystal plated to form a parallel plate condenser and let us suppose that the variation of charge Q on the plates (Fig. 1.1.1) with voltage V across the plates assumes the

† At a phenomenological level the ferroelectric crystals are the electrical analogue of the ferromagnetic crystals and, as a result, a good deal of the nomenclature peculiar to ferromagnetism has been adopted in the field of ferroelectricity. Thus, for example, paraelectric is the analogue of paramagnetic and the small field or linear dielectric permittivity in the paraelectric phase obeys a Curie–Weiss law.

‡ It has become customary to use the word "anomalous" to describe any quantity with a temperature dependence which is much greater than the corresponding temperature dependence in normal or non-ferroelectric crystals, such as, for example, sodium chloride. The word "true" sometimes serves as an antonym to anomalous; because of the overtones associated with true we do not subscribe to this usage.

form sketched in Fig. 1.1.2. This is a simple example of a *hysteresis loop*. We shall assume that the lower branch is obtained by an inversion of the upper branch through the origin O. Thus, in particular the two branches cut the Q-axis at points equidistant from O, namely $(O, +Q^0)$

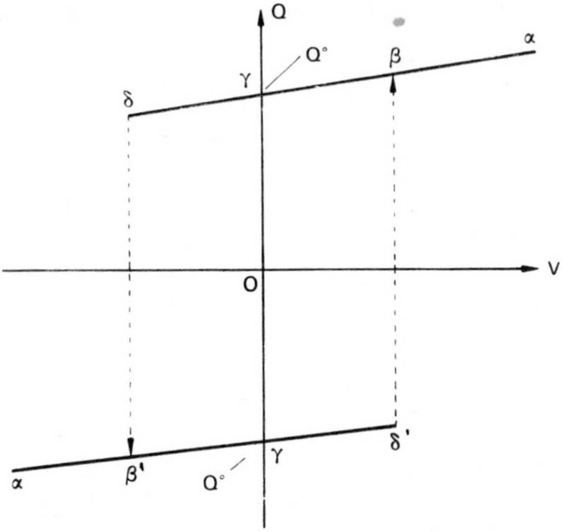

FIG. 1.1.2. Ideal hysteresis loop. As the voltage V is reduced below the point δ the system switches from the state δ to β' as indicated by the vertical dotted line. The switchover to the upper branch occurs at δ'.

and $(O, -Q^0)$. Q is a double valued function of V. For V large and positive, the values of Q fall on the upper branch $\alpha\delta$. A switchover to the lower branch only occurs if V is reduced below its value at δ. Similarly, a switchover to the upper branch occurs if V is increased above its value at δ'. It is evident that when the applied potential lies between these switchover values the charge on a given plate is determined by the past history of the system in the sense that the sign of the last applied voltage which lies outside the switchover values determines the branch describing the response Q. This effect is referred to as *dielectric hysteresis.*

If the electric field and the electric displacement field are uniform (fringing effects at the condenser edges are assumed negligible) and normal to the plates, we may immediately relate their components E

and D respectively along the axis X_3 (see Fig. 1.1.1) to V and Q; thus

$$E = V/z, \quad D = Q/a, \tag{1.1.1}$$

where z is the plate separation and a the plate area. In these circumstances the dielectric equation of state E versus D can be obtained from Fig. 1.1.2 by a simple scaling $V/z \to E, Q/a \to D$. In the shorted crystal, $E = 0$ and $D = P^0 = \pm Q^0/a$, where P^0 is the polarization and the signs refer to the upper and lower branches respectively. Thus with $E = 0$ there is a non-zero value for the polarization—the *spontaneous polarization*—and hence the slab possesses a spontaneous dipole moment. The direction of this dipole moment is parallel (anti-parallel) to X_3 for the point $E = 0$ on the upper (lower) branch. Clearly the spontaneous dipole moment can be reversed in direction with no net change in magnitude by applying a potential difference which causes a switchover from one branch to the other. Thus by definition the slab under discussion is made up of ferroelectric material. A direction in the crystal along which the spontaneous polarization vector may point is called a *ferroelectric axis.* The value of the electric field required to cause a switch from one branch to the other is referred to as the *coercive field.* Thus in Fig. 1.1.2 the magnitude of the coercive field is given by the magnitude of the potential at δ divided by the plate separation.

The dielectric behaviour described in the previous paragraph is the ideal response of a ferroelectric crystal. Observed hysteresis loops are more commonly continuous. A simple example is sketched in Fig. 1.1.3. The deviations from the ideal behaviour are attributed to the occurrence of a domain structure in the slab. The *domains* are small crystalline regions, presumably with equations of state E versus D obtained by scaling Fig. 1.1.2. The dipole moments of adjacent domains are not parallel (they are anti-parallel if the material has one ferroelectric axis and more complicated configurations occur if there is more than one ferroelectric axis) and hence the electric field and the displacement field in the slab are not uniform. The net result is a rounding off of the ideal hysteresis loop. In particular it is clear that because of the pattern of parallel and anti-parallel domains the net charge on the plates of the shorted specimen must be less than the charge in the equivalent

homogeneous system with uniform fields. Thus the intercept of the rounded loop (Fig. 1.1.3) with the Q-axis provides a poor measure of the spontaneous polarization. With large applied potential differences the domains tend to coalesce into a single domain and hence the tail of the loops provide a better representation of the single crystal behaviour than the remaining portions. Because of this, it is customary

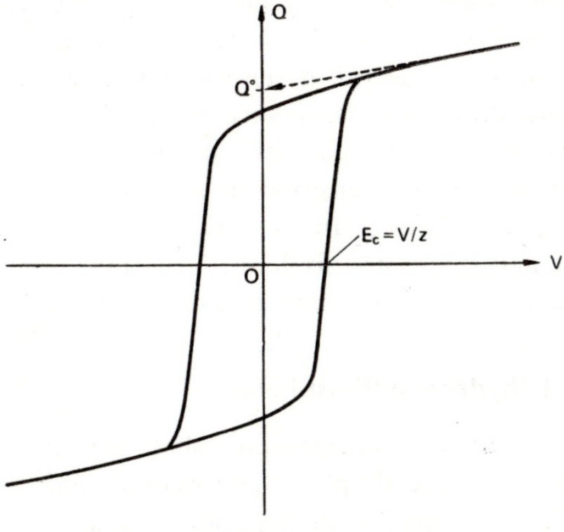

FIG. 1.1.3. Hysteresis loop.

to extrapolate the tail on to the Q-axis and use this intercept Q^0 as a measure of the spontaneous polarization, namely $P^0 = Q^0/a$ (see Fig. 1.1.3).† It is difficult to estimate how accurate this method is. The concept of a coercive field becomes vague when the hysteresis loop has the appearance sketched in Fig. 1.1.3. In these circumstances the "coercive field" E_c is usually taken as that determined by the value of V where the loop cuts the V-axis. Clearly this value is some average electric field in the crystal and is connected in some complicated fashion with the single crystal properties. The incidence and growth of domains

† This method fails when the E versus D curve of the single crystal has a large curvature close to the D-axis, as can occur in the region close to the transition to the paraelectric phase. In these circumstances the electrocaloric method (§ 5.2.5) offers a more suitable method of determining the spontaneous polarization.

is a complex topic which lies outside the scope of the present text. As we have mentioned above, it is a difficult experimental problem to obtain and keep specimens in single crystal form in the ferroelectric phase. The presence of domains clearly affects the measured values of all properties of the specimen to some extent and as in the case of the spontaneous polarization it is difficult to give a quantitative estimate of their effect on any given property. Many of the numerical discrepancies which appear when thermodynamic identities are checked with ferroelectric phase data are attributed in an *ad hoc* fashion to the effect of domains. The dielectric hysteresis loop is a function of temperature, and generally, though not invariably (as, for example, Rochelle salt), the area of the loop shrinks with increasing temperature; unless melting intervenes, a phase transition or *ferroelectric transition*† to a paraelectric state takes place. In this state there is no spontaneous dipole moment and no hysteresis loop. The transition is called the *Curie point* and the transition temperature θ_c the *Curie temperature*.

1.2 Potassium Dihydrogen Phosphate

The crystal potassium dihydrogen phosphate KH_2PO_4, abbreviated to KDP, has probably the simplest macroscopic behaviour of all the ferroelectrics. The only phase transition occurs at a 123°K. Above this Curie point the crystal is in a paraelectric phase with tetragonal symmetry‡ $\bar{4}2m$ and below in a ferroelectric phase with orthorhombic symmetry $mm2$. There is a single ferroelectric axis which is parallel to the X_3 or c-axis.§ The magnitude of the spontaneous polarization P^0 increases rapidly from zero at the Curie point and *saturates* or reaches a temperature independent value some 20 degrees below this point (Fig. 1.2.1). The slope of the spontaneous polarization curve appears to be vertical (i.e. infinite) at the Curie point.

In the paraelectric phase of KDP the zero stress and zero electric field values of any compliance matrix display point group symmetry

† Examples of transitions between two ferroelectric phases occur but we shall reserve this phrase for a ferroelectric–paraelectric transition.

‡ See page 117.

§ Unless otherwise noted we shall always use the standard crystallographic axes appropriate to the crystal of interest (Nye, 1964).

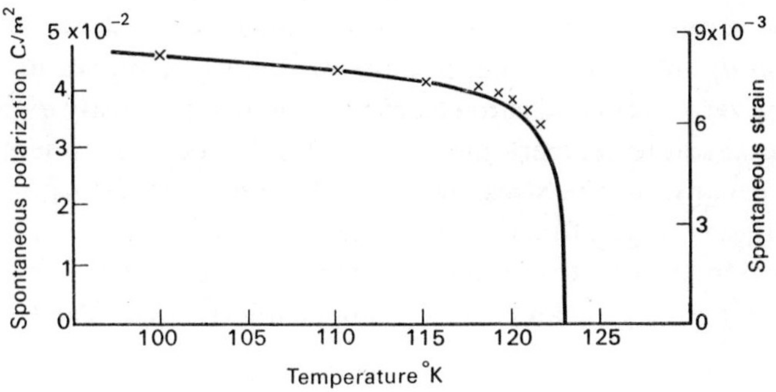

FIG. 1.2.1. Temperature dependence of the spontaneous polarization P_3^0 (full line) (von Arx and Bantle, 1944) and spontaneous strain S_6^0 (crosses) (de Quervain, 1944).

$\bar{4}2m$. From Table 1.2.1a we see that these matrices contain twenty-eight non-vanishing elements described by thirteen independent functions of temperature. In the case of the Gibbs function G, these functions are the six linear elastic compliances $s_{11}^{E,\theta}$, $s_{12}^{E,\theta}$, $s_{13}^{E,\theta}$, $s_{33}^{E,\theta}$, $s_{44}^{E,\theta}$, and $s_{66}^{E,\theta}$, the two linear piezoelectric strain coefficients d_{14}^{θ}, d_{36}^{θ}, the two linear

TABLE 1.2.1.

Arrays representing the symmetry of compliance matrices subject to the point group (a) $\bar{4}2m$ *and* (b) $mm2$ (§ 3.3.7).

Both arrays refer to the standard tetragonal crystallographic axes. The symbol ◆ represents half the difference of the two adjacent diagonal elements; thus, for example, in the case of the compliance matrix generated by the Gibbs function G, ◆ $= \frac{1}{2}(s_{44}^{E,\theta} - s_{55}^{E,\theta})$.

(a)

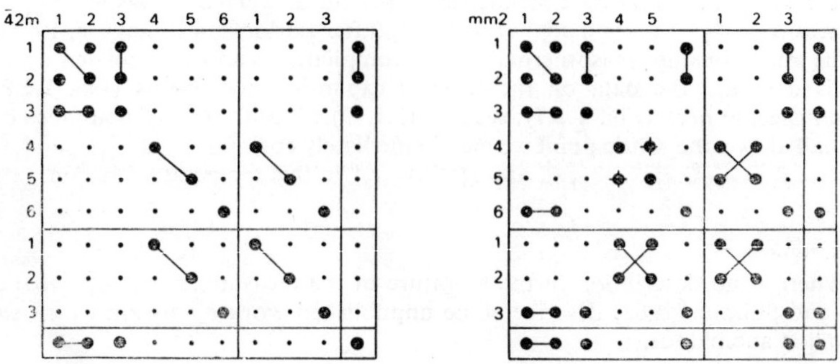

(b)

dielectric permittivities $\varepsilon_{11}^{T,\theta}$, $\varepsilon_{33}^{T,\theta}$, the two linear coefficients of thermal expansion $d\gamma_1/d\theta$, $d\gamma_3/d\theta$ and, finally, $d\sigma^0/d\theta$, the temperature derivative of the zero stress and zero electric field entropy density σ^0.† θ represents the absolute temperature and γ_α, T_α, E_i the components of the thermal expansion, the stress tensor and the electric field respectively. Each of these thirteen functions ($s_{11}^{E,\theta}$ etc.) displays anomalous behaviour n the region of the Curie point (Figs. 1.2.2 to 1.2.8).‡ The anomialies in $s_{66}^{E,\theta}$, $\varepsilon_{33}^{T,\theta}$, d_{36}^{θ} and $d\sigma^0/d\theta$ can be interrelated (see Chapter 5)

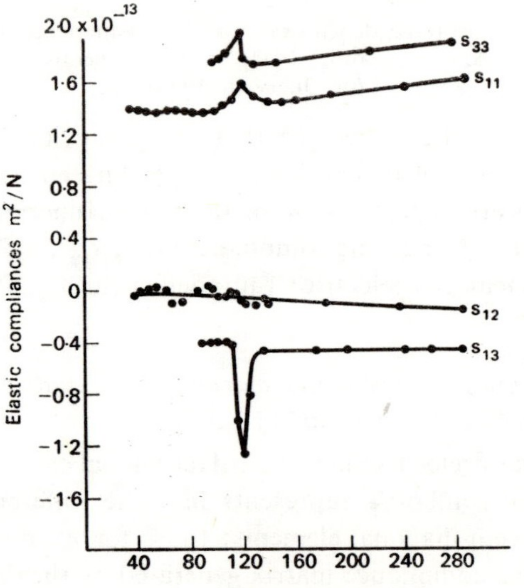

FIG. 1.2.2. Temperature dependence of the linear adiabatic elastic compliances $s_{11}^{E,\sigma}$ $s_{33}^{E,\sigma}$ $s_{12}^{E,\sigma}$ and $s_{13}^{E,\sigma}$ of KH_2PO_4 (after Barkla and Finlayson, 1953). The analogous linear isothermal elastic compliances can be calculated using this data and the data on the thermal expansion coefficients (Fig. 1.2.8) and specific heat (Fig. 1.2.7); see eqn. (3.1.19). Since the $(d\gamma_\alpha/d\theta)$ are zero or small above the Curie point we may immediately conclude that $s_{11}^{E,\sigma} \simeq s_{11}^{E,\theta}$, $s_{33}^{E,\sigma} \simeq s_{33}^{E,\theta}$, $s_{12}^{E,\sigma} \simeq s_{12}^{E,\theta}$, and $s_{13}^{E,\sigma} \simeq s_{13}^{E,\theta}$ in this temperature region.

† § 3.3.5.

‡ There is no description in the literature of the measurement of d_{14}^{θ}. However, Jona and Shirane (1962) describe some unpublished work by Essam which shows that d_{14}^{θ} is anomalous.

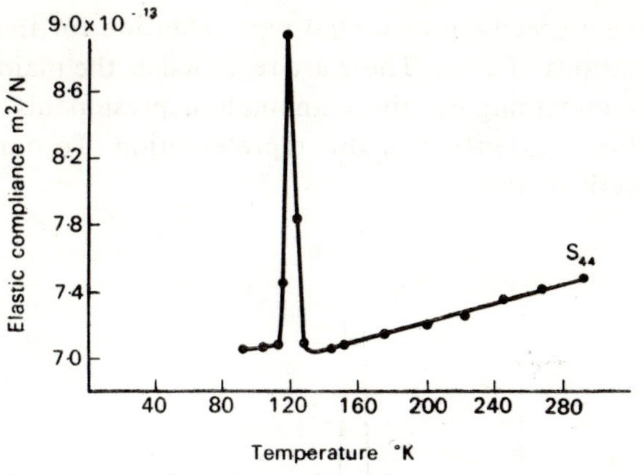

FIG. 1.2.3. Temperature dependence of the linear adiabatic elastic compliance $s_{44}^{E,\sigma}$ of KH_2PO_4 (after Barkla and Finlayson, 1953) (see remarks in caption to Fig. 1.2.2).

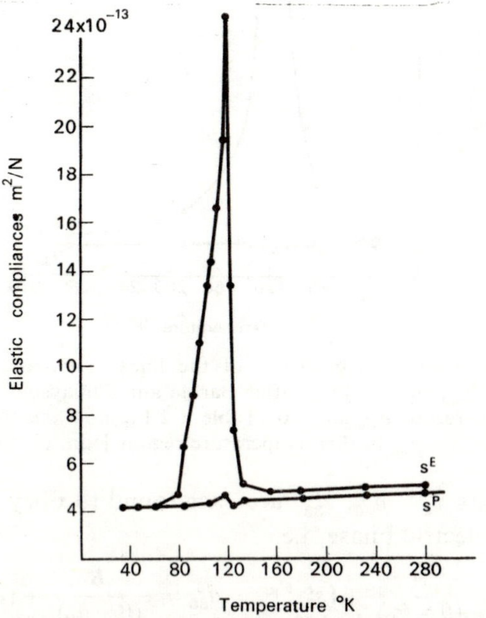

FIG. 1.2.4. Temperature dependence of the quantities $s^E = (s_{11}^{E,\sigma}+s_{12}^{E,\sigma})/2 + s_{66}^{E,\sigma}/4$ and $s^P = (s_{11}^{D,\sigma}+s_{12}^{D,\sigma})/2+s_{66}^{D,\sigma}/4$ for KH_2PO_4 (after Barkla and Finlayson, 1953). On comparing the temperature dependence of $s_{11}^{E,\sigma}$ and $s_{12}^{E,\sigma}$ (Fig. 1.2.2) with s^E we conclude that the major contribution to the peak in s^E comes from $s_{66}^{E,\sigma}$ (see remarks in caption to Fig. 1.2.2).

by introducing a specific polynomial representation for the elastic and dielectric equations of state. These are regarded as the major anomalies in KDP. The remaining or minor anomalies, presumably, can be accounted for by an extension of this representation. To our knowledge no one has verified this.

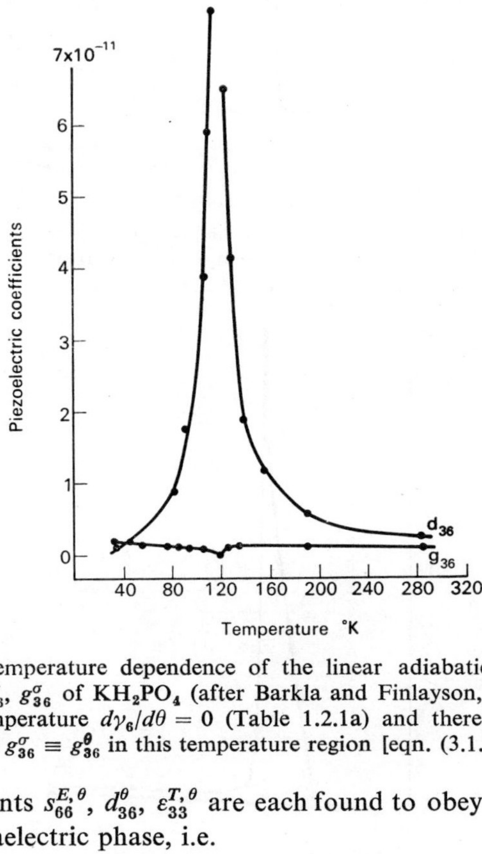

FIG. 1.2.5. Temperature dependence of the linear adiabatic piezoelectric coefficients d_{36}^{σ}, g_{36}^{σ} of KH_2PO_4 (after Barkla and Finlayson, 1953). Above the Curie temperature $d\gamma_6/d\theta = 0$ (Table 1.2.1a) and therefore $d_{36}^{\sigma} \equiv d_{36}^{\theta}$ and $g_{36}^{\sigma} \equiv g_{36}^{\theta}$ in this temperature region [eqn. (3.1.20)].

The coefficients $s_{66}^{E,\theta}$, d_{36}^{θ}, $\varepsilon_{33}^{T,\theta}$ are each found to obey a Curie–Weiss law in the paraelectric phase, i.e.

$$s_{66}^{E,\theta} = \frac{A}{(\theta-\theta_0)} + (s_{66}^{E,\theta})^0, \quad d_{36}^{\theta} = \frac{B}{(\theta-\theta_0)} + (d_{36}^{\theta})^0,$$

$$\varepsilon_{33}^{T,\theta} = \frac{C}{(\theta-\theta_0)} + (\varepsilon_{33}^{T,\theta})^0, \quad \theta > \theta_c,$$

where A, B, C, are constants—the Curie–Weiss constants with values

$A = 4{\cdot}6\times10^{-10}$ m² deg/N (Mason, 1946), $B = 4{\cdot}2\times10^{-9}$ deg C/N (Bantle and Caflisch, 1943) and $C = 2{\cdot}8\times10^{-8}$ deg F/m (Mason, 1946). θ_0 is the Curie–Weiss temperature which in the case of KDP coincides with the Curie temperature. The quantities $(s_{66}^{E,\theta})^0$, $(d_{36}^{\theta})^0$, $(\varepsilon_{33}^{T,\theta})^0$ are constants which can be neglected in the vicinity of the Curie point. These forms have been found to hold accurately over a wide range of

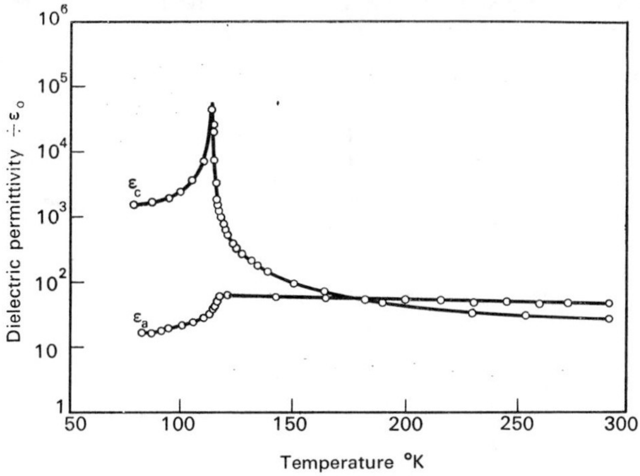

FIG. 1.2.6. Temperature dependence of the linear adiabatic permittivities $\varepsilon_a = \varepsilon_{11}^{T,\sigma}$ and $\varepsilon_c = \varepsilon_{33}^{T,\sigma}$ of KH_2PO_4 (after Busch, 1938). From Table 1.2.1 and eqn. (3.1.21) we deduce that $\varepsilon_{11}^{T,\sigma} \equiv \varepsilon_{11}^{T,\theta}$ above and below θ_c, whereas $\varepsilon_{33}^{T,\sigma} \equiv \varepsilon_{33}^{T,\theta}$ only above θ_c.

temperatures ($\sim$100 deg) starting just above the Curie temperature. The singular nature of these functions is reflected in the very large values observed close to the Curie point. Below θ_c these coefficients are also singular functions of θ, but it does not appear possible to fit their behaviour to a simple functional form valid over a wide range of temperatures. The specific heat $C^0_{E,T} = (\theta/\varrho_0)(d\sigma^0/d\theta)$ has a logarithmic singularity† of the form

$$C^0_{E,T} = -A \ln|\theta-\theta_c|, \quad \theta_c - 1°\text{K} < \theta < \theta_c + 0{\cdot}2°\text{K}$$

with $A = 170$ J/mole deg (Reese and May, 1967).

† There is also evidence for logarithmic singularities in the ferroelectrics KH_2AsO_4 (an isomorph of KDP) and TGS (Grindlay, 1965a, b). The factor common to these three ferroelectrics is the apparent order–disorder nature of their transitions.

Before there was evidence of a logarithmic singularity (Grindlay, 1965a), it appeared that the peak in the specific heat of KDP could be related by thermodynamic arguments to the dielectric properties (§ 5.2.4). However, on the basis of present experimental data the dielectric properties can only account for a finite peak. Thus at this time it seems that the peak consists of two contributions—a finite part and an infinite part, and only the former can be linked thermodynamically with the isothermal dielectric properties (§ 5.2.4).

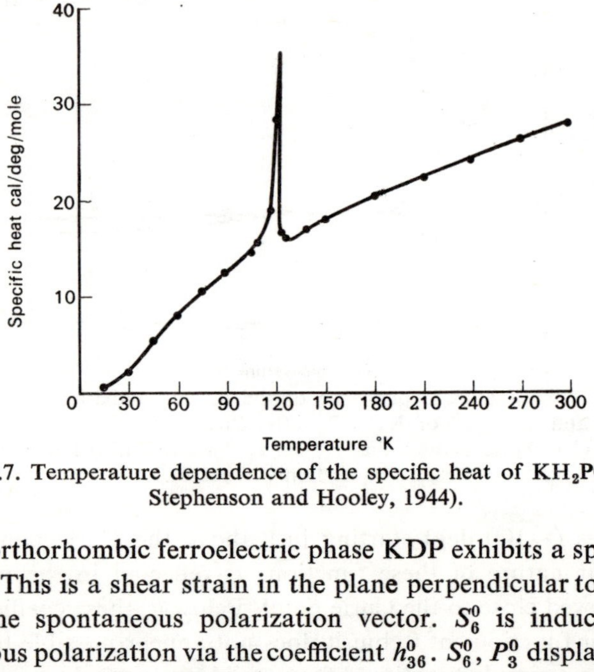

FIG. 1.2.7. Temperature dependence of the specific heat of KH_2PO_4 (after Stephenson and Hooley, 1944).

In the orthorhombic ferroelectric phase KDP exhibits a spontaneous strain S_6^0. This is a shear strain in the plane perpendicular to the direction of the spontaneous polarization vector. S_6^0 is induced by the spontaneous polarization via the coefficient h_{36}^0. S_6^0, P_3^0 display the same temperature dependence close to θ_c (Fig. 1.2.1), which suggests that h_{36}^θ is a slowly varying function of θ in this region. As KDP is lowered in symmetry, $\bar{4}2m$ to $mm2$, through the transition, additional elements appear in any zero stress and zero electric field compliance matrix. These elements can be accounted for, at least qualitatively, in terms of the spontaneous polarization (see end of § 3.3.8). No experimental quantitative study of these new elements appears to have been carried out.

The three coefficients $s_{66}^{E,\theta}$, d_{36}^{θ}, $\varepsilon_{33}^{T,\theta}$ occur in only two of the linear equations of state, namely

$$S_6 = s_{66}^{E,\theta} T_6 + d_{36}^{\theta} E_3, \tag{1.2.1}$$

$$D_3 = d_{36}^{\theta} T_6 + \varepsilon_{33}^{T,\theta} E_3. \tag{1.2.2}$$

These two equations are central to Mueller's theory described in § 1.5. Based on the experimental data available at the present time it appears that KDP is an example of the model (B) with $\xi^T > 0$ which is described in Chapter 5.

A number of crystals isomorphous to KDP, e.g. KD_2PO_4, KH_2AsO_4, RbH_2PO_4, etc., are ferroelectric with behaviour qualitatively similar to that of KDP.

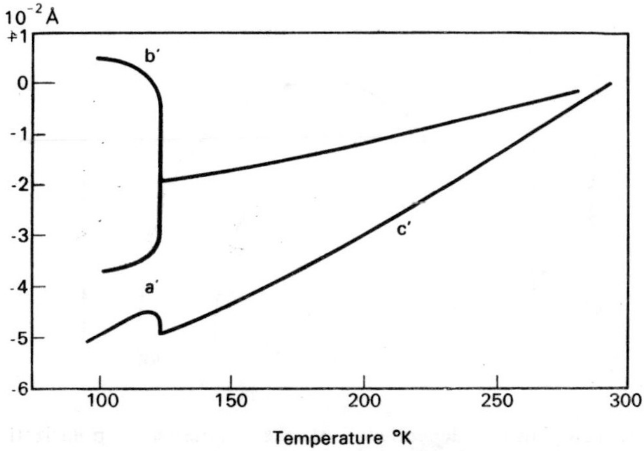

FIG. 1.2.8. Temperature dependence of the lattice constants of KH_2PO_4 measured in the standard orthorhombic crystallographic axes (after de Quervain, 1944). The ordinates represent the deviations of the lattice constants from their room temperature values. The thermal expansion coefficients in this set of axes are $\gamma_1' = (1/a_0')(da'/d\theta)$, $\gamma_2' = (1/b_0')(db'/d\theta)$, $\gamma_3' = (1/c_0')(dc'/d\theta)$, $\gamma_4' = \gamma_5' = \gamma_6' = 0$, whereas in the tetragonal crystallographic axes they read $\gamma_1 = \gamma_2 = \frac{1}{2}[(1/a_0')(da'/d\theta) + (1/b_0')(db'/d\theta)]$, $\gamma_3 = (1/c_0')(dc'/d\theta)$, $\gamma_4 = \gamma_5 = 0$, $\gamma_6 = [(1/a_0')(da'/d\theta) - (1/b_0')(db'/d\theta)]$. $a_0' = b_0'$, c_0' represent the room temperature values of the lattice constants. By inspection γ_3 has a strong temperature dependence at θ_c. $da'/d\theta$ and $db'/d\theta$ have opposite signs; however, the sum of these two quantities shows a strong temperature dependence; hence γ_1, γ_2 are anomalous.

1.3 Rochelle Salt

Rochelle salt was the first crystal to be designated a ferroelectric. The salt is a double tartrate of sodium and potassium; the chemical formula is $NaKC_4H_4O_6 \cdot 4H_2O$.

Rochelle salt has two Curie temperatures—297°K and 255°K. Above 297°K the phase is paraelectric with orthorhombic symmetry 222. Between 255°K and 297°K the phase is ferroelectric with monoclinic symmetry 2 and below 255°K the crystal reverts to the paraelectric state with symmetry 222. The sole ferroelectric axis is parallel to the orthorhombic X_1- or a-axis. The temperature dependence of the spontaneous polarization is shown in Fig. 1.3.1. The major anomalies in

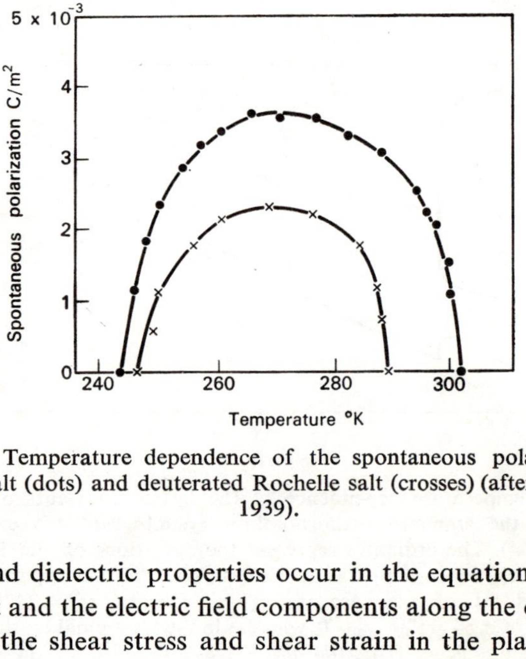

FIG. 1.3.1. Temperature dependence of the spontaneous polarization of Rochelle salt (dots) and deuterated Rochelle salt (crosses) (after Hablützel, 1939).

the elastic and dielectric properties occur in the equations relating the displacement and the electric field components along the orthorhombic X_1-axis and the shear stress and shear strain in the plane normal to this axis. Relative to the orthorhombic crystallographic axes these equations read

$$S_4 = s_{44}^{E,\theta} T_4 + d_{14}^{\theta} E_1, \tag{1.3.1}$$

$$D_1 = d_{14}^{\theta} T_4 + \varepsilon_{11}^{T,\theta} E_1, \tag{1.3.2}$$

In the upper paraelectric phase each of these coefficients satisfy a Curie–Weiss law with the Curie–Weiss temperature coinciding with the upper Curie point. The Curie–Weiss constants have the values $A = 6{\cdot}7\times10^{-10}$ deg m²/N; $B = 2{\cdot}9\times10^{-9}$ deg C/N; $C = 2{\cdot}8\times10^{-8}$ deg F/m (Mueller, 1940).

In the lower paraelectric phase each of these coefficients has a temperature dependence of the form (const.)/$(\theta_0^l - \theta)$, where θ_0^l is the lower Curie temperature. The complete temperature dependence of $\varepsilon_{11}^{T,\sigma}$ is shown in Fig. 1.3.2. The spontaneous strain in this crystal is the shear strain S_4^0 in the plane normal to the ferroelectric axis.

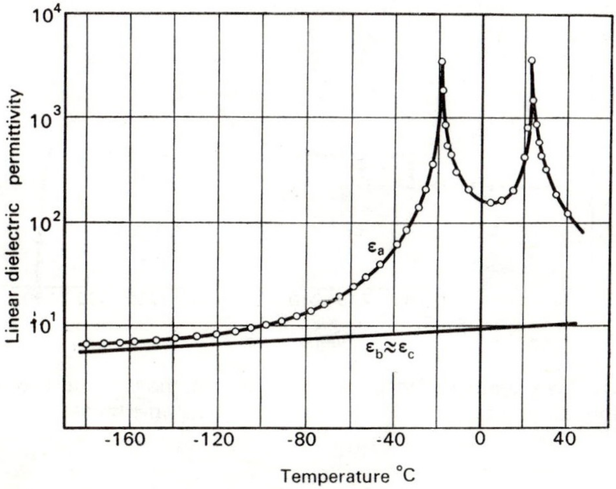

FIG. 1.3.2. Temperature dependence of the linear adiabatic permittivity of Rochelle salt (after Hablützel, 1939).

There appears to be a small anomaly in the specific heat of Rochelle salt at each of the Curie points, but the effect is not as dramatic as in the case of KDP.

The behaviour of Rochelle salt in the vicinity of the upper Curie point provides an example of model (B) with $\xi^T > 0$, described in Chapter 5.

1.4 Barium Titanate

Barium titanate undergoes three phase changes. The transition temperatures are 393°K, 278°K and 180°K; the symmetries in the four phases are, in order of decreasing temperature, cubic *m*3*m*, tetragonal 4*mm*, orthorhombic *mm*2 and rhombohedral 3*m*. The cubic phase is paraelectric and the remaining ferroelectric. As the crystal is cooled the spontaneous polarization vector successively points along the [1, 0, 0],† the [1, 1, 0] and the [1, 1, 1] directions (relative to the cubic axes) in the tetragonal, orthorhombic and rhombohedral phases respectively. The temperature dependence of the component of spontaneous polarization along the [1, 0, 0] axis is illustrated in Fig. 1.4.1. More recent work (Merz, 1953) has shown that P^0 drops discontinuously to zero and not continuously as indicated in this figure. Barium titanate has one independent linear dielectric permitivity component $\varepsilon_{11}^{T,\theta}$ in the cubic phase and two, $\varepsilon_{11}^{T,\theta}$, $\varepsilon_{22}^{T,\theta}$, in the tetragonal phase. In the paraelectric phase $\varepsilon_{11}^{T,\theta}$ obeys a Curie–Weiss law with $C = 1{\cdot}5\times 10^{-6}$ deg F/m and $\theta_0 = 383$°K (Drougard and Young, 1954), i.e. θ_0,

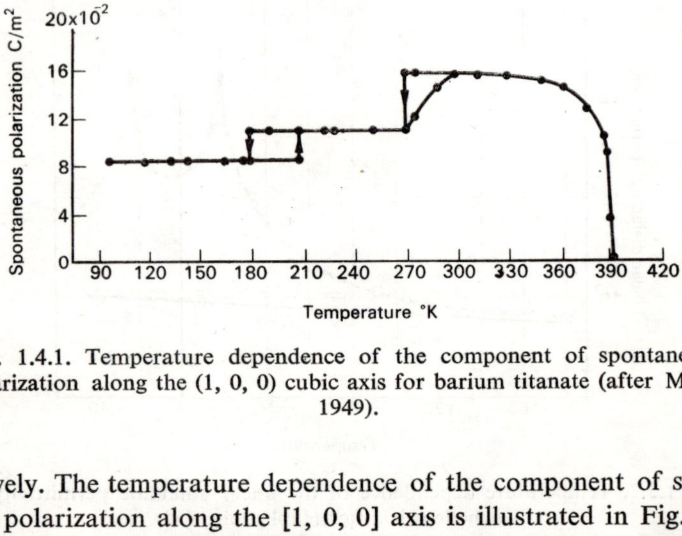

FIG. 1.4.1. Temperature dependence of the component of spontaneous polarization along the (1, 0, 0) cubic axis for barium titanate (after Merz, 1949).

† By this we mean that the spontaneous polarization vector may point along any one of the six directions [±1, 0, 0], [0, ±1, 0], [0, 0, ±1].

is lower than the Curie temperature by 10 degrees. There is no piezoelectric effect in the cubic phase, but in the lower symmetry phases there is an induced piezoelectric effect. The spontaneous strain in the tetragonal phase corresponds to an expansion in the direction parallel to the spontaneous polarization vector and a contraction at right angles to it. For example, at room temperatures with P^0 in the X_3-direction, $S_1^0 = S_2^0 = -3{\cdot}4\times10^{-3}$ and $S_3^0 = 7{\cdot}5\times10^{-3}$ (Kay and Vousden, 1949). The specific heat exhibits anomalies at each of the transitions. Thermal hysteresis effects are observed in the spontaneous polarization and in the dielectric permittivity measurements.

Based on the experimental data available at present it appears that the transition at 393°K is described by an extension of the model (A) with $\xi^T < 0$ (Chapter 5). The analysis of the other transitions is somewhat more complicated.

1.5 Mueller's Theory

In 1935 Mueller published the first of a series of papers in which he analysed the behaviour of the elastic, piezoelectric and dielectric coefficients of Rochelle salt. From the results of this work he drew two important conclusions which have since proved to be applicable to other ferroelectric crystals. In this section we describe Mueller's results.

Consider the high temperature paraelectric phase of Rochelle salt. The major anomalies in the linear, elastic and dielectric properties of this phase occur in the coefficients appearing in the following equations:

$$S_4 = s_{44}^{E,\theta} T_4 + d_{14}^{\theta} E_1, \tag{1.5.1}$$

$$D_1 = d_{14}^{\theta} T_4 + \varepsilon_{11}^{T,\theta} E_1. \tag{1.5.2}$$

These may be rewritten in the form

$$S_4 = s_{44}^{D,\theta} T_4 + g_{14}^{\theta} D_1, \tag{1.5.3}$$

$$E_1 = -g_{14}^{\theta} T_4 + \beta_{11}^{T,\theta} D_1, \tag{1.5.4}$$

with

$$s_{44}^{E,\theta} = s_{44}^{D,\theta} + \frac{(g_{14})^2}{\beta_{11}^{T,\theta}}, \quad d_{14}^{\theta} = \frac{g_{14}^{\theta}}{\beta_{11}^{T,\theta}}, \quad \varepsilon_{11}^{T,\theta} = \frac{1}{\beta_{11}^{T,\theta}}. \qquad (1.5.5)^\dagger$$

It follows immediately that, since $\varepsilon_{11}^{T,\theta}$ obeys a Curie–Weiss law, $\beta_{11}^{T,\theta} = (\theta - \theta_0)/C$, i.e. $\beta_{11}^{T,\theta}$ is a linear function of θ vanishing at θ_0. Thus the dielectric permeability must be classed as anomalous. In contrast, however, when Mueller used the experimental values of $s_{44}^{E,\theta}$, d_{14}^{θ}, $\varepsilon_{11}^{T,\theta}$ to determine $s_{44}^{D,\theta}$ and g_{14}^{θ}, he found that the latter two quantities were slowly varying functions of θ, that is to say not anomalous.‡ Thus referring to eqn. (1.5.5) we conclude that the singularities appearing in the coefficients of the equations of state (1.5.1) and (1.5.2) are associated with the anomalous behaviour of one coefficient, namely $\beta_{11}^{T,\theta}$ in (1.5.3) and (1.5.4). The expressions in (1.5.3) and (1.5.4) represent the leading terms in a power series expansion for S_4 and E_1 in terms of T_4 and D_1. Thus when $\beta_{11}^{T,\theta}$ is vanishingly small ($\theta \sim \theta_0$) the relationship between E_1 and D_1, for fixed T_4, is determined by the next higher order term in the expansion. Since the plane normal to the X_1-axis is a plane of symmetry the coefficient of the term quadratic in D_1 is identically zero. Hence in the next approximation eqn. (1.5.4) takes the form

$$E_1 = g_{14}^{\theta} T_4 + \beta_{11}^{T,\theta} D_1 + \xi_{11}^{T,\theta} D_1^3. \qquad (1.5.6)$$

$\xi_{11}^{T,\theta}$ proves to be positive in Rochelle salt. If $\xi_{11}^{T,\theta}$ were negative, it would

† These equations can be combined to give the thermodynamic identities

$$(s_{44}^{E,\theta} - s_{44}^{D,\theta})\varepsilon_{11}^{T,\theta} = \left(\frac{g_{14}}{\beta_{11}^{T,\theta}}\right)^2 = (d_{14}^{\theta})^2.$$

Now close to the Curie point $(s_{44}^{E,\theta} - s_{44}^{D,\theta}) = A/(\theta - \theta_0)$, $d_{14}^{\theta} \cong B/(\theta - \theta_0)$, $\varepsilon_{11}^{T,\theta} \cong C/(\theta - \theta_0)$, and therefore A, B, C are related, $B^2 = AC$. To test this prediction we use the results quoted in § 1.3. Thus

$$B^2 = 8{\cdot}4 \times 10^{-18} \text{ MKS}, \quad AC = 13{\cdot}4 \times 10^{-18} \text{ MKS}.$$

Similarly for KDP (§ 1.2), we find

$$B^2 = 18{\cdot}4 \times 10^{-18} \text{ MKS}, \quad AC = 13 \times 10^{-18} \text{ MKS}.$$

The discrepancies here provide an indication of the degree of internal consistency of the experimental data.

‡ The analogous coefficients in the case of KDP are $s_{66}^{D,\theta}$ and g_{36}^{θ}. Referring to Figs. 1.2.4 and 1.2.5 we see that there is virtually no anomaly in either of these functions.

be necessary to include further powers of D_1 to ensure local stability.† Setting $T_4 = 0$ and $\beta_{11}^{T,\theta} = A(\theta-\theta_0)$, where $A = 1/C$, we get

$$E_1 = A(\theta-\theta_0)\,D_1 + \xi_{11}^{T,\theta} D_1^3. \tag{1.5.7}‡$$

In particular, when $E = 0$, eqn. (1.5.7) has three roots

$$D_1 = 0 \quad D_1 = \pm \frac{\sqrt{-A(\theta-\theta_0)}}{\sqrt{\xi_{11}^T}}.$$

If $\xi_{11}^T > 0$, the non-zero roots are real only for $\theta < \theta_0$. Thus when $E_1 = 0$, eqn. (1.5.7) has two non-vanishing real roots if $\theta < \theta_0$. It is natural to identify the magnitude of these roots with the spontaneous polarization, namely

$$P^\circ = \frac{\sqrt{A(\theta_0-\theta)}}{\sqrt{\xi_{11}^T}}. \tag{1.5.8}$$

It is important to realize that this is not a thermodynamic identity in the sense of, for example, the Maxwell relation

$$\left(\frac{\partial S_4}{\partial D_1}\right)_T = \left(\frac{\partial E_1}{\partial T_4}\right)_D. \tag{1.5.9}$$

Equation (1.5.9) is based on the conservation of energy whereas eqn. (1.5.8) is based on the assumption that the equation of state (1.5.6) has the same form in the ferroelectric phase as it has in the paraelectric phase. Equation (1.5.8) is a prediction which serves primarily as a test of this assumption and not as a test of the laws of thermodynamics.

† Local stability requires $(\partial E_1/\partial D_1) > 0$ (§ 3.2.2). From eqn. (1.5.6)

$$(\partial E_1/\partial D_1)_T = \beta_{11}^{T,\theta} + 3\xi_{11}^{T,\theta} D_1^2;$$

when $\beta_{11}^{T,\theta} = 0$, $(\partial E_1/\partial D_1)_T \gtrless 0$ according as $\xi_{11}^T \gtrless 0$.

‡ Close to the Curie point of Rochelle salt large values of D_1 are observed for quite modest applied values of E_1 and the statement is sometimes made that this is a direct consequence of the large value of $\varepsilon_{11}^{T,\theta}$. However, this statement is inaccurate. More correctly, when $\varepsilon_{11}^{T,\theta}$ is large, $\beta_{11}^{T,\theta}$ is small, and D_1 is related to E_1 by the equation $D_1 \simeq \sqrt[3]{E_1/\xi_{11}^T}$; hence the values of D_1 are determined by ξ_{11}^T (for a given E_1) and are independent of $\varepsilon_{11}^{T,\theta}$ (provided $\varepsilon_{11}^{T,\theta}$ is sufficiently large).

Other predictions of a similar nature can be made; for example, from eqn. (1.5.7)

$$\left(\frac{\partial E_1}{\partial D_1}\right)_{T_4} = -A(\theta_0-\theta)+3\xi_{11}^{T,\theta}D_1^2;$$

when $D_1 = P^0$,

$$(\partial E_1/\partial D_1) = -2A(\theta-\theta_0) \quad (\theta < \theta_0). \tag{1.5.10}$$

This is the zero field permeability. Further, having assumed that eqn. (1.5.6) is valid in the ferroelectric phase it is natural to assume that the associated equation of state (1.5.3) is also valid in this phase [we note that, since neither $s_{44}^{D,\theta}$ nor g_{14}^{θ} vanish, we are not necessarily obliged to add higher order terms to this equation when $\theta \sim \theta_0$, as we were in the case of eqn. (1.5.4)]. In particular, when $T_4 = 0$ and $D_1 = P^0$, then eqn. (1.5.3) predicts a spontaneous shear strain

$$S_4^0 = g_{14}P^0. \tag{1.5.11}$$

Equations (1.5.10) and (1.5.11) serve as additional tests of the assumption discussed above. Mueller was able to show that these and other predictions are borne out in Rochelle salt and thus he concluded that the equations of state in the ferroelectric phase are described by the same functions as those describing the equations of state in the paraelectric phase.

The following observation on the choice of independent variables is appropriate at this point. It is clear from eqn. (1.5.7) that D_1 is a multi-valued function of E_1 when $\theta < \theta_0$. As a result when one is analysing the dielectric properties of Rochelle salt in particular and other ferroelectrics in general it is more convenient to work with this function rather than one of the branches of the inversion $D_1 = D_1(T_4, E_1)$. S_4 is a single valued function of T_4 and D_1, and these considerations do not apply.

Bibliography

The book *Ferroelectric Crystals* by Jona and Shirane (1962) gives a very complete description of the experimental data on ferroelectric crystals and the theories—both macroscopic and microscopic—devised

for these crystals. The review article by Devonshire (1964) serves as a supplement to this book, bringing the references up to date (1963).

Other review articles and books on ferroelectricity are (Cady, 1946), (von Hippel, 1950), (Jaynes, 1953), (Devonshire, 1954), (Forsbergh, 1956), (Megaw, 1957), (Merz, 1962), (Fatuzzo and Merz, 1967), (Burfoot, 1967).

CHAPTER 2

The Linear Elastic Dielectric

2.1 Introduction

The phenomenological theory of the electroelastic behaviour of an insulator (or dielectric) aims at providing a mathematical apparatus with which one can calculate the macroscopic response (e.g. change of shape, dipole moment, etc.) of the dielectric to mechanical and electrical stimuli (e.g. surface tractions, nearby charges, etc.). The adjectives phenomenological and macroscopic imply that the dielectric material is treated as a continuum without regard to the underlying atomic structure.

A common approach to the development of such a phenomenological theory may be summarized in the following way.

(i) One considers some arbitrary volume v of the material.

(ii) Fields are postulated to describe (a) the behaviour of the volume (position, density, dipole moment, etc.), (b) the interaction of the volume with the rest of the system (stresses, electric fields, etc.) and (c) the energy stored in the volume.

(iii) The laws of classical mechanics are applied to the volume v and there results a set of integral equations or their equivalent, a set of partial differential equations and boundary conditions for the fields.

(iv) A certain number of functional relationships (the equations of state) characteristic of the material are assumed to exist between

the fields at each point in the dielectric. These relationships are sufficient in number so that when they are combined with the field equations one may calculate (not necessarily trivially) a unique response of the dielectric for a given stimulus.

An alternative development of the theory consists of postulating fields and constraints to be used in conjunction with a variational principle for the energy. From this one obtains field equations, boundary conditions and equations of state. The choice of approach is a matter of taste and of one's confidence in postulating either appropriate fields or appropriate fields and constraints. The two approaches, when applied to a common system, must, of course, yield the same results, i.e. the same field equations, boundary conditions and equations of state. In the present chapter we describe the standard linear time-dependent electroelastic theory using the first mentioned approach, and in Chapter 4 we base our discussion of the non-linear static properties of an elastic dielectric on a variational principle.

In the first three sections of this chapter we introduce the infinitesimal strain, the stress tensor, the electric and the electric displacement fields and derive a few of their properties. These fields are then applied in the derivation of the standard linear time-dependent theory of the elastic dielectric (§ 2.5). This strictly linear theory has been extended in an *ad hoc* fashion to describe non-linear effects; the result is the quasi-linear theory also described in § 2.5.

The results obtained in this chapter have been used extensively in the measurement and analysis of the properties of ferroelectric crystals.

2.2 Infinitesimal Strain

2.2.1 Deformation

Consider a body in an initial configuration occupying some region V_0 of space. Suppose the body is *deformed*, i.e. altered in position and/or shape, to occupy a region V. As a result of the deformation a typical material point of the body is carried from a point with position vector $\boldsymbol{X}$, say, in V_0 to a point with position vector $\boldsymbol{x}$ in V. Thus to each point $\boldsymbol{X}$ in V_0 there corresponds a unique point $\boldsymbol{x}$ in V and conversely. We

can regard the deformation as a mapping of the points in V_0 on to the points in V. We refer all points in V_0 and V to a rectangular Cartesian coordinate system F with axes X_1, X_2, X_3. The components of $\boldsymbol{X}$ in this frame are denoted by X_i ($i = 1, 2, 3$) and the components of $\boldsymbol{x}$ by x_i ($i = 1, 2, 3$). The mapping referred to above is described by three functions $x_i = x_i(X_j)$ ($i, j = 1, 2, 3$). Further, because we intend to treat the body as a continuum, it is physically reasonable to suppose that these functions are continuous and as many times differentiable as may be required. Since the mapping describes a deformation, the

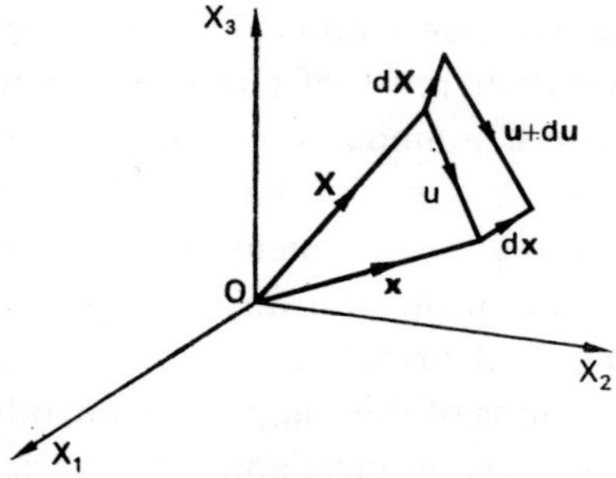

FIG. 2.2.1.

$x_i(X_j)$ will be single valued and invertable. The deformation is completely specified by these three functions. The nine quantities $(\partial x_i/\partial X_j)$ are the *deformation gradients*. A deformation is *uniform* if all the deformation gradients are constant throughout the body. The *displacement vector* is defined $\boldsymbol{u} = \boldsymbol{x} - \boldsymbol{X}$; the components of $\boldsymbol{u}$ in F are denoted by u_i; clearly, $u_i = x_i - X_i$ ($i = 1, 2, 3$); $\partial u_i/\partial X_j$ are the *displacement gradients*. From the definition of u_i, $\partial u_i/\partial x_j = \partial x_i/\partial X_j - \delta_{ij}$, where δ_{ij} is the Kronecker delta ($\delta_{ij} = 1$, $i = j$; $\delta_{ij} = 0$, $i \neq j$). The deformation is *infinitesimal* or vanishingly small if $|\partial u_i/\partial X_j| \ll 1$ ($i, j = 1, 2, 3$).

Consider the line element $d\boldsymbol{X}$ at $\boldsymbol{X}$. In a general deformation $d\boldsymbol{X}$ is translated, rotated and altered in length to become a line element $d\boldsymbol{x}$ at $\boldsymbol{x}$. We are principally concerned with the changes in length. From the definition of u_i, we have $du_i = dx_i - dX_i$ (Fig. 2.2.1) and hence

$$dx_i = dX_i + \sum_{j=1}^{3} \left(\frac{\partial u_i}{\partial X_j}\right) dX_j.$$

On squaring both sides of this equation and summing over i, we get

$$|d\boldsymbol{x}|^2-|d\boldsymbol{X}|^2 = 2\sum_{i,j}\left(\frac{\partial u_i}{\partial X_j}\right) dX_j\, dX_i+\sum_{i,j,k}\left(\frac{\partial u_i}{\partial X_j}\right)\left(\frac{\partial u_i}{\partial X_k}\right) dX_j\, dX_k,$$

where $|d\boldsymbol{x}|$ is the magnitude of $d\boldsymbol{x}$, etc.

In the case of infinitesimal deformations the second term on the right is vanishingly small compared with the first; hence for such deformations we get

$$|d\boldsymbol{x}|^2-|d\boldsymbol{X}|^2 = 2\sum_{i,j}\left(\frac{\partial u_i}{\partial X_j}\right) dX_j\, dX_i$$

or

$$|d\boldsymbol{x}|^2-|d\boldsymbol{X}|^2 = 2\sum_{i,j} S_{ij}\, dX_i\, dX_j, \tag{2.2.1}$$

where $S_{ij} = \frac{1}{2}(\partial u_i/\partial X_j+\partial u_j/\partial X_i)$ is the *infinitesimal strain tensor.* S_{ij} is a symmetric tensor of rank two.† The diagonal ($i = j$) and off-diagonal ($i \neq j$) components of S_{ij} are referred to as *normal* and *shear* respectively. S_{ij} provides a measure of the local infinitesimal deformation at $\boldsymbol{x}$ in the sense that the changes in length of all line elements through $\boldsymbol{x}$ are determined by the S_{ij} [see eqn. (2.2.1)].

For some purposes it is convenient to replace the infinitesimal strain tensor S_{ij} by the closely related quantity S_α ($\alpha = 1, 2, 3, 4, 5, 6$),‡ where $S_1 = S_{11}$, $S_2 = S_{22}$, $S_3 = S_{33}$, $S_4 = 2S_{23}$, $S_5 = 2S_{13}$, $S_6 = 2S_{12}$. This is the *Voigt notation* for subscripts. We shall refer to the S_α as *infinitesimal strain components.* It should be emphasized that the S_α (a) are not tensor components, i.e. do not transform as tensor components under changes of coordinate axes, and (b) do not provide a suitable measure of the local deformation when the deformation is finite as opposed to infinitesimal. This last remark applies also to the S_{ij}.

Consider now the effect of the deformation on a mass element dm at $\boldsymbol{X}$. Let dm occupy a volume dV at $\boldsymbol{X}$ and a volume dv at $\boldsymbol{x}$, where $dV = dX_1\, dX_2\, dX_3$ and $dv = dx_1\, dx_2\, dx_3$. We set

$$dm = \varrho_0 dV = \varrho\, dv, \tag{2.2.2}$$

† The transformation properties of the strain tensor are described in § 3.3.6.

‡ To prevent confusion we reserve italic letters for indices taking the values 1, 2, 3, and Greek letters for indices taking the values 1, 2, 3, 4, 5, 6.

where ϱ_0, ϱ are the *mass densities* in the initial and final states respectively. A standard result from calculus gives

$$dv = \frac{\partial(x_1, x_2, x_3)}{\partial(X_1, X_2, X_3)} dV, \tag{2.2.3}$$

where $\partial(x_1, x_2, x_3)/\partial(X_1, X_2, X_3)$ is the Jacobian describing the mapping $x_i(X_j)$. Now $\partial x_i/\partial X_j = \delta_{ij}+(\partial u_i/\partial X_j)$ and therefore the Jacobian can be written in the form

$$\frac{\partial(x_1, x_2, x_3)}{\partial(X_1, X_2, X_3)} = \begin{vmatrix} 1+\dfrac{\partial u_1}{\partial X_1} & \dfrac{\partial u_1}{\partial X_2} & \dfrac{\partial u_1}{\partial X_3} \\ \dfrac{\partial u_2}{\partial X_1} & 1+\dfrac{\partial u_2}{\partial X_2} & \dfrac{\partial u_2}{\partial X_3} \\ \dfrac{\partial u_3}{\partial X_1} & \dfrac{\partial u_3}{\partial X_2} & 1+\dfrac{\partial u_3}{\partial X_3} \end{vmatrix},$$

where the vertical bars denote a three-by-three determinant. When $|\partial u_i/\partial X_j| \ll 1$,

$$\partial(x_1, x_2, x_3)/\partial(X_1, X_2, X_3) \cong 1+\partial u_1/\partial X_1+\partial u_2/\partial X_2+\partial u_3/\partial X_3.$$

Thus for infinitesimal deformations

$$dv = (1+S_{11}+S_{22}+S_{33})\, dV, \tag{2.2.4}$$

and hence, by eqn. (2.2.2),

$$\varrho = (1-S_{11}-S_{22}-S_{33})\, \varrho_0. \tag{2.2.5}$$

We conclude that infinitesimal shear strains change the shape but not the size of volume elements.

2.2.2 Physical Significance of the S_{ij}

In this section we illustrate the physical significance of the infinitesimal strain tensor by means of two simple examples.

(i) Suppose that S_{11} is the only non-vanishing component of S_{ij}. Then a line element with components (dX_1, dX_2, dX_3) is, by definition,

deformed to a line element with components $[dX_1(1+S_{11}), dX_2, dX_3]$. Thus the normal strain S_{11} is the fractional elongation suffered by line elements parallel to the X_1-axis.

(ii) Suppose that $\partial u_1/\partial X_2$ and $\partial u_2/\partial X_1$ are the only non-vanishing components of $\partial u_i/\partial X_j$. Then a line element with components $[dX_1, dX_2, dX_3]$ is deformed to the line element $[dX_1+(\partial u_1/\partial X_2)dX_2, dX_2+(\partial u_2/\partial X_1)dX_1, dX_3]$. In particular the line elements $[dX_1, 0, 0]$ and $[0, dX_2, 0]$ are deformed to $[dX_1, (\partial u_2/\partial X_1)dX_1, 0]$ and $[(\partial u_1/\partial X_2)dX_2, dX_2, 0]$ respectively. If we denote the angle between the latter two vectors by $\frac{1}{2}\pi+\theta$ then, since $|\partial u_1/\partial X_2|, |\partial u_2/\partial X_1| \ll 1$, we have immediately that $\theta = -\partial u_1/\partial X_2 - \partial u_2/\partial X_1 = -2S_{12}$. Thus the shear strain S_{12} is one-half the decrease in angular separation, occurring between two initially perpendicular line elements, one parallel to the X_1-axis and the other parallel to the X_2-axis.

2.2.3 Compatibility Relations

Given a displacement field $u_i(X_j)$, $(i = 1, 2, 3)$, we can always determine the associated strains. However, the converse problem is not so trivial; that is, given six functions $S_{ij}(X_k)$ determine the associated displacement field $u_i(X_k)$. In general a solution does not exist. It is easy to see why this is so. The problem is equivalent to the following one. Given six partial-differential equations $\partial u_i/\partial X_j + \partial u_j/\partial X_i = 2S_{ij}(X_k)$, $(i, j = 1, 2, 3)$, determine the three functions u_1, u_2, u_3. Clearly there can be no general solution, for the functions are over-determined. To insure that solutions for the u_i do exist, the S_{ij} must satisfy a set of equations—the *compatibility relations*. We may think of these equations as field equations for the strain components. However, in practice, when carrying out measurements on a crystal one normally chooses the shape and orientation of the crystal in such a way that these relations are trivially satisfied. For a more detailed discussion see, for example, Pearson (1959).

2.3 Stress

2.3.1 Stress and Body Forces

Consider the case of a finite body of volume V. We direct our attention to an arbitrary volume v of this body, bounded by an imaginary surface s. We shall suppose for the present that s does not include any part of the surface of V.

We assume that the net force acting on the matter within v can be represented by a body force of density $\boldsymbol{f}$† and a surface traction field $\boldsymbol{t}$. The fields $\boldsymbol{f}$ and $\boldsymbol{t}$ are defined by writing down expressions for (a) the net force $\boldsymbol{F}$ and net torque $\boldsymbol{\Gamma}$ acting on v and (b) the work done δW when v is subjected to an infinitesimal displacement. Thus $\boldsymbol{f}$ and $\boldsymbol{t}$ are defined by the following expressions:

$$\boldsymbol{F} = \int_v \boldsymbol{f}\,dv + \oint_v \boldsymbol{t}\,dS, \qquad (2.3.1)$$

$$\boldsymbol{\Gamma} = \int_v \boldsymbol{x}\times\boldsymbol{f}\,dv + \oint_v \boldsymbol{x}\times\boldsymbol{t}\,dS, \qquad (2.3.2)$$

$$\delta W = \int_v \boldsymbol{f}\cdot\delta\boldsymbol{x}\,dv + \oint_v \boldsymbol{t}\cdot\delta\boldsymbol{x}\,dS, \qquad (2.3.3)$$

where the field $\delta\boldsymbol{x}$ describes the infinitesimal displacement of the elementary volume dv at $\boldsymbol{x}$. We have arbitrarily chosen to take moments about the origin of coordinates.

The component of $\boldsymbol{t}$ perpendicular to the outward normal at a point on s is called the *shear stress* and the component parallel to the normal the *normal stress*. By convention the normal stress is taken to be positive for tension on the volume v and negative for compression.‡

† In deriving the field equations of the standard treatment of the linear dielectric (§ 2.5), we put $\boldsymbol{f} \equiv 0$. However, we feel it is worth while to retain the field $\boldsymbol{f}$ explicitly in the present section, largely to show that the properties deduced for the stress vector $\boldsymbol{t}$ are not predicated on vanishing body forces.

‡ The opposite convention also occurs in the literature (see, for example, Cady, 1946).

2.3.2 Conservation of Momentum

Applying the laws of classical mechanics to describe the motion of the matter within v we have

$$\frac{d}{dt}\int\limits_{v} \varrho\dot{\boldsymbol{x}}\,dv = \boldsymbol{F} \quad \text{conservation of linear momentum,} \tag{2.3.4}$$

$$\frac{d}{dt}\int\limits_{v} \varrho\boldsymbol{x}\times\dot{\boldsymbol{x}}\,dv = \boldsymbol{\Gamma} \quad \text{conservation of angular momentum,} \tag{2.3.5}$$

where $\varrho\,dv$ is the element of mass within dv situated at $\boldsymbol{x}$ and moving with velocity $\dot{\boldsymbol{x}}$. The quantities on the left of these equations represent respectively the time rate of change of the linear momentum and angular momentum of the material in v at the time t and hence the limits of the integrals on the left are time dependent. However, we may differentiate under the integral sign in a special sense which we now introduce.

Consider the function $H = \int\limits_{v} h(x_i, t)\,\varrho\,dv$ *associated with the mass* $m = \int\limits_{v} \varrho\,dv$ which occupies the volume v at the time t; $\boldsymbol{x}$ represents the position vector of the mass element $dm = \varrho\,dv$ and the x_i are the components of $\boldsymbol{x}$ in some rectangular frame fixed in the laboratory. In general both $\boldsymbol{x}$ and v are functions of t. We wish to calculate an expression for the time rate of change of H. Suppose at time $t = t_0$ the mass dm is at $\boldsymbol{x}^0$ in a volume dv^0 and the total mass occupies a volume v^0 with density ϱ^0, i.e. $dm = \varrho\,dv = \varrho^0\,dv^0$. Clearly the solutions to the equations of motion for dm yield some relationship

$$x_i = x_i(x_j^0, t^0, t) \qquad (i, j = 1, 2, 3).$$

Using this relationship we transform the variables of integration $x_i \to x_i^0$; hence

$$H \parallel \int\limits_{v_0} h[x_i(x_j^0, t^0, t), t]\,\varrho^0\,dv^0.$$

Since the limits of the integral are independent of t,

$$\frac{dH}{dt} = \int_{v_0} \dot{h}\varrho^0 \, dv^0.$$

The reverse transformation $x_i^0 \rightarrow x_i$ now yields

$$\frac{dH}{dt} = \int_{v} \dot{h}\varrho \, dv,$$

where $\dot{h}$ is the so-called *material derivative*. Explicitly,

$$\dot{h} = \left(\frac{\partial h}{\partial t}\right)_x + \sum_{i=1}^{3} \left(\frac{\partial h}{\partial x_i}\right)_t \dot{x}_i$$

$\left(\frac{\partial h}{\partial t}\right)_x$ is the *local time derivative* and $\sum_{i=1}^{3} \left(\frac{\partial h}{\partial x_i}\right)_t \dot{x}_i$ the *convection* of h. We note that the components of $\boldsymbol{x}$ are functions only of t and thus the convection of x_i vanishes; i.e. $\dot{x}_i = (\partial x_i/\partial t)$. However, $\ddot{x}_i \not\equiv \partial(\dot{x}_i)/\partial t$.

In terms of material derivatives, eqns. (2.3.4) and (2.3.5) now read

$$\int_v \varrho \ddot{\boldsymbol{x}} \, dv = \int_v \boldsymbol{f} \, dv + \oint_v \boldsymbol{t} \, dS, \quad \text{by (2.3.1)}, \tag{2.3.6}$$

$$\int_v \varrho \boldsymbol{x} \times \ddot{\boldsymbol{x}} \, dv = \int_v \boldsymbol{x} \times \boldsymbol{f} \, dv + \oint_v \boldsymbol{x} \times \boldsymbol{t} \, dS, \quad \text{by (2.3.2)}, \tag{2.3.7}$$

The demands of the principle of conservation of momentum, as expressed by eqns. (2.3.6) and (2.3.7), put restrictions on the permissible forms of the vector field $\boldsymbol{t}$ but not on the vector field $\boldsymbol{f}$. This point is dealt with in the following section. We reserve the discussion of the analogous constraints imposed by the form of the work δW and by the principle of conservation of energy for later sections (§§ 2.3.5 and 2.3.6).

2.3.3 Stress Tensor

In general, the infinitesimal force $\boldsymbol{t}\,dS$ which acts on the surface element dS of the volume v is a function of both the position and orientation of dS, i.e. $\boldsymbol{t} = \boldsymbol{t}(\boldsymbol{x}, \boldsymbol{n})$, where $\boldsymbol{n}$ is the unit vector parallel to

the outward normal to the surface of v at $\boldsymbol{x}$. To explore the restrictions on the physically permissible functions, t_i, let us first of all consider the case where v is a small tetrahedron with three mutually perpendicular faces, each of which is perpendicular to one of the coordinate axes. Let a_i be the area of the face perpendicular to the X_i-axis ($i = 1$, 2, 3) and a the area of the fourth face (Fig. 2.3.1). If $\boldsymbol{n} = (n_1, n_2, n_3)$

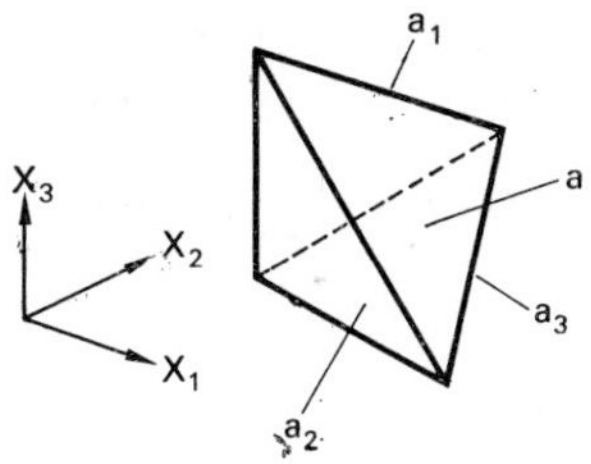

FIG. 2.3.1.

is the unit vector normal to this fourth face and h is the length of the altitude parallel to $\boldsymbol{n}$, standard geometric arguments give the following results: (i) $a_1/n_1 = a_2/n_2 = a_3/n_3 = a$, and (ii) $v = \frac{1}{3}ha$. For a small tetrahedron we may neglect, as a first approximation, the variation of the stress across the faces and the variation of the body force throughout the volume; then the net force acting on the tetrahedron is

$$F_i = \tfrac{1}{3}(haf_i) + at_i - \sum T_{ji}\, a_j \qquad (i = 1, 2, 3),$$

where F_i and t_i are the components of $\boldsymbol{F}$ and $\boldsymbol{t}$ respectively and T_{ji} is the ith component of the stress vector acting on the small face of area a_j. The conservation of linear momentum [eqn. (2.3.6)] now reads in this case

$$\tfrac{1}{3}(\varrho ha\ddot{x}_i) = \tfrac{1}{3}(haf_i) + a\Big[t_i - \sum_j T_{ji}\, a_j/a\Big].$$

Divide both sides of this equation by a and let h tend to zero; thus

$$0 = \Big[t_i - \sum T_{ji}\, a_j/a\Big].$$

Finally, using (i) above, we get

$$t_i = \sum_j T_{ji} n_j. \qquad (2.3.8)$$

Since $\boldsymbol{t}$ and $\boldsymbol{n}$ are vectors, T_{ji} is a second-rank tensor—the so-called *stress tensor*. To summarize, (a) the components of a stress vector acting on any element of area are linear combinations of the nine components of the stress tensor, with the direction cosines of the normal to the elementary area acting as coefficients, and (b) the stress vector is an implicit function of position through the T_{ji}, i.e.

$$t_i(x_k, n_l) = \sum_j T_{ji}(x_k) n_j.$$

We now show that if the force $\boldsymbol{F}$ and torque $\boldsymbol{\Gamma}$ are fully represented by the two fields $\boldsymbol{f}$ and $\boldsymbol{t}$, the stress tensor is symmetric, $T_{ji} = T_{ij}$.† The proof of this statement is considerably simplified if we use the Levi–Civita density e_{ijk} to represent the vector cross product; this density is defined $e_{ijk} = +1\ (-1)$ if i, j, k is an even (odd) permutation of 1, 2, 3 and zero otherwise.

In terms of the stress tensor and the Levi–Civita density, F_i and Γ_i are

$$F_i = \int_v f_i\, dv + \sum_j \oint_v T_{ji} n_j\, dS,$$

$$\Gamma_i = \sum_{j,k} e_{ijk} \left[\int_v x_j f_k\, dv + \sum_l \left(\oint_v T_{lk} x_j n_l\, dS \right) \right].$$

The surface integrals in each of these expressions can be transformed to volume integrals by means of Gauss' theorem; thus

$$F_i = \int_v \left[f_i + \sum_j \left(\frac{\partial T_{ji}}{\partial x_j} \right) \right] dv, \qquad (2.3.9)$$

$$\left.\begin{aligned} \Gamma_i &= \sum_{j,k} e_{ijk} \int_v \left[x_j \left(f_k + \sum_l \frac{\partial T_{lk}}{\partial x_l} \right) + \sum_l T_{lk}\, \delta_{jl} \right] dv \\ &= \sum_{j,k} e_{ijk} \int_v \left[x_j \left(f_k + \sum_l \frac{\partial T_{lk}}{\partial x_l} \right) + T_{jk} \right] dv, \end{aligned}\right\} \qquad (2.3.10)$$

† In the case where volume and/or surface couples are necessary this statement is not valid (Truesdell and Toupin, 1960).

where δ_{jl} is the Kronecker delta. The conservation conditions now read

$$\int_v \varrho \ddot{x}_i \, dv = \int_v \left(f_i + \sum_j \frac{\partial T_{ji}}{\partial x_j} \right) dv,$$

$$\sum_{j,k} e_{ijk} \int_v \varrho x_j \ddot{x}_k \, dv = \sum_{j,k} e_{ijk} \int_v \left[x_j \left(f_k + \sum_l \frac{\partial T_{lk}}{\partial x_l} \right) + \sum_l T_{lk} \delta_{jl} \right] dv.$$

Since the volume v is arbitrary, the necessary and sufficient conditions that these conservations conditions hold for any volume in the body are

$$\varrho \ddot{x}_i = f_i + \sum_j \frac{\partial T_{ji}}{\partial x_j} \tag{2.3.11}$$

and

$$\sum_{j,k} e_{ijk} \varrho x_j \ddot{x}_k = \sum_{j,k} e_{ijk} \left[x_j \left(f_k + \sum_l \frac{\partial T_{lk}}{\partial x_l} \right) + T_{jk} \right]. \tag{2.3.12}$$

These field equations are to be satisfied at each point in the body. When $\varrho \ddot{x}_i$ is substituted from eqn. (2.3.11) into eqn. (2.3.12), the latter equation reduces to the form

$$\sum_{j,k} e_{ijk} T_{jk} = 0. \tag{2.3.13}$$

The field equations (2.3.11) and (2.3.13) are the differential forms of the conservation of linear momentum and conservation of angular momentum respectively. The solution to eqn. (2.3.13) is trivial, viz. $T_{ji} = T_{ij}$. Thus the stress tensor must be symmetric. However, if additional fields are required to describe the net force on v, the field equations derived from the conservation of linear momentum and the conservation of angular momentum may not yield such a simple solution (Truesdell and Toupin, 1960.)

2.3.4 Boundary Conditions

The arguments in § 2.3.3, which led to the introduction of the stress tensor, also yield the conditions to be satisfied by the stress tensor field at the boundary of the body. Suppose the body is subjected to surface tractions represented by the stress field $\boldsymbol{t}^0$. Let the large face of the tetrahedron, introduced in § 2.3.3, lie wholly on the surface

of the body. Applying the argument of that section we get immediately [see eqn. (2.3.8)]

$$\sum_j T_{ji} n_j = t_i^0 \qquad (i = 1, 2, 3). \tag{2.3.14}$$

This is the sole boundary condition for the stress tensor field.

2.3.5 Mechanical Work

With the introduction of the stress tensor, eqn. (2.3.8), the expression for the mechanical work δW assumes the form

$$\delta W = \sum_i \left[\int_v f_i \, \delta x_i \, dv + \sum_j \oint_v T_{ji} \, \delta x_i \, n_j \, dS \right]. \tag{2.3.15}$$

Using Gauss' theorem we can put δW in the form

$$\delta W = \int_v \left[\sum_i \left(f_i + \sum_j \frac{\partial T_{ji}}{\partial x_j} \right) \delta x_i + \sum_{i,j} T_{ji} \frac{\partial}{\partial x_j} (\delta x_i) \right] dv \tag{2.3.16}$$

or, since $T_{ji} = T_{ij}$,

$$\delta W = \int_v \left[\sum_i \left(f_i + \sum_j \frac{\partial T_{ji}}{\partial x_j} \right) \delta x_i + \frac{1}{2} \sum_{i,j} T_{ji} \left\{ \frac{\partial}{\partial x_j} (\delta x_i) + \frac{\partial}{\partial x_i} (\delta x_j) \right\} \right] dv. \tag{2.3.17}$$

The net work done on the body by the surface tractions and the body forces is

$$\delta \mathcal{W} = \int_v \boldsymbol{f} \cdot \delta \boldsymbol{x} \, dv + \oint_v \boldsymbol{t}^0 \cdot \delta \boldsymbol{x} \, dS. \tag{2.3.18}$$

Since, at each point on the surface of V, $t_i^0 = \sum_j T_{ji} n_j$, eqn. (2.3.14),

$$\delta \mathcal{W} = \sum \int_V f_i \, \delta x_i \, dv + \sum_{i,j} \oint_V T_{ji} \, \delta x_i \, n_j \, dS.$$

Using the arguments which enabled us to replace the expressions in

(2.3.15) by those in (2.3.17), we obtain

$$\delta\mathcal{W} = \int_V \left[\sum_i \left(f_i + \sum_j \frac{\partial T_{ji}}{\partial x_j}\right)\delta x_i + \frac{1}{2}\sum_{i,j} T_{ji}\left\{\frac{\partial}{\partial x_j}(\delta x_i) + \frac{\partial}{\partial x_i}(\delta x_j)\right\}\right] dV. \tag{2.3.19}$$

We shall use these expressions for δW and $\delta\mathcal{W}$ in later sections.

2.3.6 Inequivalence of Body Forces and Stresses

The expression for the work [eqn. (2.3.3)], also imposes a constraint on the fields $\boldsymbol{f}$ and $\boldsymbol{t}$. However, this constraint is somewhat different in nature from the constraints imposed by the balance of linear momentum and the balance of angular momentum. Consider first the expressions for F_i and Γ_i [eqns. (2.3.9) and (2.3.10)],

$$F_i = \int_v \left(f_i + \sum_j \frac{\partial T_{ji}}{\partial x_j}\right) dv \qquad (i = 1, 2, 3),$$

$$\Gamma_i = \sum_{j,k} e_{ijk} \int_v x_j \left(f_k + \frac{\partial T_{lk}}{\partial x_l}\right) dv \qquad (i = 1, 2, 3),$$

since T_{ji} is symmetric. It is clear from these two expressions that we can fully represent the effect of the stress tensor in both $\boldsymbol{F}$ and $\boldsymbol{\Gamma}$ by a body force of density $\sum_j \partial T_{ji}/\partial x_j$. Conversely, if we obtain a symmetric tensor T'_{ji} such that $f_i = \sum_j \frac{\partial T'_{ji}}{\partial x_j}$ (and this can always be done for any arbitrary vector), then we can fully represent the effect of the body force by a stress field $T'_{ji} n_j$. These results suggest that it is redundant to introduce two fields $\boldsymbol{f}$ and $\boldsymbol{t}$ as we have done in § 2.3, and certainly—were we only considering force and torque—this would be the case. However, the introduction of an expression for the infinitesimal work δW is sufficient to distinguish between these two fields. To show this, let us suppose that we may subtract off part of the body force density $\boldsymbol{f}$ and replace it by the gradient of a stress tensor T'_{ji}, i.e. the original system of fields f_i, T_{ji} is fully equivalent to the new system

$$f_i - \sum_j \frac{\partial T'_{ji}}{\partial x_j}, \qquad T_{ji} + T'_{ji}.$$

As discussed above, the expressions for F_i and Γ_i in the old and new systems are equivalent. For the case of the infinitesimal work, the new system of fields yields the expression

$$\delta W = \sum_i \int_v \Big(f_i - \sum_j \partial T'_{ji}/\partial x_i\Big)\, \delta x_i\, dv + \oint_v \sum_{i,j} (T_{ji} + T'_{ji})\, n_j\, \delta x_i\, dS$$

$$= \sum_i \int_v f_i\, \delta x_i\, dv + \sum_{i,j} \oint_v T_{ji}\, n_j\, \delta x_i\, dS + \sum_{i,j} \int_v T'_{ji} \frac{\partial(\delta x_i)}{\partial x_j}\, dv.$$

If this is to be equivalent to the expression in eqn. (2.3.15), we must have

$$\int_v \sum_{i,j} T'_{ji} \frac{\partial}{\partial x_j} (\delta x_i)\, dv \equiv 0.$$

Since δx_i is arbitrary, $T'_{ji} \equiv 0$. Therefore we cannot replace part or all of the body force by an equivalent stress field. Conversely, we cannot replace part or all of a stress field by an equivalent body force.

2.4 Electric and Displacement Fields

2.4.1 Electric Field

The standard analysis of the macroscopic electrostatic properties of systems of conductors and insulators is based on the assumption that each conductor can be regarded as an equipotential region with an associated surface charge density and each insulator or *dielectric* a region of dipole-moment density.†

Consider an isolated system consisting of a charged conductor and an insulator. Let the charge density on the conductor surface be ω and the dipole-moment density (or *polarization*) of the insulator be $\boldsymbol{P}$; by introducing the concept of an infinitesimal test charge we may define in the usual way the electrostatic potential ϕ and the static electric

† We shall not be concerned with systems in which it is necessary to associate with the insulator either a charge density or a multipole density higher than the first order.

field E at points x in free space; for this system we have (in rationalized units)

$$\phi(x) = \frac{1}{4\pi\varepsilon_0}\left[\oint_C \frac{\omega(x')}{r} dS' + \int_V P \cdot \frac{(x-x')}{r^3} dv'\right], \quad (2.4.1)$$

$$E(x) = -\operatorname{grad} \phi, \quad (2.4.2)$$

where ε_0 is the permittivity of free space, the subscripts C and V refer to the regions occupied by the conductor and dielectric respectively (Fig. 2.4.1), $r = |x-x'|$ and the primes denote the variables of integration. Using eqns. (2.4.1) and (2.4.2) we *define*† fields ϕ and E for points

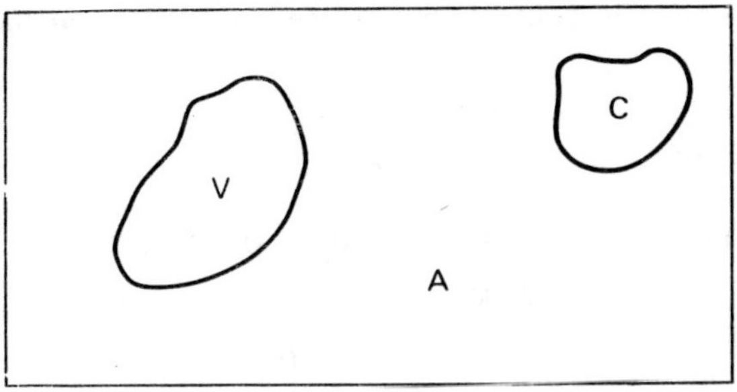

FIG. 2.4.1. Diagram of the regions referred to in the text. C, V, A are the regions occupied by the conductor, the dielectric and free space respectively.

within the dielectric. Furthermore, it is natural to retain the names potential and electric field‡ in these circumstances, but, of course, the fields ϕ, E cannot be interpreted, except in special circumstances as

† The integrand of the volume integral appearing in eqn. (2.4.1) has a singularity when x lies within the dielectric; thus in defining ϕ for points within the dielectric we must use the principal value of this integral. Further, since the gradient operator occurring in the definition of E does not commute with the principal value operator, the order of these operations, indicated by eqns. (2.4.1) and (2.4.2), must be preserved. If the order is reversed, i.e. the gradient operator applied to the integrand before the integration is carried out, then the value of the resulting integral is a function of the shape of the so-called excluded volume used in the limiting process (Kellogg, 1926). These shape-dependent values give rise to the *cavity* definition of the electric field (Fuller-Brown, 1956).

‡ The alternative expressions "macroscopic electric field", "Maxwell electric field" and "Maxwell–Faraday electric field" are also current in the literature.

either the work done per unit charge or the force per unit charge respectively.

From eqn. (2.4.1), the electric field $\boldsymbol{E}$ can be written in the form $\boldsymbol{E} = \boldsymbol{E}^s + \mathcal{E}$, where

$$\boldsymbol{E}^s = -\operatorname{grad}\left[\frac{1}{4\pi\varepsilon_0}\int_V \frac{\boldsymbol{P}\cdot(\boldsymbol{x}-\boldsymbol{x}')}{r^3}\,dv'\right],$$

$$\mathcal{E} = -\operatorname{grad}\left[\frac{1}{4\pi\varepsilon_0}\oint_C \frac{\omega(\boldsymbol{x}')\,dS'}{r}\right],$$

where $\boldsymbol{E}^s$ is the *self-field* of the dielectric and $\mathcal{E}$ the *applied field.* We note that $\mathcal{E}$ is the electric field of the charged conductor *in the presence* of the dielectric.

The definitions given above are easily extended to the case of systems of several conductors and dielectrics.

2.4.2 Field Equations and Boundary Conditions

Since $\boldsymbol{E} = -\operatorname{grad}\phi$ everywhere,

$$\operatorname{curl}\boldsymbol{E} = 0 \tag{2.4.3}$$

at all points in space. To derive the other field equation for $\boldsymbol{E}$ we first write ϕ in the form

$$\phi(\boldsymbol{x}) = \frac{1}{4\pi\varepsilon_0}\left[\oint_C \frac{\omega(\boldsymbol{x}')\,dS'}{r} + \oint_V \frac{\boldsymbol{P}(\boldsymbol{x}')\cdot\boldsymbol{n}dS'}{r} + \int_V \frac{-\operatorname{div}\boldsymbol{P}(\boldsymbol{x}')\,dv'}{r}\right] \tag{2.4.4}$$

by means of the identity

$$\oint_v \frac{\boldsymbol{P}\cdot\boldsymbol{n}dS}{r} = \int_v \frac{1}{r}\operatorname{div}\boldsymbol{P}\,dv + \int_v \boldsymbol{P}\cdot\operatorname{grad}\left(\frac{1}{r}\right)dv;$$

$\boldsymbol{n}$ is the unit vector parallel to the outward normal at the point $\boldsymbol{x}'$ on the surface of $\boldsymbol{V}$. From eqn. (2.4.4) we see immediately that ϕ is equivalent to the potential of surface distributions of charge ω and $\boldsymbol{P}\cdot\boldsymbol{n}$ on the conductor and dielectric surfaces respectively and a volume distribution of charge $-\operatorname{div}\boldsymbol{P}$ in the region occupied by the dielectric. These are the so-called *Poisson–Kelvin equivalent charge distributions.*

Standard electrostatic theory then yields the results

$$\text{div grad } \phi = \frac{1}{\varepsilon_0} \text{div } \boldsymbol{P}$$

at all points within the dielectric and div grad $\phi = 0$ at all points outside the dielectric. Since $\boldsymbol{E} = -\text{grad } \phi$ and since $\boldsymbol{P}$ vanishes everywhere outside the dielectric, the above field equations can be written in the form

$$\text{div } \boldsymbol{E} = -\frac{1}{\varepsilon_0} \text{div } \boldsymbol{P} \tag{2.4.5}$$

for points both inside and outside the dielectric. A third vector field the *displacement* $\boldsymbol{D}$ proves useful, where by definition $\boldsymbol{D} = \varepsilon_0 \boldsymbol{E} + \boldsymbol{P}$. From eqn. (2.4.5)

$$\text{div } \boldsymbol{D} = 0. \tag{2.4.6}$$

Standard results from electrostatic theory applied to the field equations (2.4.3) and (2.4.6) give the following boundary conditions:

(i) At a dielectric-vacuum boundary the normal component of $\boldsymbol{D}$, viz. $\boldsymbol{D} \cdot \boldsymbol{n}$, and the tangential component $\boldsymbol{E}$, viz. $\boldsymbol{E} \times \boldsymbol{n}$, are continuous across the boundary.

(ii) At a dielectric-conductor boundary $\boldsymbol{D} \cdot \boldsymbol{n} = \omega$ and $\boldsymbol{E} \times \boldsymbol{n} = 0$, where $\boldsymbol{n}$ is a unit vector parallel to the outward normal to the conductor surface.

When $\boldsymbol{P}$ and ω are functions of time, equation (2.4.6) continues to hold (provided, of course, that there is no free charge density) but equations (2.4.3) is valid only if the associated frequencies are much smaller than $2\pi c/L$, where c is the speed of light *in vacuo* and L is the maximum dimension of the system. This is the *electrodynamic quasi-static approximation.* (See, for example, Fano, Chu and Adler, 1960.)

2.4.3 Electrostatic Work

Consider again the isolated system consisting of the conductor occupying a region C and a dielectric occupying a region V. The net charge on the conductor $Q = \oint_C \omega \, dS$. Let the potential of the conductor be

ϕ_0; then the work required to place an additional infinitesimal charge δQ on the conductor is $\phi_0\,\delta Q$. Provided the boundaries of the conductor and dielectric are held fixed, the addition of an infinitesimal charge δQ to the conductor results in an infinitesimal† change $\delta \boldsymbol{D}$ in the field $\boldsymbol{D}$ at all points in $A+V$ (see Fig. 2.4.1). Let $\delta\omega$ be the corresponding change in ω. The boundary condition at the conductor surface, § 2.4.2, demands that $\delta\omega = \boldsymbol{n}\cdot\delta\boldsymbol{D}$, where $\boldsymbol{n}$ is a unit vector parallel to the outward normal to the conductor surface. Hence the expression for the electrostatic work becomes

$$\delta\mathcal{W} = \phi_0\,\delta Q = \oint_C \phi\,\delta\omega\,dS = +\oint_C \phi\,\delta\boldsymbol{D}\cdot\boldsymbol{n}\,dS. \qquad (2.4.7)$$

Using Gauss' theorem, we obtain

$$\delta\mathcal{W} = -\int_{A+V} \mathrm{div}\,(\phi\,\delta\boldsymbol{D})\,dv + \oint \phi\,\delta\boldsymbol{D}\cdot\boldsymbol{n}\,dS,$$

where the second integral is over the surface at infinity and hence vanishes since we are dealing with an isolated system. Differentiating by parts and using the relation grad $\phi = -\boldsymbol{E}$ and the field equation div $\boldsymbol{D} = 0$, we get

$$\delta\mathcal{W} = \int_{A+V} \boldsymbol{E}\cdot\delta\boldsymbol{D}\,dv \qquad (2.4.8)$$

In the case of a system of several conductors and insulators, the work done is of the form

$$\delta\mathcal{W} = \sum_t \phi_{0,\,t} \oint_{C_t} \delta\omega_t\,dS \qquad (2.4.9)$$

with an obvious extension of the notation used previously. The analysis leading to eqn. (2.4.8) applies to eqn. (2.4.9) and gives the same result, i.e. a volume integral of $\boldsymbol{E}\cdot\delta\boldsymbol{D}$ throughout all space outside the conductors.

† If, for example, the boundaries of the dielectric were moved, then at points either previously in the vacuum and now in the dielectric or vice versa the displacement field suffers a discontinuous and not an infinitesimal change. A specific example is discussed in § 4.3.

Consider now the system consisting only of the isolated conductor occupying the region C. Let $\mathscr{E}'$† be the electric field in the system when the conductor bears a net charge Q. From the analysis of the previous paragraph the work required to bring a charge δQ from infinity to the conductor is $\varepsilon_0 \int \mathscr{E}' \cdot \delta\mathscr{E}' dv$, where $\delta\mathscr{E}'$ is the change in $\mathscr{E}'$ resulting from the additional charge δQ; the integral is over all space outside the conductor. Returning to the expression for the work $\delta\mathcal{W}$, from eqn. (2.4.8) we write the identity

$$\delta\mathcal{W} = \int_{A+V} (\boldsymbol{E} \cdot \delta\boldsymbol{D} - \varepsilon_0 \mathscr{E}' \cdot \delta\mathscr{E}')\, dv + \varepsilon_0 \int_{A+V} \mathscr{E}' \cdot \delta\mathscr{E}'\, dv.$$

The first integral may be rewritten in the following manner

$$\int_{A+V} (\boldsymbol{E} \cdot \delta\boldsymbol{D} - \varepsilon_0 \mathscr{E}' \cdot \delta\mathscr{E}')\, dv = -\int_{A+V} (\boldsymbol{D} - \varepsilon_0 \boldsymbol{E}) \cdot \delta\mathscr{E}'\, dv$$
$$+ \int_{A+V} (\boldsymbol{D} - \varepsilon_0 \mathscr{E}') \cdot \delta\mathscr{E}'\, dv + \int_{A+V} \boldsymbol{E} \cdot (\delta\boldsymbol{D} - \varepsilon_0\, \delta\mathscr{E}')\, dv. \quad (2.4.10)$$

The second and third integrals on the right-hand side vanish identically; to prove this set $\mathscr{E}' = -\text{grad}\, \phi'$, where ϕ' is the potential associated with $\mathscr{E}'$. Let $\delta\phi'$ be the change in ϕ' resulting from the additional charge δQ; hence $\delta\mathscr{E}' = -\text{grad}\, \delta\phi'$. Consider first the second integral. From the identity $(\boldsymbol{D} - \varepsilon_0 \mathscr{E}') \cdot \text{grad}\, \delta\phi' = \text{div}[(\boldsymbol{D} - \varepsilon_0 \mathscr{E}')\, \delta\phi'] - \delta\phi'\, \text{div}(\boldsymbol{D} - \varepsilon_0 \mathscr{E}')$ and Gauss' theorem, we have

$$\int_{A+V} (\boldsymbol{D} - \varepsilon_0 \mathscr{E}') \cdot \delta\mathscr{E}'\, dv = -\oint_C \delta\phi' (\boldsymbol{D} - \varepsilon_0 \mathscr{E}') \cdot \boldsymbol{n}\, dS + \int_{A+V} \delta\phi'\, \text{div}(\boldsymbol{D} - \varepsilon_0 \mathscr{E}')\, dv$$
$$= -\delta\phi_0'(Q - Q) + \int_{A+V} \delta\phi' [\text{div}\, \boldsymbol{D} - \varepsilon_0\, \text{div}\, \mathscr{E}']\, dv \equiv 0.$$

The third integral on the right-hand side of eqn. (2.4.10) can be transformed in a similar analytical sequence; thus

$$\int_{A+V} \boldsymbol{E} \cdot (\delta\boldsymbol{D} - \varepsilon_0\, \delta\mathscr{E}')\, dv = -\oint_C \phi (\delta\boldsymbol{D} - \varepsilon_0\, \delta\mathscr{E}') \cdot \boldsymbol{n}\, dS$$
$$+ \int_{A+V} \phi\, \text{div}\, (\delta\boldsymbol{D} - \varepsilon_0\, \delta\mathscr{E}')\, dv \equiv 0.$$

† In general the macroscopic electric field $\boldsymbol{E}$, the applied electric field $\mathscr{E}$ (§ 2.4.1) and this field $\mathscr{E}'$ are quite distinct.

Hence

$$\delta W = -\int_V \boldsymbol{P}\cdot\delta\mathscr{E}'\,dv + \varepsilon_0 \int_{A+V} \mathscr{E}'\cdot\delta\mathscr{E}'\,dv, \tag{2.4.11}$$

where $\boldsymbol{P}$ is the polarization field in V. The second term in this expression represents a part of the work which does not depend on the nature or properties of the dielectric.

We note that $\varepsilon_0\mathscr{E}'\cdot\delta\mathscr{E}' = \delta\left(\frac{\varepsilon_0}{2}\mathscr{E}'\cdot\mathscr{E}'\right)$. The quantity in brackets is referred to as the *free space energy density* of the electrostatic field.

2.4.4 Depolarization Factor

We now show that if an ellipsoid of dielectric material is uniformly polarized (i.e. $\boldsymbol{P}$ = constant throughout the body), the self-field within the ellipsoid is uniform. Consider the case of an ellipsoid with uniform polarization $\boldsymbol{P}$. The potential of the self-field $\boldsymbol{E}^s$ can then be written

$$\phi^s(x_i) = \frac{\boldsymbol{P}}{4\pi\varepsilon_0}\cdot\int_V \frac{(\boldsymbol{x}-\boldsymbol{x}')\,dv'}{r^3} = -\boldsymbol{P}\cdot\operatorname{grad}\left[\frac{1}{4\pi\varepsilon_0}\int_V \frac{dv'}{r}\right].$$

The quantity in square brackets is the potential of an ellipsoid with a constant charge density, of value unity. Suppose the coordinate axes lie parallel to the principal axes of the ellipsoid, then the surface of the ellipsoid is defined by an equation of the form $(x_1/a_1)^2+(x_2/a_2)^2+(x_3/a_3)^2 = 1$, where a_i $(i = 1, 2, 3)$ are lengths of the principal axes. A standard calculation from electrostatic theory gives (Becker, 1964) for positions $\boldsymbol{x}$ within the ellipsoid,

$$\frac{1}{4\pi\varepsilon_0}\int_V \frac{dv}{r} = -\sum_{i=1}^{3}\frac{x_i^2 L_i}{2\varepsilon_0} + \text{constant},$$

where L_i, the depolarization factor, is defined

$$L_i = \frac{a_1a_2a_3}{2}\int_0^\infty \frac{1}{(a_i^2+\lambda)}\,\frac{d\lambda}{D(\lambda)} \tag{2.4.12}$$

with $D(\lambda) = \sqrt{(a_1^2+\lambda)(a_2^2+\lambda)(a_3^2+\lambda)}$. In the case of a sphere, for example, $L_i = 1/3$.

Hence

$$\phi^s(\boldsymbol{x}) = +\sum_{i=1}^{3} \frac{P_i x_i L_i}{\varepsilon_0} + \text{constant}$$

and

$$E_i^s = -(\partial\phi^s/\partial x_i) = -\frac{L_i P_i}{\varepsilon_0} \quad (i = 1, 2, 3),$$

i.e. the self-field of the dielectric is uniform within the ellipsoid. In this case the electric field is given by the expression

$$E_i = \mathscr{E}_i - \frac{L_i P_i}{\varepsilon_0} \quad (i = 1, 2, 3). \tag{2.4.13}$$

Thus if $\mathscr{E}$ and $\boldsymbol{P}$ are uniform within the ellipsoid, so also is $\boldsymbol{E}$.

2.5 The Linear Elastic Dielectric†

2.5.1 Assumptions

The standard theory of the linear time-dependent behaviour of a homogeneous dielectric is based on four main assumptions which we list below. These assumptions, together with the conservation laws for linear momentum, angular momentum and energy, yield the field equations, boundary conditions and equations of state characteristic of this theory.

(i) The behaviour of the dielectric is fully specified by the following four fields: the displacement vector $\boldsymbol{u}$ describing the deformation of the dielectric relative to some reference configuration (§ 2.2); the stress field $\boldsymbol{t}$ (§ 2.3); the electric field $\boldsymbol{E}$ and the electric displacement field $\boldsymbol{D}$ (§ 2.4).

Thus in particular it is assumed that (a) the force on any arbitrary volume of the dielectric is fully represented by the stress field $\boldsymbol{t}$ and no additional fields, such as the body force density $\boldsymbol{f}$, are required, and (b) the electric self-field (§ 2.4) arises only from a dipole moment distribution in the dielectric (see footnote on p. 36).

† In this section we continue to measure all vector components in a single rectangular coordinate system F.

(ii) The electrodynamic quasi-static approximation holds. Thus the fields $\boldsymbol{E}$ and $\boldsymbol{D}$ are assumed to be at most slowly varying functions of time.

(iii) The dielectric possesses a state in which $\boldsymbol{t}$, $\boldsymbol{E}$ and $\boldsymbol{D}$ are identically zero throughout the dielectric.

The configuration of the dielectric in this state is used as the reference configuration and the state is referred to as the *initial state*. Thus the initial state is characterized by the vanishing of the four fields $\boldsymbol{t}$, $\boldsymbol{u}$, $\boldsymbol{D}$ and $\boldsymbol{E}$; in particular since $\boldsymbol{t} \equiv 0$ this state is an equilibrium state. By a "homogeneous dielectric" we shall mean a dielectric which is homogeneous in the initial state, i.e. local properties of the dielectric (mass density, energy density, etc.) do not vary from point to point in the dielectric when it is in the initial state.

(iv) The energy (kinetic plus potential) of an arbitrary volume $\boldsymbol{v}$ of the dielectric is of the form

$$\int_{v} \left[\frac{|\dot{\boldsymbol{u}}|^2}{2} + \psi(u_i,\ \partial u_i/\partial X_j,\ D_i) \right] \varrho\, dv,$$

where ϱ is the mass density and ψ the potential energy density of the dielectric; u_i are the components of the displacement vector $\boldsymbol{u}$; $\partial u_i/\partial X_j$ are the displacement gradients, and D_i the components of the electric displacement vector $\boldsymbol{D}$.

Since the dielectric is assumed to be homogeneous in the initial state, neither ψ nor ϱ are explicit functions of the coordinates X_i describing the initial state. This means in particular that the mass density ϱ_0 in the initial state is a constant, independent of position. The form† of the function ψ (i.e. the parameters describing the function) is a function

† It is also assumed, usually tacitly, that ψ is a local property in the sense that the energy density function at a point $\boldsymbol{x} = \boldsymbol{X} + \boldsymbol{u}$ in the dielectric is characteristic only of the material in the immediate vicinity of $\boldsymbol{x}$ and not, for example, how the material is distributed elsewhere in the dielectric. Recent work (Grindlay and Wong, 1968) on the statistical mechanics of dielectrics shows that this assumption may be valid.

of (a) the choice of coordinate axes (see § 3.3.6), (b) the material making up the dielectric and (c) temperature (see Chapter 3).

Only the case of vanishingly small fields $\boldsymbol{u}$, $\boldsymbol{t}$, $\boldsymbol{E}$ and $\boldsymbol{D}$ is considered in the linear theory. As a result, in developing this theory we merely retain the lowest order non-vanishing terms in these quantities in any expression we may require.

2.5.2 Field Equations and Boundary Conditions

For definiteness we shall consider the isolated system consisting of a stressed dielectric and charged conductor. In the initial state the conductor is uncharged, and the dielectric is free of stress and polarization. Let the regions occupied by the dielectric, conductor and free space

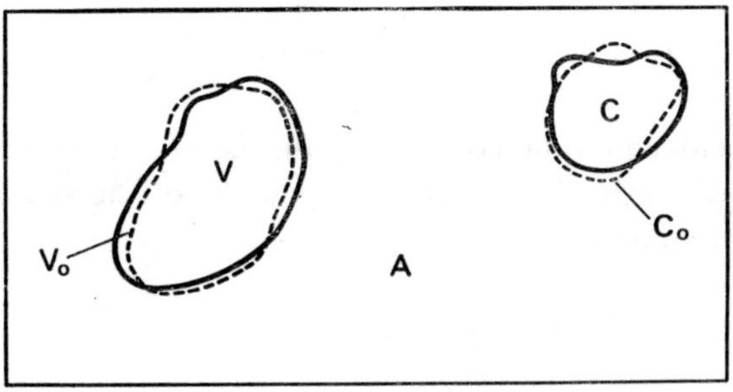

FIG. 2.5.1. Diagram of the regions referred to in the text. C_0 and V_0 are the regions occupied in the initial state by the conductor and dielectric respectively (dashed lines). C, V and A are the regions occupied in some arbitrary state by the conductor, dielectric and free space respectively (full lines).

in the initial state be denoted by V_0, C_0 and A_0 respectively and the corresponding regions in any other state by V, C and A (Fig. 2.5.1). In the linear theory we are concerned only with small deformations of the dielectric and conductor relative to their configurations in the initial state and hence $V \sim V_0$ and $C \sim C_0$. Extending the notation introduced in § 2.2.1, we denote the position vectors of *all points* in space by $\boldsymbol{X}$ when the system is in the initial state and by $\boldsymbol{x}$ otherwise.

Consider the system in some state other than the initial state. Since the force on any volume of the dielectric is assumed to be fully represented by the stress field $\boldsymbol{t}$ [assumption (i)] we have, from the conser-

vation of linear momentum and the conservation of angular momentum (§ 2.3), that

$$t_i = \sum_{j=1}^{3} T_{ji} n_j \quad \text{in } V \tag{2.5.1}$$

and

$$\varrho \ddot{x}_i = \sum_{j=1}^{3} \frac{\partial T_{ji}}{\partial x_j} \quad \text{in } V, \tag{2.5.2}$$

where T_{ji} is a symmetric tensor.

The statement "in V" means that the associated results holds at all points in the region V. Since $x_i = u_i + X_i$ and the X_i are independent of time, $\ddot{x}_i = \ddot{u}_i$; hence eqn. (2.5.2) may be replaced by

$$\varrho \ddot{u}_i = \sum_{j=1}^{3} \frac{\partial T_{ji}}{\partial x_j} \quad \text{in } V. \tag{2.5.3}$$

This is a field equation connecting the acceleration components $\ddot{\boldsymbol{u}}_i$ with the space derivatives of the components of the stress tensor. The T_{ji} satisfy the boundary condition

$$t_i^0 = \sum_{j=1}^{3} T_{ji} n_j \quad \text{on } V, \tag{2.5.4}$$

where $\boldsymbol{t}^0$ is the mechanical stress applied to the surface V and the n_i are the direction cosines of the outward normal to V [eqn. (2.3.14)].

In the electrodynamic quasi-static approximation [assumption (ii)] the fields $\boldsymbol{E}$ and $\boldsymbol{D}$ satisfy the field equations

$$\nabla_x \times \boldsymbol{E} = 0, \quad \nabla_x \cdot \boldsymbol{D} = 0 \quad \text{in } A \text{ and } V \tag{2.5.5}$$

and the boundary conditions

$$\boldsymbol{n} \times (\boldsymbol{E}^V - \boldsymbol{E}^A) = 0, \quad \boldsymbol{n} \cdot (\boldsymbol{D}^V - \boldsymbol{D}^A) = 0 \quad \text{on } V, \tag{2.5.6}$$

$$\boldsymbol{n} \times \boldsymbol{E} = 0, \quad \boldsymbol{n} \cdot \boldsymbol{D} = \omega \quad \text{on } C, \tag{2.5.7}$$

where ∇_x is the gradient operator for the coordinates x_i, $\boldsymbol{n}$ a unit vector parallel to the outward normal at the surface in question and ω the charge density on the conductor surface. The superscripts V and A

are used to distinguish, where necessary, field quantities in the dielectric from those in free space.

The set of field equations and boundary conditions listed above hold for the regions V, A and C. In applications of the theory it is convenient to use the transformed set which holds for the regions V_0, A_0 and C_0 of the initial state. To effect this transformation we introduce a coordinate transformation $x_i = f_i(X_j) + X_i$ to map the points of V on to V_0, C on to C_0 and A on to A_0. We choose

$$\begin{aligned} f_i &= u_i(X_j), \quad X_j \ \text{in } V_0, \\ f_i &= u_i'(X_j), \quad X_j \ \text{in } A_0 \text{ and } C_0. \end{aligned}$$

$\boldsymbol{u}(\boldsymbol{X})$ is the displacement vector describing the deformation of the dielectric; clearly this part of the transformation maps the boundary of V on to V_0. $\boldsymbol{u}'(\boldsymbol{X})$ is some vector function with the following properties: (i) $\boldsymbol{u}'(\boldsymbol{X}) \equiv \boldsymbol{u}(\boldsymbol{X})$ on V_0, (ii) the surface C maps on to C_0 and (iii) $|\partial u_i'/\partial X_j| \ll 1$. Otherwise $\boldsymbol{u}'$ is arbitrary. Consider first the field equation (2.5.3). If we regard T_{ji} as a function of the X_j, then by the chain rule we have

$$\frac{\partial T_{ji}}{\partial x_j} = \sum_{k=1}^{3} \frac{\partial T_{ji}}{\partial X_k}\frac{\partial X_k}{\partial x_j} \cong \sum_{k=1}^{3} \frac{\partial T_{ji}}{\partial X_k}(\delta_{kj} - \partial u_k/\partial X_j).$$

T_{ji} and $\partial u_j/\partial X_k$ are small and we retain only the leading term to write

$$\frac{\partial T_{ji}}{\partial x_j} \cong \frac{\partial T_{ji}}{\partial X_j}.$$

Similarly the mass density in the initial state

$$\begin{aligned} \varrho_0 &= \varrho\, \partial(x_1, x_2, x_3)/\partial(X_1, X_2, X_3) \quad \text{[by eqn. (2.2.3)]} \\ &\cong \varrho, \end{aligned}$$

and the material acceleration component

$$\begin{aligned} \ddot{u}_i &= \left(\frac{\partial^2 u_i}{\partial t^2}\right) + \sum_{j=1}^{3} \left(\frac{\partial \dot{u}_i}{\partial x_j}\right) \dot{u}_j \quad \text{(from § 2.3.2)} \\ &\cong \left(\frac{\partial^2 u_i}{\partial t^2}\right). \end{aligned}$$

It is thus evident that in the linear theory there is no distinction between the material derivative and the local time derivative of small fields. We choose to retain the dot notation to represent time derivatives. Equation (2.5.2) now reads

$$\varrho_0 \ddot{u}_i = \sum_{j=1}^{3} \left(\frac{\partial T_{ji}}{\partial X_j} \right) \quad \text{in } V_0. \tag{2.5.8}$$

The boundary condition for T_{ji} transforms to

$$t_i^0[u_k(X_j)+X_k] = \sum_{j=1}^{3} T_{ji}[u_k(X_l)+X_k]\, n_j[u_k(X_l)+X_k] \quad \text{on } V_0,$$

where each square bracket encloses the arguments of the function on the immediate left. Again, since T_{ji} and u_i are small quantities, the leading term on the right of this equation is $\sum_{j=1}^{3} T_{ji}[u_k(X_l)+X_l]n_j^0$; the n_j^0 are the direction cosines of the normal to the surface V_0 at X; hence we can write this boundary condition in the form

$$t_i^0 = \sum_{j=1}^{3} T_{ji} n_j^0. \tag{2.5.9}$$

We note that in eqns. (2.5.8) and (2.5.9) ϱ_0, $\boldsymbol{n}^0$ are the fields in the initial state and $\boldsymbol{t}^0$, T_{ji}, $\boldsymbol{u}$ are the fields in the deformed state. Similar arguments using the full mapping, when applied to the remaining field equations and boundary conditions yield

$$\nabla_X \times \boldsymbol{E} = 0, \qquad \nabla_X \cdot \boldsymbol{D} = 0 \quad \text{in } A_0 \text{ and } V_0, \tag{2.5.10}$$

$$\boldsymbol{n}^0 \times (\boldsymbol{E}^V - \boldsymbol{E}^A) = 0, \qquad \boldsymbol{n}^0 \cdot (\boldsymbol{D}^V - \boldsymbol{D}^A) = 0 \quad \text{on } V_0, \tag{2.5.11}$$

$$\boldsymbol{n}^0 \times \boldsymbol{E} = 0, \qquad \boldsymbol{n}^0 \cdot \boldsymbol{D} = \omega \quad \text{on } C_0, \tag{2.5.12}$$

where ∇_X is the gradient operator for the coordinates X_j. In eqn. (2.5.12) $\boldsymbol{n}^0$ is the unit vector parallel to the outward normal at a point on the surface C_0. Because the fields $\boldsymbol{E}$, $\boldsymbol{D}$, ω and $\boldsymbol{u}'$ are all small, these results are independent of the quantity $\boldsymbol{u}'$.

We note that this linear theory also applies when the dielectric is in contact with the conductor. In this case the boundary condition at

this dielectric-conductor interface reads

$$\boldsymbol{n}^0 \times \boldsymbol{E}^V = 0, \quad \boldsymbol{n}^0 \cdot \boldsymbol{D}^V = \omega.$$

Equations (2.5.8) to (2.5.12) are the field equations and boundary conditions characteristic of the standard linear theory of the elastic dielectric. To complete the theory we require a set of equations of state.

2.5.3 Equations of State I

At each point in the dielectric the values of the fields T_{ji} and E_i are related to the values of the fields $\partial u_i/\partial X_j$ and D_i. The form of these relationships or *equations of state*† is dictated by the law of conservation of energy. The equations of state for the linear theory form the main topic of this and the following subsection.

Consider an arbitrary volume v of the dielectric. From eqn. (2.3.16) the density of mechanical work when there are no body forces is

$$\sum_{i,j} [(\partial T_{ji}/\partial x_j)\, \delta x_i + T_{ji}\{\partial(\delta x_i)/\partial x_j\}],$$

where $\delta \boldsymbol{x}$ is an infinitesimal displacement; from eqn. (2.4.8) the density of electric work is $\boldsymbol{E} \cdot \delta \boldsymbol{D}$. Thus if the volume v subjected to a deformation $\delta \boldsymbol{x} = \delta \boldsymbol{u}$ and a change in $\boldsymbol{D}$, $\delta \boldsymbol{D}$, then the work done on v is

$$\delta W = \int_v \left[\sum_{i,j} \left\langle \left(\frac{\partial T_{ji}}{\partial x_j} \right) \delta u_i + T_{ji} \left\{ \frac{\partial (\delta u_i)}{\partial x_j} \right\} \right\rangle + \boldsymbol{E} \cdot \delta \boldsymbol{D} \right] dv.$$

Consider the term in curly brackets. For the linear theory we retain only the leading term and write

$$T_{ji} \left[\frac{\partial (\delta u_i)}{\partial x_j} \right] \cong T_{ji} \frac{\partial (\delta u_i)}{\partial X_j} = T_{ji} \delta \left(\frac{\partial u_i}{\partial X_j} \right)^{\ddagger}.$$

$\partial u_i/\partial X_j$ are the deformation gradients introduced in § 2.2.1.

† Sometimes referred to as constitutive relations.

‡ To prove that the operations δ and $\partial/\partial X_j$ commute we treat the u_i as functions of some parameter λ and define the variation in the u_i as $\delta u_i = (\partial u_i/\partial \lambda)_X \delta\lambda$. It follows immediately that

$$\frac{\partial}{\partial X_j}(\delta u_i) = \frac{\partial^2 u_i}{\partial \lambda \partial X_j} \delta\lambda = \delta \left(\frac{\partial u_i}{\partial X_j} \right).$$

Thus

$$\delta W = \int_v \left\{ \sum_{i,j} \left[\left(\frac{\partial T_{ji}}{\partial x_j} \right) \delta u_i + T_{ji}\, \delta \left(\frac{\partial u_i}{\partial X_j} \right) \right] + \boldsymbol{E} \cdot \delta \boldsymbol{D} \right\} dv. \qquad (2.5.13)$$

Hence

$$\frac{dW}{dt} = \int_v \left\{ \sum_{i,j} \left[\left(\frac{\partial T_{ji}}{\partial x_j} \right) \dot{u}_i + T_{ji} \left(\frac{\partial \dot{u}_i}{\partial X_j} \right) + \boldsymbol{E} \cdot \dot{\boldsymbol{D}} \right\} dv, \qquad (2.5.14)$$

where dW/dt is the rate of work done on the volume v and $\dot{\boldsymbol{u}}$, $\dot{\boldsymbol{D}}$ and $\partial \dot{u}_i/\partial X_j$ are the time derivatives of the fields $\boldsymbol{u}$, $\boldsymbol{D}$ and $\partial u_i/\partial X_j$ respectively. Strictly these should be written as local time derivatives, but the form assumed for δW already contains the small field assumption, § 4.1.4, and hence at this stage it is pointless to maintain the distinction between local and material derivatives.

The law of conservation of energy demands that

$$\frac{dU}{dt} = \frac{dW}{dt},$$

where U is the energy stored in v. From assumption (iv)

$$\frac{dU}{dt} = \frac{d}{dt} \int_v \left[\frac{|\dot{\boldsymbol{u}}|^2}{2} + \psi \left(u_i, \frac{\partial u_i}{\partial X_j}, D_i \right) \right] \varrho \, dv. \qquad (2.5.15)$$

At this point it is convenient to introduce the notation $\partial u_i/\partial X_j = u_{ij}$. Then

$$\frac{dU}{dt} = \int_v \left[\sum_{i=1}^{3} \left\{ \dot{u}_i \ddot{u}_i + \left(\frac{\partial \psi}{\partial u_i} \right) \dot{u}_i + \left(\frac{\partial \psi}{D_i} \right) \dot{D}_i \right\} + \sum_{i,j} \left(\frac{\partial \psi}{\partial u_{ij}} \right) \dot{u}_{ij} \right] \varrho \, dv. \qquad (2.5.16)$$

Substituting for dW/dt and dU/dt from eqns. (2.5.13) and (2.5.16) into the expression for the conservation of energy, we get

$$\int_v \left[\sum_i \left\{ \dot{u}_i \ddot{u}_i + \left(\frac{\partial \psi}{\partial u_i} \right) \dot{u}_i + \left(\frac{\partial \psi}{\partial D_i} \right) \dot{D}_i \right\} + \sum_{i,j} \left(\frac{\partial \psi}{\partial u_{ij}} \right) \dot{u}_{ij} \right] \varrho \, dv$$

$$= \int_v \left[\sum_{i,j} \left\{ \left(\frac{\partial T_{ji}}{\partial x_j} \right) \dot{u}_i + T_{ji} \dot{u}_{ij} \right\} + \sum_i E_i \dot{D}_i \right] dv$$

or

$$\int\limits_v \left\{ \sum_i \dot{u}_i \left(\varrho \ddot{u}_i + \varrho \frac{\partial \psi}{\partial u_i} - \sum_j \frac{\partial T_{ji}}{\partial x_j} \right) + \dot{D}_i \left(\varrho \frac{\partial \psi}{\partial D_i} - E_i \right) \right.$$
$$\left. + \sum_{i,j} \left(\varrho \frac{\partial \psi}{\partial u_{ij}} - T_{ji} \right) \dot{u}_{ij} \right\} dv = 0.$$

The volume v and the time derivatives $\dot{u}_i$, $\dot{D}_j$ and u_{ij} are arbitrary and hence energy is conserved if and only if the quantities in round brackets vanish; thus

$$\varrho \ddot{u}_i = \sum_j \left(\frac{\partial T_{ji}}{\partial x_j} \right) - \varrho \left(\frac{\partial \psi}{\partial u_i} \right) \qquad \text{in } V, \tag{2.5.17}$$

$$E_i = \varrho \left(\frac{\partial \psi}{\partial D_i} \right) \qquad \text{in } V, \tag{2.5.18}$$

$$T_{ji} = \varrho \left(\frac{\partial \psi}{\partial u_{ij}} \right) \qquad \text{in } V. \tag{2.5.19}$$

Now from the conservation of linear momentum $\varrho \ddot{u}_i = \sum_j \partial T_{ji}/\partial x_j$; hence eqn. (2.5.17) means that $\partial \psi/\partial u_i = 0$, i.e. ψ is not an explicit function of $u_i (i = 1, 2, 3)$. We note the contrast; ψ is not an explicit function of X because the dielectric is assumed to be homogeneous in the initial state whereas ψ is not an explicit function of $\boldsymbol{u}$ because of the demands of the principles of conservation of linear momentum and energy. Consider eqn. (2.5.19); from the law of conservation of angular momentum $T_{ij} = T_{ji}$ and hence

$$\left(\frac{\partial \psi}{\partial u_{ij}} \right) = \left(\frac{\partial \psi}{\partial u_{ji}} \right) \qquad (i, j = 1, 2, 3).$$

Thus the energy density ψ must satisfy the three partial differential equations

$$\left(\frac{\partial \psi}{\partial u_{12}} \right) - \left(\frac{\partial \psi}{\partial u_{21}} \right) = 0, \quad \left(\frac{\partial \psi}{\partial u_{23}} \right) - \left(\frac{\partial \psi}{\partial u_{32}} \right) = 0,$$

$$\left(\frac{\partial \psi}{\partial u_{13}} \right) - \left(\frac{\partial \psi}{\partial u_{31}} \right) = 0.$$

By inspection, any function of the six arguments $S_{ij} = \frac{1}{2}(u_{ij}+u_{ji})$ is a solution to these equations. S_{ij} is the infinitesimal strain tensor (§ 2.2.1). Therefore to conserve both energy and angular momentum, ψ can only be a function of the twelve arguments u_{ij}, D_i through the nine arguments S_{ij}, D_i. It is sometimes convenient to use the strain components S_α as arguments in preference to S_{ij}. In the linear theory we may replace ϱ by ϱ_0 in the expression for T_{ji} given by eqn. (2.5.19). Hence on using the convention that $\psi = \psi(S_\alpha, D_i)$, we get

$$T_\alpha = \varrho_0 \left(\frac{\partial \psi}{\partial S_\alpha}\right) \qquad (\alpha = 1, 2, 3, 4, 5, 6) \qquad \text{in } V_0. \qquad (2.5.20)$$

Similarly, from eqn. (2.5.18)

$$E_i = \varrho_0 \left(\frac{\partial \psi}{\partial D_i}\right) \qquad (i = 1, 2, 3) \qquad \text{in } V_0. \qquad (2.5.21)$$

Thus at each point in the dielectric the fields T_α, E_i are determined by the fields S_α, D_i and the determining relations are generated from a single function $\varrho_0\psi$ which is an explicit function only of S_α, D_i and not of u_i, X_i or $\partial u_i/\partial X_j - \partial u_j/\partial X_i$. Equations (2.5.20) and (2.5.21) are the equations of state. From these equations we get the following identities:

$$\left(\frac{\partial T_\alpha}{\partial S_\beta}\right) = \left(\frac{\partial T_\beta}{\partial S_\alpha}\right), \quad \left(\frac{\partial T_\alpha}{\partial D_i}\right) = \left(\frac{\partial E_i}{\partial S_\alpha}\right),$$

$$\left(\frac{\partial E_i}{\partial D_j}\right) = \left(\frac{\partial E_j}{\partial D_i}\right). \qquad (2.5.22)$$

Thus not only are the fields T_α, E_i coupled to the fields S_α, D_i but the couplings, as measured by the partial derivatives appearing in eqn. (2.5.22), are interrelated.

The equation of state for $\boldsymbol{E}$ in free space is

$$E_i = \left(\frac{\partial \chi}{\partial D_i}\right)$$

with

$$\chi = \left(\sum_{i=1}^{3} D_i^2\right) \Big/ 2\varepsilon_0 .$$

2.5.4 EQUATIONS OF STATE II

It is customary to assume† that we may represent the energy density ψ by a polynomial (i.e. a truncated Taylor series) in the arguments S_α, D_i. Consider the following quadratic representation:

$$\psi = \sum_{\alpha=1}^{6}\left(\frac{\partial\psi}{\partial S_\alpha}\right)^0 S_\alpha + \sum_{i=1}^{3}\left(\frac{\partial\psi}{\partial D_i}\right)^0 D_i + \frac{1}{2}\sum_{\alpha,\,\beta}\left(\frac{\partial^2\psi}{\partial S_\alpha\,\partial S_\beta}\right)^0 S_\alpha S_\beta + \sum_{\alpha,\,i}\left(\frac{\partial^2\psi}{\partial S_\alpha\,\partial D_i}\right)^0 S_\alpha D_i + \frac{1}{2}\sum_{i,\,j}\left(\frac{\partial^2\psi}{\partial D_i\,\partial D_j}\right)^0 D_i D_j$$

where the superscript 0 indicates that the quantities enclosed in the brackets are the values of the functions for $S_\alpha = 0$, $D_i = 0$. From eqns. (2.5.20) and (2.5.21)

$$T_\alpha = \varrho_0\left(\frac{\partial\psi}{\partial S_\alpha}\right)^0 + \sum_{\beta=1}^{6}\varrho_0\left(\frac{\partial^2\psi}{\partial S_\alpha\,\partial S_\beta}\right)^0 S_\beta + \sum_{j=1}^{3}\varrho_0\left(\frac{\partial^2\psi}{\partial S_\alpha\,\partial D_j}\right)^0 D_j, \tag{2.5.23}$$

$$E_i = \varrho_0\left(\frac{\partial\psi}{\partial D_i}\right)^0 + \sum_{\beta=1}^{6}\varrho_0\left(\frac{\partial^2\psi}{\partial S_\beta\,\partial D_i}\right)^0 S_\beta + \sum_{j=1}^{3}\varrho_0\left(\frac{\partial^2\psi}{\partial D_i\partial D_j}\right) D_j, \tag{2.5.24}$$

By assumption (iii), $T_\alpha = 0$, $E_i = 0$ when $S_\alpha = 0$ and $D_i = 0$; hence

$$\left(\frac{\partial\psi}{\partial S_\alpha}\right)^0 = 0, \quad \left(\frac{\partial\psi}{\partial D_i}\right)^0 = 0.$$

Thus the leading non-vanishing terms in the expressions for T_α, E_i are linear functions of S_α, D_i and the associated energy density is a quadratic function of S_α, D_i. In a strictly linear theory, higher order terms must be rejected. The coefficients appearing in eqns. (2.5.23) and (2.5.24) characterize the electroelastic properties of the material at the point of interest in the dielectric. The following notation and

† It appears that this may not be a valid assumption close to a transition (see discussion in § 5.1.10).

nomenclature are commonly used in the literature:

$$c^D_{\alpha\beta} = \varrho_0 \left(\frac{\partial^2 \psi}{\partial S_\alpha\, \partial S_\beta}\right)^0 \quad \text{linear elastic stiffness at constant displacement.} \tag{2.5.25}$$

$$h_{i\alpha} = -\varrho_0 \left(\frac{\partial^2 \psi}{\partial S_\alpha\, \partial D_i}\right)^0 \quad \text{linear piezoelectric coefficient} \tag{2.5.26}$$

$$\beta^S_{ij} = \varrho_0 \left(\frac{\partial^2 \psi}{\partial D_i\, \partial D_j}\right)^0 \quad \text{linear dielectric permeability at constant strain.} \tag{2.5.27}$$ †

The superscripts D and S are used as a reminder that, for example, the elastic stiffness $c^D_{\alpha\beta}$ is defined, regarding the displacement components D_i as independent variables. Following from these definitions,

$$c^D_{\alpha\beta} = \varrho_0 \left(\frac{\partial^2 \psi}{\partial S_\alpha\, \partial S_\beta}\right)^0 = c^D_{\beta\alpha}$$

and

$$\beta^S_{ij} = \varrho_0 \left(\frac{\partial^2 \psi}{\partial D_i\, \partial D_j}\right)^0 = \beta^S_{ji}.$$

Thus the matrices $[c^D_{\alpha\beta}]$ and $[\beta^S_{ij}]$ are symmetric with at most twenty-one and six independent components respectively. $[h_{i\alpha}]$ has no analogous property and hence has at most eighteen independent components. This gives a total of forty-five independent coefficients appearing in the linear equations of state. For a given crystal this number may be reduced by the demands of material symmetry.‡ Dielectrics for which there is at least one non-vanishing coefficient $h_{i\alpha}$ are referred to as *piezoelectric*. That is to say, in piezoelectric materials there is linear coupling between the fields describing the mechanical behaviour and the fields describing the electrical behaviour.

With the notation introduced above the equations of state now read

$$T_\alpha = \sum_\beta c^D_{\alpha\beta} S_\beta - \sum_j h_{j\alpha} D_j, \tag{2.5.28}$$

$$E_i = -\sum_\beta h_{i\beta} S_\beta + \sum_j \beta^S_{ij} D_j, \tag{2.5.29}$$

† $c^D_{\alpha\beta}$, $h_{i\alpha}$ and β^S_{ij} transform as tensors of rank four, three and two respectively (see § 3.3.6).

‡ This point is discussed in § 3.3.7.

and $\varrho_0\psi$ has the form

$$\varrho_0\psi = \tfrac{1}{2}\sum_{\alpha,\beta} c^D_{\alpha\beta} S_\alpha S_\beta - \sum_{i,\alpha} h_{i\alpha} S_\alpha D_i + \tfrac{1}{2}\sum_{i,j} \beta^S_{ij} D_i D_j. \qquad (2.5.30)$$

These equations of state provide nine linear equations between the eighteen components T_α, S_α, E_i, D_i. We may solve these equations and express any nine of the components in terms of the remaining nine. The coefficients appearing in such a set of equations are of course determined by the $c^D_{\alpha\beta}$, $h_{i\alpha}$, β^S_{ij} appearing in eqns. (2.5.28) and (2.5.29). The following three sets of equations are used extensively in the literature:

$$T_\alpha = \sum_\beta c^E_{\alpha\beta} S_\beta - \sum_j e_{j\alpha} E_j, \qquad (2.5.31)$$

$$D_i = \sum_\beta e_{i\beta} S_\beta + \sum_j \varepsilon^S_{ij} E_j, \qquad (2.5.32)$$

$$S_\alpha = \sum_\beta s^D_{\alpha\beta} T_\beta + \sum_j g_{j\alpha} D_j, \qquad (2.5.33)$$

$$E_i = -\sum_\beta g_{i\beta} T_\beta + \sum_j \beta^T_{ij} D_j, \qquad (2.5.34)$$

$$S_\alpha = \sum_\beta s^E_{\alpha\beta} T_\beta + \sum_j d_{j\alpha} E_j, \qquad (2.5.35)$$

$$D_i = \sum_\beta d_{i\beta} T_\beta + \sum_j \varepsilon^T_{ij} E_j. \qquad (2.5.36)$$

The names and dimensions of these coefficients are given in Table 2.5.1. The arrays of coefficients $c^E_{\alpha\beta}$, $s^D_{\alpha\beta}$, $s^E_{\alpha\beta}$, ε^S_{ij}, ε^T_{ij}, β^T_{ij} are invariant under the interchange of indices α, β or i, j, whichever is appropriate. To prove this we use the technique of Legendre transformations. For example, consider the quantity $\varrho_0\psi' = \varrho_0\psi - \sum_\alpha T_\alpha S_\alpha$. From eqns. (2.5.20) and (2.5.21) the differential $\varrho_0\, d\psi$ satisfies the relation

$$\varrho_0\, d\psi = \sum_\alpha T_\alpha\, dS_\alpha + \sum_i E_i\, dD_i.$$

Hence

$$\varrho_0\, d\psi' = -\sum_\alpha S_\alpha\, dT_\alpha + \sum_\beta E_i\, dD_i.$$

In particular,

$$\varrho_0(\partial\psi'/\partial T)_D = -S_\alpha,$$

and therefore

$$(\partial S_\alpha/\partial T_\beta)_D = (\partial S_\beta/\partial T_\alpha)_D.$$

TABLE 2.5.1.

Nomenclature, units and notation for the coefficients appearing in the linear representation of the elastic and dielectric equations of state.

Quantity	Units (MKS)	Symbol
Linear elastic compliance (*E* constant)	m^2/N	$s^E_{\alpha\beta}$
Linear elastic compliance (*D* constant)	m^2/N	$s^D_{\alpha\beta}$
Linear elastic stiffness (*E* constant)	N/m^2	$c^E_{\alpha\beta}$
Linear elastic stiffness (*D* constant)	N/m^2	$c^D_{\alpha\beta}$
Linear piezoelectric strain coefficient	C/N	$d_{i\alpha}$
Linear piezoelectric strain coefficient	m^2/C	$g_{i\alpha}$
Linear piezoelectric stress coefficient	C/m^2	$e_{i\alpha}$
Linear piezoelectric stress coefficient	N/C	$h_{i\alpha}$
Linear dielectric permeability (*T* constant)	m/F	β^T_{ij}
Linear dielectric permeability (*S* constant)	m/F	β^S_{ij}
Linear dielectric permittivity (*T* constant)	F/m	ε^T_{ij}
Linear dielectric permittivity (*S* constant)	F/m	ε^S_{ij}

Substituting for S_α and S_β from eqn. (2.5.33) we obtain $s^D_{\alpha\beta} = s^D_{\beta\alpha}$. It is clear that a similar argument applies in the case of each of the other coefficients listed above.

It is easy to obtain explicit relations between the coefficients of one set and those of another. As an illustration consider the relationship between $c^D_{\alpha\beta}$, $h_{j\alpha}$, β^S_{ij} and $c^E_{\alpha\beta}$, $e_{j\alpha}$, ε^S_{ij}. We denote the elements of the matrix reciprocal to the matrix $[\beta^S_{ij}]$ by $(\beta^S_{ij})^{-1}$. Thus

$$\sum_{j=1}^{3} (\beta^S_{ij})^{-1}\beta^S_{jk} = \delta_{ik} \qquad (i, k = 1, 2, 3). \tag{2.5.37}$$

Multiply both sides of eqn. (2.5.29) by $(\beta^S_{ki})^{-1}$ and sum over i,

$$\sum_{i=1}^{3} (\beta^S_{ki})^{-1} E_i = -\sum_{\beta,\, i} (\beta^S_{ki})^{-1} h_{i\beta} S_\beta + \sum_{i,\, j} (\beta^S_{ki})^{-1}\beta^S_{ij} D_j.$$

Using eqn. (2.5.37) and rearranging we get

$$D_i = \sum_{\beta,\, j} (\beta^S_{ij})^{-1} h_{j\beta} S_\beta + \sum_{j=1}^{3} (\beta^S_{ij})^{-1} E_j. \tag{2.5.38}$$

Substitute for D_i from eqn. (2.5.38) in eqn. (2.5.28),

$$T_\alpha = \sum_\beta \Big[c^D_{\alpha\beta} - \sum_{j,\,k} h_{j\alpha}(\beta^S_{jk})^{-1} h_{k\beta}\Big] S_\beta - \sum_{j,\,k} h_{j\alpha}(\beta^S_{jk})^{-1} E_k. \quad (2.5.39)$$

Comparing eqns. (2.5.31) and (2.5.32) with eqns. (2.5.38) and (2.5.39), we get the following identities:

$$c^E_{\alpha\beta} = c^D_{\alpha\beta} - \sum_{j,\,k} h_{j\alpha}(\beta^S_{jk})^{-1} h_{k\beta}, \quad (2.5.40)$$

$$e_{j\alpha} = \sum_k h_{k\alpha}(\beta^S_{jk})^{-1}, \quad (2.5.41)$$

$$\varepsilon^S_{ij} = (\beta^S_{ij})^{-1}. \quad (2.5.42)$$

These are the identities we seek. Any other relations which may be required can be obtained in a similar fashion.

The various coefficients introduced above have simple, obvious interpretations. Thus, for example, if at some point in the dielectric, stress vanishes then the linear relation between D_i and the E_j is $D_i = \sum_j \varepsilon^T_{ij} E_j$ [see eqn. (2.5.36)]. Whereas if the dielectric is free of strain at that point, the relation is $D_i = \sum_j \varepsilon^S_{ij} E_j$. Clearly if the crystal is piezoelectric, $\varepsilon^T_{ij} \neq \varepsilon^S_{ij}$. We say that $\varepsilon^T_{ij}(\varepsilon^S_{ij})$ is the dielectric coefficient of the *free* (*clamped*) dielectric. The following alternative notation is sometimes used:

$$\varepsilon^T_{ij} \to \varepsilon'_{ij} \quad \text{and} \quad \varepsilon^S_{ij} \to \varepsilon''_{ij}.$$

2.5.5 Quasi-linear Theory

Non-linear effects have been observed in the electroelastic behaviour of ferroelectric crystals; the non-linear theory commonly used to analyse these effects is obtained by (a) adopting unchanged the field equations and boundary conditions of the linear theory, and (b) introducing cubic and higher order terms (i.e. "non-linear" terms) in the polynomial representation for the energy density. We shall refer to this theory as the *quasi-linear theory*. For the non-linear effects so far studied experimentally it has never proved necessary to introduce any "non-linear" terms in addition to the ones appearing in the following repre-

sentation for the energy density:

$$\varrho_0\psi = \tfrac{1}{2}\sum_{\alpha,\beta} c^D_{\alpha\beta} S_\alpha S_\beta - \sum_{i,\alpha} h_{i\alpha} S_\alpha D_i + \tfrac{1}{2}\sum_{i,j} \beta^S_{ij} D_i D_j$$
$$+ \sum_{i,j,\alpha} q_{ij\alpha} S_\alpha D_i D_j + \tfrac{1}{3}\sum_{i,j,k} \eta^S_{ijk} D_i D_j D_k + \tfrac{1}{4}\sum_{i,j,k,l} \xi^S_{ijkl} D_i D_j D_k D_l$$
$$+ \tfrac{1}{5}\sum_{i,j,k,l,m} \mu^S_{ijklm} D_i D_j D_k D_l D_m + \tfrac{1}{6}\sum_{i,j,k,l,m,n} \zeta^S_{ijklmn} D_i D_j D_k D_l D_m D_n. \tag{2.5.43}$$

The equations of state generated by this representation are:

$$T_\alpha = \varrho_0\left(\frac{\partial\psi}{\partial S_\alpha}\right)_D = \sum_\beta c^D_{\alpha\beta} S_\beta - \sum_j h_{j\alpha} D_j + \sum_{i,j} q_{ij\alpha} D_i D_j, \tag{2.5.44}$$

$$E_i = \varrho_0\left(\frac{\partial\psi}{\partial D_i}\right)_S = -\sum_\beta h_{i\beta} S_\beta + \sum_j \beta^S_{ij} D_j + 2\sum_{j,\beta} q_{ij\beta} S_\beta D_j + \sum_{j,k} \eta^S_{ijk} D_j D_k$$
$$+ \sum_{j,k,l} \xi^S_{ijkl} D_j D_k D_l + \sum_{j,k,l,m} \mu^S_{ijklm} D_j D_k D_l D_m$$
$$+ \sum_{j,k,l,m,n} \zeta^S_{ijklmn} D_j D_k D_l D_m D_n. \tag{2.5.45}$$

The following symmetries under subscript interchanges are easily deduced from these equations:

$q_{ij\alpha}$	is invariant under the interchange	$i \rightleftharpoons j$
η^S_{ijk}	is invariant under the interchange	$i \rightleftharpoons j \rightleftharpoons k$
ξ^S_{ijkl}	is invariant under the interchange	$i \rightleftharpoons j \rightleftharpoons k \rightleftharpoons l$
μ^S_{ijklm}	is invariant under the interchange	$i \rightleftharpoons j \rightleftharpoons k \rightleftharpoons l \rightleftharpoons m$
ζ^S_{ijklmn}	is invariant under the interchange	$i \rightleftharpoons j \rightleftharpoons k \rightleftharpoons l \rightleftharpoons m \rightleftharpoons n$

Hence the arrays of coefficients q_{ij}, η^S_{ijk}, ξ^S_{ijkl}, μ^S_{ijklm}, ζ^S_{ijklmn} have at most 36, 10, 12, 21 and 28 independent components respectively. In a given material these numbers may be further reduced by the demands of material symmetry (§ 3.3.7).

The $q_{ij\alpha}$ describe the lowest order non-linear coupling between the electric and elastic variables. This coupling is referred to as *electrostriction*† and the $q_{ij\alpha}$ as *electrostrictive coefficients*. The so-called piezo-

† Electrostriction is sometimes used to describe the non-linear coupling to all orders.

electric coupling (see § 3.3.2) between the electric and elastic variables may be described by the derivatives $\partial T_\alpha/\partial D_i$. For the representation given above, we find

$$\left(\frac{\partial T_\alpha}{\partial D_i}\right)_S \equiv \left(\frac{\partial E_i}{\partial S_\alpha}\right)_D = -h_{i\alpha}+2\sum_{j=1}^{3} q_{ij\alpha}D_j\,.$$

This consists of a linear term, the $-h_{i\alpha}$ and a non-linear electrostrictive term $2\sum_j q_{ij\alpha}D_j$. Electrostriction has been observed only in non-piezoelectric crystals ($h_{i\alpha} \equiv 0$) presumably because of the masking effect of the linear piezoelectric coupling in the piezoelectric crystals.

The coefficients η^S_{ijk}, ξ^S_{ijkl}, μ^S_{ijklm}, ζ^S_{ijklmn} describe the non-linear dielectric equations of state of the clamped crystal. If we choose to consider the components T_α, D_i as the independent variables, we write the equations of state in the form

$$S_\alpha = \sum_\beta s^D_{\alpha\beta}T_\beta+\sum_j g_{j\alpha}D_j+\sum_{j,\,k} Q_{jk\alpha}D_jD_k\,, \tag{2.5.46}$$

$$\left.\begin{aligned} E_i = &-\sum_\beta g_{i\beta}T_\beta+\sum_j \beta^T_{ij}D_j-2\sum_{j,\,\beta} Q_{ij\beta}D_jT_\beta+\sum_{j,\,k}\eta^T_{ijk}D_jD_k \\ &+\sum_{j,\,k,\,l}\xi^T_{ijkl}D_jD_kD_l+\sum_{j,\,k,\,l,\,m}\mu^T_{ijklm}D_jD_kD_lD_m \\ &+\sum_{j,\,k,\,l,\,m,\,n}\zeta^T_{ijklmn}D_jD_kD_lD_mD_n \end{aligned}\right\} \tag{2.5.47}$$

To obtain the relationships between the non-linear coefficients appearing in eqns. (2.5.44) and (2.5.45) and those in eqns. (2.5.46) and (2.5.47) we first introduce the matrix $[(c^D_{\alpha\beta})^{-1}]$ which is reciprocal to the matrix $[c^D_{\alpha\beta}]$, i.e.

$$\sum_\beta (c^D_{\alpha\beta})^{-1}c^D_{\beta\gamma} = \delta_{\alpha\gamma}\,.$$

We may solve eqn. (2.5.44) for the S_α in terms of the T_α, D_i. Thus

$$S_\alpha = \sum_\beta (c^D_{\alpha\beta})^{-1}T_\beta+\sum_{\beta,\,j}(c^D_{\alpha\beta})^{-1}h_{j\beta}D_j-\sum_{\beta,\,i,\,j}(c^D_{\alpha\beta})^{-1}q_{ij\beta}D_iD_j\,. \tag{2.5.48}$$

Inserting this expression in eqn. (2.5.45) we get

$$\left.\begin{aligned} E_i = & -\sum_{\gamma,\beta} h_{i\gamma}(c^D_{\gamma\beta})^{-1}T_\beta + \sum_j [\beta^S_{ij} - \sum_{\gamma,\beta} h_{i\gamma}(c^D_{\gamma\beta})^{-1}h_{j\beta}]D_j \\ & +2\sum_{\gamma,\beta,j} q_{ij\gamma}(c^D_{\gamma\beta})^{-1}T_\beta D_j + \sum_{j,k}\Big\{\eta^S_{ijk} + \sum_{\gamma,\beta}[h_{i\gamma}(c^D_{\gamma\beta})^{-1}q_{jk\beta} \\ & +2q_{ij\gamma}(c^D_{\gamma\beta})^{-1}h_{k\beta}]\Big\}D_j D_k + \sum_{j,k,l}\Big[\xi^S_{ijkl} - 2\sum_{\gamma,\beta} q_{ij\gamma}(c^D_{\gamma\beta})^{-1}q_{kl\beta}\Big]\times \\ & \times D_j D_k D_l + \sum_{j,k,l,m} \mu^S_{ijklm} D_j D_k D_l D_m \\ & + \sum_{j,k,l,m,n} \zeta^S_{ijklmn} D_j D_k D_l D_m D_n. \end{aligned}\right\} \quad (2.5.49)$$

Comparing eqns. (2.5.48) and (2.5.49) with eqns. (2.5.46) and (2.5.47) we find that

$$s^D_{\alpha\beta} = (c^D_{\alpha\beta})^{-1}, \quad (2.5.50)$$

$$g_{i\beta} = \sum_\gamma h_{i\gamma}(c^D_{\gamma\beta})^{-1}, \quad (2.5.51)$$

$$\beta^T_{ij} = \beta^S_{ij} - \sum_{\gamma,\beta} h_{i\gamma}(c^D_{\gamma\beta})^{-1}h_{j\beta}, \quad (2.5.52)$$

$$Q_{ij\beta} = -\sum_\gamma q_{ij\gamma}(c^D_{\gamma\beta})^{-1}, \quad (2.5.53)$$

$$\eta^T_{ijk} = \eta^S_{ijk} + \sum_{\gamma,\beta}[h_{i\gamma}(c_{\gamma\beta})^{-1}q_{jk\beta} + 2h_{k\beta}(c^D_{\gamma\beta})^{-1}q_{ij\gamma}], \quad (2.5.54)$$

$$\xi^T_{ijkl} = \xi^S_{ijkl} - 2\sum_{\gamma,\beta} q_{ij\gamma}(c^D_{\gamma\beta})^{-1}q_{kl\beta}, \quad (2.5.55)$$

$$\mu^T_{ijklm} = \mu^S_{ijklm}, \quad (2.5.56)$$

$$\zeta^T_{ijklmn} = \zeta^S_{ijklmn}. \quad (2.5.57)$$

The first three equations provide relationships between the sets of linear coefficients and the remainder the required relationships between the non-linear coefficients.

We make no attempt to invert these non-linear equations of state to obtain the D_i as functions of the E_i. In the case of the ferroelectric crystals in the ferroelectric phase the D_i are not single valued functions of E_i and as a result the inverted forms are formidable, complicated expressions.

2.5.6 Summary

The field equations, boundary conditions and equations of state characteristic of the standard linear theory of the elastic dielectric are

$$\varrho_0 \ddot{u}_i = \sum_{j=1}^{3} \frac{\partial T_{ji}}{\partial X_j} \quad \text{in } V_0, \tag{2.5.58}$$

$$\nabla_X \times \boldsymbol{E} = 0, \quad \nabla_X \cdot \boldsymbol{D} = 0 \quad \text{in } V_0 \text{ and } A_0, \tag{2.5.59}$$

$$t_i^0 = \sum_j T_{ji} n_j^0, \quad \boldsymbol{n}^0 \times (\boldsymbol{E}^V - \boldsymbol{E}^A) = 0, \quad \boldsymbol{n}^0 \cdot (\boldsymbol{D}^V - \boldsymbol{D}^A) = 0 \text{ on } V_0, \tag{2.5.60}$$

$$\boldsymbol{n}^0 \times \boldsymbol{E} = 0, \quad \boldsymbol{n}^0 \cdot \boldsymbol{D} = \omega \quad \text{on} \quad C_0, \tag{2.5.61}$$

$$T_\alpha = \varrho_0 \left(\frac{\partial \psi}{\partial S_\alpha}\right) = \sum_\beta c_{\alpha\beta}^D S_\beta - \sum_j h_{j\alpha} D_j \tag{2.5.62}$$

in V_0 or a variation thereof,

$$E_i = \varrho_0 \left(\frac{\partial \psi}{\partial D_i}\right) = -\sum_\beta h_{i\beta} S_\beta + \sum_j \beta_{ij}^S D_j \tag{2.5.63}$$

$$E_i = D_i/\varepsilon_0 \quad \text{in } A_0, \tag{2.5.64}$$

where

$$S_\beta = \partial u_i/\partial X_j + \partial u_j/\partial X_i \quad (i \neq j)$$
$$= \partial u_i/\partial X_j \quad (i = j).$$

The strains S_β are measured relative to the configuration of the dielectric in the initial state in which $T_\alpha = 0$, $D_i = 0$ and $E_i = 0$. The $c_{\alpha\beta}^D$, β_{ij}^S satisfy the equations $c_{\alpha\beta}^D = c_{\beta\alpha}^D$ and $\beta_{ij}^S = \beta_{ji}^S$. This theory has to date provided a consistent description of the linear non-dissipative behaviour of elastic dielectrics in all applications. There have been suggestions that there are fundamental inconsistencies in the theory described above, but these criticisms have proved groundless (see, for example, the discussion by Lax, 1963).

The field equations and boundary conditions of the quasilinear theory are those described by eqns. (2.5.58) to (2.5.61) and the equations of state are of the form

$$T_\alpha = \sum_\beta c_{\alpha\beta}^D S_\beta - \sum_j h_{j\alpha} D_j + \sum_{j,k} q_{jk\alpha} D_j D_k, \tag{2.5.65}$$

$$\left.\begin{aligned} E_i = & -\sum_{\beta} h_{i\beta} S_\beta + \sum_{j} \beta^S_{ij} D_j + 2\sum_{j,\beta} q_{ij\beta} S_\beta D_j + \sum_{j,k} \eta^S_{ijk} D_j D_k \\ & + \sum_{j,k,l} \xi^S_{ijkl} D_j D_k D_l + \sum_{j,k,l,m} \mu^S_{ijklm} D_j D_k D_l D_m \\ & + \sum_{j,k,l,m,n} \zeta^S_{ijklmn} D_j D_k D_l D_m D_n . \end{aligned}\right\} \quad (2.5.66)$$

The body of experimental data in support of the quasi-linear theory is considerably more meagre and is subject to considerably larger experimental uncertainty than that for the linear theory. In Chapter 4 we present results for the case of barium titanate which suggest that the range of validity of the quasi-linear theory is considerably less than has hitherto been supposed.

Modifications are required in both these theories to include thermal and dissipative effects; we discuss the former in Chapter 3 and refer the reader to the article by Truesdell and Toupin (1960) for a discussion of the latter.

Bibliography

A general reference for the topics treated in this and the following chapter is the encyclopedic article by Truesdell and Toupin (1960) on the "Classical theory of fields". This promises to be the standard reference for some years to come.

A very clear introductory account of theoretical elasticity is given by Pearson (1959). Landau and Lifshitz (1960) and Fuller-Brown, Jr. (1956) deal with the macroscopic description of dielectrics.

The definition and measurement of the linear coefficients are discussed by Cady (1946) and Mason (1950). The latter reference gives an extensive discussion of the application of the Christoffel wave equation, based on eqn. (2.5.58), to the experimental determination of the linear coefficients.

CHAPTER 3

Thermodynamics of the elastic dielectric

IN THIS chapter we develop the phenomenological theory of the static thermal, elastic and dielectric properties of a homogeneous material. The elastic and dielectric properties will be assumed to be fully described by the fields introduced in the linear theory of Chapter 2. For the most part the discussion is restricted to the case in which these fields are arbitrarily small. The theory described below has been used extensively to analyse the properties of the ferroelectrics. The results fall into two categories: (a) thermodynamic identities, §§ 3.1, 3.3, 3.4, and (b) criteria for stability, § 3.2.

3.1 Thermodynamic Identities

3.1.1 THE FIRST LAW

For definiteness we consider the isolated system consisting of a homogeneous dielectric and a conductor. Let us assume that the thermal, elastic and dielectric behaviour of the system is fully described by six fields $\boldsymbol{u}$, $\boldsymbol{t}$, $\boldsymbol{D}$, $\boldsymbol{E}$, σ' and θ. The first four are defined in Chapter 2 and σ' and θ are, respectively, the entropy density (entropy per unit mass) and the absolute temperature. The fields $\boldsymbol{D}$, $\boldsymbol{E}$ are defined at all points in space whereas $\boldsymbol{u}$, $\boldsymbol{t}$, σ', θ are defined only at points within the dielectric. We shall restrict the discussion to the states (a) which are in thermodynamic equilibrium, (b) in which the fields, $\boldsymbol{t}$, $\boldsymbol{D}$, $\boldsymbol{E}$, σ', θ

and the displacement gradients $\partial u_i/\partial X_j$ are uniform throughout the dielectric, and (c) in which $\boldsymbol{u}$, $\boldsymbol{t}$, $\boldsymbol{D}$ and $\boldsymbol{E}$ are small. By adopting restriction (c) we are assuming that the system possesses at least one state in which $\boldsymbol{t}$, $\boldsymbol{D}$ and $\boldsymbol{E}$ vanish simultaneously. We use this state as the initial state, $\boldsymbol{u} \equiv 0$, and denote its temperature by θ_i. This assumption does not hold for all materials but it is apparently valid for the ferroelectrics.

Let the energy stored in some volume v of the dielectric be of the form $\int_v \psi(S_\alpha, D_i, \sigma')\varrho \, dv$, where ψ is the energy density (energy per unit mass), S_α the infinitesimal strain and ϱ the mass density. Suppose that the system is transformed quasi-statically from a state described by the field values $\boldsymbol{u}$, $\boldsymbol{t}$, $\boldsymbol{D}$, $\boldsymbol{E}$, σ', θ to another state described by the field values, $\boldsymbol{u}+\delta\boldsymbol{u}$, $\boldsymbol{t}+\delta\boldsymbol{t}$, $\boldsymbol{D}+\delta\boldsymbol{D}$, $\boldsymbol{E}+\delta\boldsymbol{E}$, $\sigma'+\delta\sigma'$, $\theta+\delta\theta$, where $\delta\boldsymbol{u}$, etc., are infinitesimals. If δW denotes the electric and mechanical work and δQ the heat supplied to this volume during the transformation, then by the law of conservation of energy

$$\delta \int_v \psi(S_\alpha, D_i, \sigma')\varrho \, dv = \delta W + \delta Q$$

or

$$\int_v [\delta\psi(S_\alpha, D_i, \sigma')]\varrho \, dv = \delta W + \delta Q.$$

Since the displacement gradients are small we may use eqn. (2.5.13) and write

$$\delta W = \int_v \left\{ \sum_{i,j} \left[\left(\frac{\partial T_{ji}}{\partial x_j} \right) \delta u_i + T_{ji}\, \delta \left(\frac{\partial u_i}{\partial X_j} \right) \right] + \sum_{i=1}^{3} E_i \, \delta D_i \right\} dv.\dagger$$

From the assumptions listed above the stress tensor field is both uniform and symmetric, hence

$$\delta W = \int_v \left\{ \sum_{\alpha=1}^{6} T_\alpha \delta S_\alpha + \sum_{i=1}^{3} E_i \, \delta D_i \right\} dv. \qquad (3.1.1)$$

† We recall that in deriving this expression we assumed that the work done by the Maxwell stress is negligible (see discussion in § 4.1.4).

By definition

$$\delta Q = \int_v \theta\, \delta\sigma' \varrho\, dv. \tag{3.1.2}$$

Expanding $\delta\psi$ to terms linear in δS_α, δD_i, $\delta\sigma'$ and substituting the expressions for δW, δQ given in eqns. (3.1.1) and (3.1.2) we may write the condition for conservation of energy in the form

$$\int_v \left[\sum_{\alpha=1}^{6}\left(\frac{\partial\psi}{\partial S_\alpha}\right)_{D,\,\sigma'}\delta S_\alpha + \sum_{i=1}^{3}\left(\frac{\partial\psi}{\partial D_i}\right)_{\sigma',\,S}\delta D_i + \left(\frac{\partial\psi}{\partial\sigma'}\right)_{S,\,D}\delta\sigma'\right]\varrho\, dv$$

$$= \int_v \left[\sum_{\alpha=1}^{6} T_\alpha\,\delta S_\alpha + \sum_{i=1}^{3} E_i\,\delta D_i + \theta\varrho\,\delta\sigma'\right] dv.$$

There are no convective terms of the form $(\partial\psi/\partial D_i)(\partial D_i/\partial x_j)\,\delta x_j$ or $(\partial\psi/\partial\sigma')(\partial\sigma'/\partial x_i)\,\delta x_i$ on the left side because the fields $\boldsymbol{D}$, σ' are assumed to be uniform within the dielectric. Treating the quantities S_α, D_i, σ' as independent variables, we deduce that

$$\varrho\left(\frac{\partial\psi}{\partial S_\alpha}\right)_{D,\,\sigma'} = T_\alpha, \quad \varrho\left(\frac{\partial\psi}{\partial D_i}\right)_{S,\,\sigma'} = E_i, \quad \left(\frac{\partial\psi}{\partial\sigma'}\right)_{S,\,D} = \theta \tag{3.1.3}$$

at each point in v. The subscripts on the partial differentials serve as a reminder that S_α, D_i, σ' are being treated as independent variables. It rarely proves necessary to retain the component subscripts on these subscripts. Assuming that the strains are small† compared with unity, we replace ϱ by ϱ_0 the mass density in the initial state; eqns. (3.1.3) now take the form

$$\left(\frac{\partial U}{\partial S_\alpha}\right)_{D,\,\sigma'} = T_\alpha, \quad \left(\frac{\partial U}{\partial D_i}\right)_{S,\,\sigma'} = E_i, \quad \left(\frac{\partial U}{\partial\sigma}\right)_{D,\,S} = \theta, \tag{3.1.4}$$

where $U = \varrho_0\psi$ is the energy (in the state of interest) per unit volume (in the initial state) and $\sigma = \varrho_0\sigma'$ is the entropy (in the state of interest) per unit volume (in the initial state). These three equations describe the *elastic*, *dielectric* and *calorimetric* equations of state respectively.

† Typically $|S_\alpha| < 10^{-3}$.

These equations are strictly valid only for small strains and displacement fields.†

Let us consider the differential dU. With $U = U(S_\alpha, D_i, \sigma)$

$$dU = \sum_\alpha \left(\frac{\partial U}{\partial S_\alpha}\right)_{D,\sigma} dS_\alpha + \sum_i \left(\frac{\partial U}{\partial D_i}\right)_{\sigma, S} dD_i + \left(\frac{\partial U}{\partial \sigma}\right)_{S, D} d\sigma. \quad (3.1.5)$$

Combining eqns. (3.1.4) and (3.1.5) we have

$$dU = \sum_\alpha T_\alpha \, dS_\alpha + \sum_i E_i \, dD_i + \theta \, d\sigma. \quad (3.1.6)$$

This is the form the first law of thermodynamics takes in the case of the elastic dielectric. The first two sums represent the density of mechanical and electric work respectively and the last term the density of heat input [cf. eqns. (3.1.1) and (3.1.2)]. (T_α, S_α), (E_i, D_i) and (θ, σ) are conjugate pairs of variables in the usual thermodynamic sense. According to eqn. (3.1.6), states which differ only by a strain-free infinitesimal deformation, i.e. $dS_\alpha \equiv 0$ $(\alpha = 1, 2, 3, 4, 5, 6)$, have the same energy. The transformations between such states are of little interest and hence we consider only thermodynamics of transformations involving changes in at least one pair of the components of the following fields: S_α, T_α, D_i, E_i, σ, θ. Under these circumstances these twenty components serve as thermodynamic variables or *thermodynamic coordinates* describing the thermodynamic states of the system.

3.1.2 Thermodynamic Potentials

In particular applications it is not always convenient to use S_α, D_i and σ as the independent variables. A transformation of variables is effected most easily by means of a Legendre transformation. The scalar function associated with the transformation is called a *thermodynamic potential.* The following four potentials prove especially useful:

$$J = U - \sum_{\alpha=1}^{6} T_\alpha S_\alpha, \qquad F = U - \theta\sigma,$$

$$H = U - \theta\sigma - \sum_{\alpha=1}^{6} T_\alpha S_\alpha, \qquad G = U - \theta\sigma - \sum_{\alpha=1}^{6} T_\alpha S_\alpha - \sum_{i=1}^{3} D_i E_i.$$

† The assumption of small strains and small electric displacement fields is used explicitly in writing down the expression for electric and mechanical work, eqn. (3.1.1), and is used again in passing from eqn. (3.1.3) to eqn. (3.1.4).

We shall refer to F, H and G as the *free energy*, *Helmholtz energy* and *Gibbs function* respectively; J has not been named.

The corresponding differentials are [cf. eqn. (3.1.6)]:

$$dJ = \theta\, d\sigma - \sum_{\alpha=1}^{6} S_\alpha\, dT_\alpha + \sum_{i=1}^{3} E_i\, dD_i, \tag{3.1.7}$$

$$dF = -\sigma\, d\theta + \sum_{\alpha=1}^{6} T_\alpha\, dS_\alpha + \sum_{i=1}^{3} E_i\, dD_i, \tag{3.1.8}$$

$$dH = -\sigma\, d\theta - \sum_{\alpha=1}^{6} S_\alpha\, dT_\alpha + \sum_{i=1}^{3} E_i\, dD_i, \tag{3.1.9}$$

$$dG = -\sigma\, d\theta - \sum_{\alpha=1}^{6} S_\alpha\, dT_\alpha - \sum_{i=1}^{3} D_i\, dE_i. \tag{3.1.10}$$

Thus, for example, given $H = H(\theta, T_\alpha, D_i)$ we may generate the following equations of state:

$$\sigma = -\left(\frac{\partial H}{\partial \theta}\right)_{T,\,D}, \quad S_\alpha = -\left(\frac{\partial H}{\partial T_\alpha}\right)_{D,\,\theta}, \quad E_i = \left(\frac{\partial H}{\partial D_i}\right)_{\theta,\,T}. \tag{3.1.11}$$

Each thermodynamic potential contains a complete description of the thermodynamic properties of the homogeneous dielectric of interest. The potentials also possess extremal properties. This point is discussed in the following section.

From eqn. (3.1.11) we generate the following *Maxwell relations*:

$$\left(\frac{\partial \sigma}{\partial T_\alpha}\right)_{\theta,\,D} = +\left(\frac{\partial S_\alpha}{\partial \theta}\right)_{D,\,T} \qquad (\alpha = 1, 2, 3, 4, 5, 6), \tag{3.1.12}$$

$$\left(\frac{\partial \sigma}{\partial D_i}\right)_{\theta,\,T} = -\left(\frac{\partial E_i}{\partial \theta}\right)_{D,\,T} \qquad (i = 1, 2, 3), \tag{3.1.13}$$

$$\left(\frac{\partial S_\alpha}{\partial D_i}\right)_{\theta,\,T} = -\left(\frac{\partial E_i}{\partial T_\alpha}\right)_{D,\,T} \qquad (\alpha = 1, 2, 3, 4, 5, 6;\ i = 1, 2, 3). \tag{3.1.14}$$

For each thermodynamic potential we may, in this fashion, deduce twenty-seven Maxwell relations.

3.1.3 Thermodynamic Identities I

The Maxwell relations introduced above are examples of thermodynamic identities. Such identities are useful in practice either as a means of checking the internal consistency of two sets of measurements or deducing the properties of the system in a region which for technical reasons may be experimentally inaccessible.† In this subsection we derive a number of identities which are used frequently in analysing the properties of ferroelectric crystals.

The following two equations are standard identities satisfied by partial derivatives:

$$\left(\frac{\partial S_\alpha}{\partial T_\beta}\right)_{E,\,\sigma} = \left(\frac{\partial S_\alpha}{\partial T_\beta}\right)_{E,\,\theta} + \left(\frac{\partial S_\alpha}{\partial \theta}\right)_{T,\,E}\left(\frac{\partial \theta}{\partial T_\beta}\right)_{E,\,\sigma}, \tag{3.1.15}$$

$$\left(\frac{\partial \theta}{\partial T_\alpha}\right)_{E,\,\sigma} = -\left(\frac{\partial \sigma}{\partial T_\alpha}\right)_{E,\,\theta} \Bigg/ \left(\frac{\partial \sigma}{\partial \theta}\right)_{E,\,T}. \tag{3.1.16}$$

Eliminating $(\partial\theta/\partial T_\alpha)_{E,\,\sigma}$ we get

$$\left(\frac{\partial S_\alpha}{\partial T_\beta}\right)_{E,\,\sigma} = \left(\frac{\partial S_\alpha}{\partial T_\beta}\right)_{E,\,\theta} - \left(\frac{\partial S_\alpha}{\partial \theta}\right)_{T,\,E}\left(\frac{\partial \sigma}{\partial T_\beta}\right)_{E,\,\theta} \Bigg/ \left(\frac{\partial \sigma}{\partial \theta}\right)_{E,\,T} \tag{3.1.17}$$

Now by definition the specific heat at constant electric field and constant stress is

$$C_{E,\,T} = \theta\left(\frac{\partial \sigma'}{\partial \theta}\right)_{E,\,T}$$

or, since $\sigma' = \sigma/\varrho_0$,

$$C_{E,\,T} = \frac{\theta}{\varrho_0}\left(\frac{\partial \sigma}{\partial \theta}\right)_{E,\,T}. \tag{3.1.18}$$

Combining eqns. (3.1.12), (3.1.17) and (3.1.18) we obtain the final result

$$\left(\frac{\partial S_\alpha}{\partial T_\beta}\right)_{E,\,\sigma} = \left(\frac{\partial S_\alpha}{\partial T_\beta}\right)_{E,\,\theta} - \theta\left(\frac{\partial S_\alpha}{\partial \theta}\right)_{T,\,E}\left(\frac{\partial S_\beta}{\partial \theta}\right)_{T,\,E} \Bigg/ \varrho_0 C_{E,\,T}. \tag{3.1.19}$$

† An example of the latter situation occurs if one wishes to determine the properties of the dielectric held at fixed strain. In many cases the crystal shatters when subjected to this constraint.

This identity provides a relation between the adiabatic elastic behaviour (σ = const.) and the isothermal elastic behaviour (θ = const.) in terms of θ, the coefficients of expansion $(\partial S_\alpha/\partial\theta)_{E,T}$ (see § 3.3.4), the specific heat $C_{E,T}$ and the density of the dielectric *in the initial state.* Normally adiabatic properties are observed in dynamic measurements while isothermal properties are observed in static measurements, and hence eqn. (3.1.19) provides a check on the internal consistency of elastic measurements carried out under these two different experimental conditions. Similar mathematical procedures yield the following identities:

$$\left(\frac{\partial S_\alpha}{\partial E_i}\right)_{T,\sigma} = \left(\frac{\partial S_\alpha}{\partial E_i}\right)_{T,\theta} - \theta\left(\frac{\partial S_\alpha}{\partial\theta}\right)_{T,E}\left(\frac{\partial D_i}{\partial\theta}\right)_{T,E}\Big/\varrho_0 C_{E,T}, \tag{3.1.20}$$

$$\left(\frac{\partial D_i}{\partial E_j}\right)_{T,\sigma} = \left(\frac{\partial D_i}{\partial E_j}\right)_{T,\theta} - \theta\left(\frac{\partial D_i}{\partial\theta}\right)_{T,E}\left(\frac{\partial D_j}{\partial\theta}\right)_{T,E}\Big/\varrho_0 C_{E,T}, \tag{3.1.21}$$

$$\left(\frac{\partial E_i}{\partial D_j}\right)_{\theta,S} = \left(\frac{\partial E_i}{\partial D_j}\right)_{\theta,T} + \sum_{\alpha=1}^{6}\left(\frac{\partial E_i}{\partial T_\alpha}\right)_{\theta,D}\left(\frac{\partial T_\alpha}{\partial D_j}\right)_{\theta,S}, \tag{3.1.22}$$

$$\left(\frac{\partial\sigma}{\partial\theta}\right)_{D,T} = \left(\frac{\partial\sigma}{\partial\theta}\right)_{E,T} + \sum_{i=1}^{3}\left(\frac{\partial D_i}{\partial\theta}\right)_{T,E}\left(\frac{\partial E_i}{\partial\theta}\right)_{D,T}, \tag{3.1.23}$$

$$\left(\frac{\partial\sigma}{\partial\theta}\right)_{D,T} = \left(\frac{\partial\sigma}{\partial\theta}\right)_{D,S} - \sum_{\alpha=1}^{6}\left(\frac{\partial T_\alpha}{\partial\theta}\right)_{D,S}\left(\frac{\partial S_\alpha}{\partial\theta}\right)_{D,T}. \tag{3.1.24}$$

From eqns. (3.1.23) and (3.1.24) we immediately deduce that

$$C_{E,T} - C_{D,T} = -\frac{\theta}{\varrho_0}\sum_{i=1}^{3}\left(\frac{\partial D_i}{\partial\theta}\right)_{T,E}\left(\frac{\partial E_i}{\partial\theta}\right)_{D,T} \tag{3.1.25}$$

and

$$C_{D,S} - C_{D,T} = \frac{\theta}{\varrho_0}\sum_{\alpha=1}^{6}\left(\frac{\partial T_\alpha}{\partial\theta}\right)_{D,S}\left(\frac{\partial S_\alpha}{\partial\theta}\right)_{D,T}. \tag{3.1.26}$$

Of the four partial derivatives appearing on the right of eqns. (3.1.25) and (3.1.26), only $(\partial T_\alpha/\partial\theta)_{D,S}$ is not easily accessible experimentally.

To eliminate this derivative we use the identity†

$$\left(\frac{\partial T_\alpha}{\partial \theta}\right)_{D,S} = -\sum_{\beta=1}^{6}\left(\frac{\partial T_\alpha}{\partial S_\beta}\right)_{\theta,D}\left(\frac{\partial S_\beta}{\partial \theta}\right)_{T,D}$$

to replace eqn. (3.1.26) by

$$C_{D,S}-C_{D,T} = -\frac{\theta}{\varrho_0}\sum_{\alpha,\beta}\left(\frac{\partial T_\alpha}{\partial S_\beta}\right)_{\theta,D}\left(\frac{\partial S_\beta}{\partial \theta}\right)_{T,D}\left(\frac{\partial S_\alpha}{\partial \theta}\right)_{T,D}.$$

Another form of identity is obtained from the following considerations.

If we let $\partial(\boldsymbol{E}, \sigma)/\partial(\boldsymbol{D}, \theta)$ represent the Jacobian

$$\partial(E_1, E_2, E_3, \sigma)/\partial(D_1, D_2, D_3, \theta)$$

then

$$\left(\frac{\partial \sigma}{\partial \theta}\right)_E = \frac{\partial(\boldsymbol{E}, \sigma)}{\partial(\boldsymbol{E}, \theta)}$$

is an identity. For the present we do not specify which set of variables T_α or S_α is held fixed during the differentiation. From the standard transformation properties of Jacobians we may write

$$\left(\frac{\partial \sigma}{\partial \theta}\right)_E = \frac{\partial(\boldsymbol{E}, \sigma)}{\partial(\boldsymbol{D}, \sigma)}\frac{\partial(\boldsymbol{D}, \sigma)}{\partial(\boldsymbol{E}, \theta)} \tag{3.1.27}$$

Similarly,

$$\left(\frac{\partial \sigma}{\partial \theta}\right)_D = \frac{\partial(\boldsymbol{D}, \sigma)}{\partial(\boldsymbol{E}, \theta)}\frac{\partial(\boldsymbol{E}, \theta)}{\partial(\boldsymbol{D}, \theta)}. \tag{3.1.28}$$

† To derive this identity consider the following differentials

$$dT_\alpha = \sum_\beta\left(\frac{\partial T_\alpha}{\partial S_\beta}\right)_{\theta,D} dS_\beta + \left(\frac{\partial T_\alpha}{\partial \theta}\right)_{D,S} d\theta,$$

$$dS_\beta = \sum_\gamma\left(\frac{\partial S_\beta}{\partial T_\gamma}\right)_{\theta,D} dT_\gamma + \left(\frac{\partial S_\beta}{\partial \theta}\right)_{T,D} d\theta.$$

Combining these two equations, we have

$$dT_\alpha = \sum_{\beta,\gamma}\left(\frac{\partial T_\alpha}{\partial S_\beta}\right)_{\theta,D}\left(\frac{\partial S_\beta}{\partial T_\gamma}\right)_{\theta,D} dT_\gamma + \left[\sum_\beta\left(\frac{\partial T_\alpha}{\partial S_\beta}\right)_{\theta,D}\left(\frac{\partial S_\beta}{\partial \theta}\right)_{T,D} + \left(\frac{\partial T_\alpha}{\partial \theta}\right)_{D,S}\right] d\theta.$$

Since the differentials dT_α, $d\theta$ are independent, the coefficients of each differential in this equation must vanish. The coefficient of $d\theta$ yields the identity used above.

Combining eqns. (3.1.27) and (3.1.28) we have

$$\frac{C_E}{C_D} = \frac{\partial(\boldsymbol{E}, \sigma)}{\partial(\boldsymbol{D}, \sigma)} \bigg/ \frac{\partial(\boldsymbol{E}, \theta)}{\partial(\boldsymbol{D}, \theta)}$$

$$= \det\left[\left(\frac{\partial E_i}{\partial D_j}\right)_\sigma\right] \bigg/ \det\left[\left(\frac{\partial E_i}{\partial D_j}\right)_\theta\right].$$

We may choose to consider either constant strain or constant stress constraints; thus

$$\frac{C_{E,S}}{C_{D,S}} = \det\left[\left(\frac{\partial E_i}{\partial D_j}\right)_{\sigma,S}\right] \bigg/ \det\left[\left(\frac{\partial E_i}{\partial D_j}\right)_{\theta,S}\right] \qquad (3.1.29)$$

and

$$\frac{C_{E,T}}{C_{D,T}} = \det\left[\left(\frac{\partial E_i}{\partial D_j}\right)_{\sigma,T}\right] \bigg/ \det\left[\left(\frac{\partial E_i}{\partial D_j}\right)_{\theta,T}\right]. \qquad (3.1.30)$$

Similarly,

$$\frac{C_{S,E}}{C_{T,E}} = \det\left[\left(\frac{\partial S_\alpha}{\partial T_\beta}\right)_{\sigma,E}\right] \bigg/ \det\left[\left(\frac{\partial S_\alpha}{\partial T_\beta}\right)_{\theta,E}\right] \qquad (3.1.31)$$

and

$$\frac{C_{S,D}}{C_{T,D}} = \det\left[\left(\frac{\partial S_\alpha}{\partial T_\beta}\right)_{\sigma,D}\right] \bigg/ \det\left[\left(\frac{\partial S_\alpha}{\partial T_\beta}\right)_{\theta,D}\right]. \qquad (3.1.32)$$

3.1.4 Thermodynamic Identities II

The identities derived in the previous subsection referred to a single state of the system. It is sometimes convenient to have identities between quantities referring to different states of the system. To illustrate this point consider the case of a dielectric subjected to hydrostatic pressure P. Thus each thermodynamic state of the dielectric is described by the thermodynamic variables $P, V, \sigma, \theta, \boldsymbol{E}$ and $\boldsymbol{D}$, where V is the volume of the dielectric. For hydrostatic pressure, $T_1 = T_2 = T_3 = -P$, $T_4 = T_5 = T_6 = 0$, $S_1 = S_2 = S_3 = (V - V_0)/3V_0$ (§ 2.2.1), and $S_4 = S_5 = S_6 = 0$. Therefore the first law [eqn. (3.1.6)] now reads

$$dU = -\left(\frac{P}{V_0}\right) dV + \sum_{i=1}^{3} E_i\, dD_i + \theta\, d\sigma. \qquad (3.1.33)$$

Figure 3.1.1 represents the PV diagram of the dielectric—we shall ignore the variation of $\boldsymbol{E}$ and $\boldsymbol{D}$ for the time being. The lines ABD,

ACE are isobaric and isochoric (equal volume) paths respectively in this diagram; CB, ED are two isotherms of temperatures θ and $\theta + d\theta$ respectively. We denote the specific heat at constant pressure by C_P^B and the specific heat at constant volume by C_V^B for the state at point B; similarly for the state at point C the specific heats are denoted C_P^C and C_V^C respectively. If $d\theta$ is an infinitesimal temperature change, then

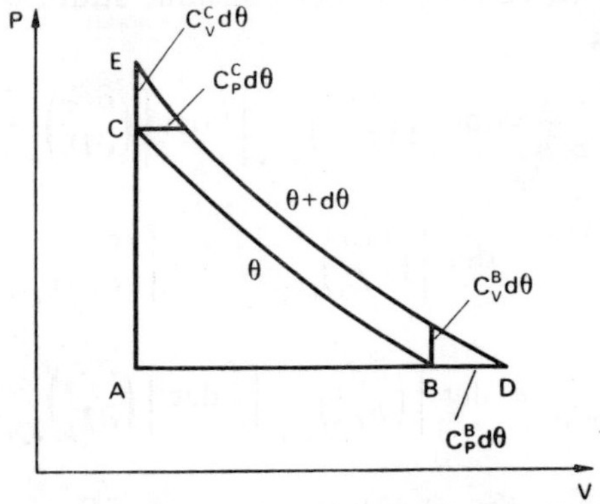

FIG. 3.1.1. The pressure–volume diagram for the dielectric, showing the two isothermal paths CB, ED, the isobaric path AD and the isochoric path AE.

to heat the system isobarically from θ to $\theta + d\theta$ requires an amount of heat $C_P^B\, d\theta$ when the system is in the state B and $C_P^C\, d\theta$ when it is in the state C. Similarly, C_V^B, C_V^C determine the heat required for infinitesimal isochoric heating processes. Clearly in general C_P^B, C_P^C, C_V^B, C_V^C exhibit different values. The techniques of the previous subsection yield relationships between C_P^B and C_V^B and between C_P^C and C_V^C (i.e. *point relations* involving only a single state).† The problem we wish to consider in this subsection is the determination of relationships

† Since

$$\frac{1}{V_0}(\partial V/\partial \theta) = [(\partial S_1/\partial \theta) + (\partial S_2/\partial \theta) + (\partial S_3/\partial \theta)]$$

eqn. (3.1.26) gives, for example, the point relationship

$$C_P^B - C_V^B = -\frac{\theta}{\varrho_0 V_0}\left(\frac{\partial V}{\partial \theta}\right)_P \left(\frac{\partial P}{\partial \theta}\right)_V.$$

As above, we have suppressed the D dependence.

between the properties of one state of a system and those of another state of the same system, i.e. *two point relations*; an example is the relationship between C_V^B and C_V^C.

Let us treat the variables V, D_i, θ as independent. Consider the free energy $F = F(V, D_i, \theta)$, where

$$\frac{P}{V_0} = -\left(\frac{\partial F}{\partial V}\right)_{D,\theta}, \quad E_i = \left(\frac{\partial F}{\partial D_i}\right)_{V,\theta}, \quad \sigma = -\left(\frac{\partial F}{\partial \theta}\right)_{D,V}$$

Then

$$F(V^C, \theta, D) - F(V^B, \theta, D) = \int_{V^B}^{V^C} \left(\frac{\partial F}{\partial V}\right)_{\theta, D} dV \qquad (3.1.34)$$

represents the difference in free energy of two states of different volumes V^B, V^C and the same temperature θ and displacement $\boldsymbol{D}$. The integral on the right is taken along a path of constant θ and constant $\boldsymbol{D}$. Equation (3.1.34) holds provided F is a smooth function of V, i.e. provided no phase transitions occur along this path between C and B. Differentiating both sides of eqn. (3.1.34) with respect to D_i, keeping V^B, V^C and θ fixed, yields

$$\frac{\partial F}{\partial D_i}(V^C, \theta, D_i) - \frac{\partial F}{\partial D_i}(V^B, \theta, D_i) = \int_{V^B}^{V^C} \left(\frac{\partial^2 F}{\partial V \partial D_i}\right)_\theta dV$$

or

$$E_i(V^C, \theta, D_i) = E_i(V^B, \theta, D_i) + \int_{V^B}^{V^C} \left(\frac{\partial E_i}{\partial V}\right)_{D,\theta} dV. \qquad (3.1.35)$$

$E_i(V, \theta, D_i)$ is the dielectric equation of state of the system clamped at the volume V; hence eqn. (3.1.35) provides a relationship for determining $E_i(V^C, \theta, D_i)$ in terms of $E_i(V^B, \theta, D_i)$.

Similarly, we have

$$P(V^C, \theta, D_i) = P(V^B, \theta, D_i) + \int_{V^B}^{V^C} \left(\frac{\partial P}{\partial V}\right)_{D,\theta} dV \qquad (3.1.36)$$

and
$$\sigma(V^C, \theta, D_i) = \sigma(V^B, \theta, D_i) + \int_{V^B}^{V^C} \left(\frac{\partial \sigma}{\partial V}\right)_{D,\theta} dV. \tag{3.1.37}$$

Equations (3.1.35) to (3.1.37) provide exact relationships between the dielectric, elastic and thermal equations of state at one state and those of another state. Clearly other forms of these relations can be derived by choosing different paths (θ = const., V = const. and $\boldsymbol{D}$ = const., V = const.), but the forms given above are especially useful in the problem of relating the dielectric properties of a system along an isobaric path to those of the system along an isochoric path. The former path is most readily accessible experimentally ($P = 0$) while the latter is the most convenient one to treat in any statistical mechanical study of the dielectric properties (in this case V^C is the equilibrium volume of the dielectric at $P = 0$, $\theta = 0$°K).

A number of identities can be obtained from the above three equations. We derive only one as an example. Differentiating both sides of eqn. (3.1.37) with respect to θ, keeping V^B, V^C and $\boldsymbol{D}$ fixed and multiplying through by θ/ϱ_0 yields
$$C_V(V^C, \theta, D_i) = C_V(V^B, \theta, D_i) + \frac{\theta}{\varrho_0} \int_{V^B}^{V^C} \left(\frac{\partial^2 \sigma}{\partial V \partial \theta}\right) dV.$$

Now $(\partial \sigma/\partial V) = 1/V_0(\partial P/\partial \theta)$, hence
$$C_V(V^C, \theta, D_i) = C_V(V^B, \theta, D_i) + \frac{\theta}{\varrho_0 V_0} \int_{V^B}^{V^C} \left(\frac{\partial^2 P}{\partial \theta^2}\right) dV. \tag{3.1.38a}$$

The exact identities given in eqns. (3.1.35) to (3.1.38) can be replaced by approximate relations.† For example, in eqn. (3.1.36), if we expand $(\partial P/\partial V)_{\theta, D}$ in a power series in $(V - V^B)$ to terms of order one in this quantity, we get
$$P(V^C, \theta, D_i) = P(V^B, \theta, D_i) + \left(\frac{\partial P}{\partial V}\right)\bigg|_{V=V^B} (V^C - V^B)$$
$$+ \frac{1}{2}\left(\frac{\partial^2 P}{\partial V^2}\right)\bigg|_{V=V^B} (V^C - V^B)^2. \tag{3.1.38b}$$

† See footnote on p. 219.

Thus if

$$V^B,\ (\partial P/\partial V)\Big|_{V=V^B} \quad \text{and} \quad (\partial^2 P/\partial V^2)\Big|_{V=V^B}$$

are known (measured) functions of θ and $\boldsymbol{D}$ along some isobaric path ($P = 0$ normally), then we may use this last equation to calculate an approximate expression for $P(V^C, \theta, D_i)$ along the isochoric path $V = V^C$.

The extension of the results listed above to the case of arbitrary stresses (non-hydrostatic) is easily accomplished.

3.1.5. Change of Initial State

It sometimes proves convenient to describe the deformation of a material relative to two different initial states (i.e. two states with $T_\alpha = 0$, $E_i = 0$ and different θ_i's); in this subsection we derive expressions relating the deformation gradients appropriate to two initial states.

Consider the position coordinates $x_j(\theta, T_\alpha, E_i)$ of some point in the material which is in an arbitrary state defined by the fields θ, T_α, E_i. We set

$$x_j = u_j(\theta, T_\alpha, E_i, X_k) + X_j \tag{3.1.39}$$

and

$$x_j = u_j'(\theta, T_\alpha, E_i, X_k') + X_j', \tag{3.1.40}$$

where

$$u_j(\theta_i, 0, 0, X_k) = 0, \quad u_j'(\theta_i', 0, 0, X_k') = 0.$$

Thus the set of coordinates X_k define one initial state with temperature θ_i and the X_k' another initial state with temperature θ_i'. Since the expressions on the right of eqns. (3.1.39) and (3.1.40) refer to the same point, we have

$$u_j(\theta, T_\alpha, E_i, X_k) + X_j \equiv u_j'(\theta, T_\alpha, E_i, X_k') + X_j', \tag{3.1.41}$$

in particular

$$X_j' = u_j(\theta_i', 0, 0, X_k) + X_j. \tag{3.1.42}$$

From eqn. (3.1.41)

$$\frac{\partial u_j}{\partial X_l}(\theta, T_\alpha, E_i, X_k) + \delta_{jl} = \sum_m \left[\frac{\partial u_j'}{\partial X_m'}(\theta, T_\alpha, E_i, X_k') + \delta_{jm}\right] \frac{\partial X_m'}{\partial X_l}.$$

The quantity $(\partial X'_m/\partial X_l)$ is determined by eqn. (3.1.42); hence

$$\frac{\partial u_j}{\partial X_l}(\theta, T_\alpha, E_i, X_k)+\delta_{jl}$$
$$=\sum_m\left[\frac{\partial u'_j}{\partial X'_m}(\theta, T_\alpha, E_i, X_k)+\delta_{jm}\right]\left[\frac{\partial u_m}{\partial X_l}(\theta'_i, 0, 0, X_k)+\delta_{ml}\right]$$

or

$$\left.\begin{aligned}\frac{\partial u_j}{\partial X_l}(\theta, T_\alpha, E_i, X_k)&=\sum_m\frac{\partial u_j}{\partial X'_m}(\theta, T_\alpha, E, X'_k)\frac{\partial u_m}{\partial X_l}(\theta'_i, 0, 0, X_k)\\&\quad+\frac{\partial u_j}{\partial X_l}(\theta'_i, 0, 0, X_k)+\frac{\partial u'_j}{\partial X_l}(\theta, T_\alpha, E_i, X'_k).\end{aligned}\right\}\quad(3.1.43)$$

For small deformations the sum of products on the right of this expression may be neglected; thus

$$\frac{\partial u'_j}{\partial X'_l}(\theta, T_\alpha, E_i, X'_k)=\frac{\partial u_j}{\partial X_l}(\theta, T_\alpha, E_i, X_k)-\frac{\partial u_j}{\partial X_l}(\theta'_i, 0, 0, X_k).\quad(3.1.44)$$

In terms of the strains this result takes the form

$$S'_\alpha(\theta, T_\beta, E_i)=S_\alpha(\theta, T_\beta, E_i)-S_\alpha(\theta'_i, 0, 0).\quad(3.1.45)$$

We see that the strain measured relative to the primed initial state is equal to the strain measured relative to the unprimed initial state minus the strain of the primed initial state measured relative to the unprimed state.

3.2 Stability

3.2.1 Stable Equilibrium

The energy density in the dielectric is $U(S_\alpha, D_i, \sigma)$ and the energy density in free space is $\sum_{i=1}^{3} D_i^2/2\varepsilon_0$; hence the total energy of the system consisting of a dielectric and a conductor is

$$\mathcal{U}=\int_V U(S_\alpha, D_i, \sigma)\,dv+\int_A\sum_{i=1}^{3}\frac{D_i^2}{2\varepsilon_0}\,dv,\quad(3.2.1)$$

where V, A denote the regions occupied by the dielectric and free space respectively. Since the mechanical and thermal properties of the conductor are of no interest to us, we assume that in any thermodynamic transformations of the system the changes in the mechanical and thermal properties of the conductor are negligible. Let us consider transformations in this system when it is subjected to the following constraints: $\boldsymbol{t}^0 = \text{const.}$, $\phi_0 = \text{const.}$ and $\theta = \text{const.}$ To achieve this last constraint we suppose that the dielectric is in thermal contact with a heat bath. Let the system suffer an infinitesimal transformation in which the various changes in fields are denoted by δT_α, δS_α, etc. The work done in this transformation is

$$\delta \mathcal{W} = \oint_V (\boldsymbol{t} \cdot \delta \boldsymbol{u})\, dS + \phi_0 \delta Q \tag{3.2.2}$$

and the corresponding energy change is

$$\delta \mathcal{U} = \delta \int_V U(S_\alpha, D_i, \sigma)\, dv + \delta \int_A \sum_{i=1}^{3} \frac{D_i^2}{2\varepsilon_0}\, dv.$$

From the conservation of energy, the heat transferred from the dielectric to the heat bath is $-(\delta\mathcal{U} - \delta\mathcal{W})$, and hence the entropy change of the heat bath is $-(\delta\mathcal{U} - \delta\mathcal{W})/\theta$. Now we may regard the heat bath, dielectric and conductor as an isolated system and by the second law of thermodynamics, the only *allowed transformations* in this system are those in which the total entropy change is either positive or zero; hence for allowed transformations of the system we have

$$\frac{-(\delta\mathcal{U} - \delta\mathcal{W})}{\theta} + \delta \int_V \sigma\, dv \geqslant 0.$$

Using eqn. (3.2.2) and the fact that θ, $\boldsymbol{t}^0$ and ϕ_0 are all held constant, we can replace this inequality by

$$\delta \mathcal{G} \leqslant 0, \tag{3.2.3}$$

where

$$\mathcal{G} = \mathcal{U} - \oint_V \boldsymbol{t}^0 \cdot \boldsymbol{u}\, dS - \phi_0 Q - \int_V \theta\sigma\, dv. \tag{3.2.4}$$

Stable and unstable equilibrium are defined in terms of the inequality (3.2.3). Before introducing these definitions we first obtain some alternative expressions for $\mathcal{G}$. We may apply the arguments, used to derive eqn. (2.3.19) from eqn. (2.3.18) and eqn. (2.4.8) from eqn. (2.4.7), to eqn. (3.2.4) and thus obtain†

$$\mathcal{G} = \mathcal{U} - \int_V \sum_\alpha T_\alpha S_\alpha \, dv - \int_{A+V} \sum_i E_i D_i \, dv - \int_V \theta\sigma \, dv. \qquad (3.2.5)$$

From the definition of the Gibbs function G and the free space relationship $D_i = \varepsilon_0 E_i$, the right-hand side of eqn. (3.2.5) can be written in the form

$$\mathcal{G} = \int_V G \, dv - \int_A \sum_{i=1}^{3} \frac{\varepsilon_0 E_i^2}{2} \, dv. \qquad (3.2.6)$$

The term $\left[-\sum_i \varepsilon_0 E_i^2\right]$ plays the role of the Gibbs function of free space‡ and hence we can regard $\mathcal{G}$ as the integral of the Gibbs function over all space. Finally, using the definition of H, the Helmholtz energy, and the free space relation $D_i = \varepsilon_0 E_i$, we can replace eqn. (3.2.6) with

$$\mathcal{G} = \int_V \left(H - \sum_i D_i E_i\right) dv - \int_A \sum_i \frac{D_i^2}{2\varepsilon_0} \, dv. \qquad (3.2.7)$$

Thermodynamic stability is defined in the following manner. Consider the system and constraints introduced above. A state of this system described by the fields S_α, T_α, D_i, E_i, D_i, θ, σ is a *stable* state if for any arbitrary set of small changes δS_α, δT_α, δD_i, δE_i, $\delta\theta$, $\delta\sigma$ which observe the constraints listed above, the corresponding $\delta\mathcal{G}$ is positive definite (i.e. $\mathcal{G}$ is a minimum); in other words the state is stable if there are no *allowed* transformations which will carry this state into an adjacent state without violating the constraints imposed: if transformations do exist for which $\delta\mathcal{G} \leqslant 0$ then the state is *unstable*. Thus the stability of a state (subject to the constraints θ = const., t^0 = const., ϕ_0 =

† In deriving eqn. (3.2.5) from eqn. (3.2.4) we have used the linear approximation $(\partial u_i/\partial X_j) \cong (\partial u_i/\partial x_j)$.

‡ We note that $\delta\left[-\sum_i \varepsilon_0 E_i^2/2\right] = -\sum_i D_i \delta E_i$ in agreement with eqn. (3.1.10).

const.) is determined by the extremal properties of $\mathcal{G}$ with respect to these virtual changes in the fields S_α, T_α, etc. In general, to each set of constraints there corresponds an integrated thermodynamic potential and the stability of the state under these constraints is determined by the extremal properties of this integral. For example, if $\theta = \text{const.}$, $t^0 = \text{const.}$ and $Q = \text{const.}$, then we can show that the state is stable if and only if $\delta\mathcal{H} > 0$, where

$$\mathcal{H} = \int\limits_V H\,dv + \int\limits_A \sum_i \frac{D_i^2}{2\varepsilon_0}\,dv.$$

H is the Helmholtz energy.

3.2.2 Stability Conditions

The demands of stability put restrictions on the behaviour of the integrands of $\mathcal{G}$, $\mathcal{H}$, etc., in the stable states. We shall describe the effects of these demands only in the case of the thermodynamic coordinates E_i and D_i. Consider virtual changes δD_i, δE_i subject to the constraint $\delta\phi_0 = 0$, then from eqn. (3.2.7)

$$\delta\mathcal{G} = \delta\int\limits_V H\,dv - \delta\int\limits_V \sum_i D_i E_i\,dv - \delta\int\limits_A \sum_i \frac{D_i^2}{2\varepsilon_0}\,dv.$$

Ignoring changes in the mechanical and thermal variables, we calculate $\delta\mathcal{G}$ correct to second order in the small quantities δD_i, δE_i:

$$\begin{aligned}\delta\mathcal{G} = &\int\limits_V \left[\sum_i \left(\frac{\partial H}{\partial D_i}\right)\delta D_i + \frac{1}{2}\sum_{i,j}\left(\frac{\partial^2 H}{\partial D_i\,\partial D_j}\right)\delta D_i\,\delta D_j\right] dv \\ &- \int\limits_V \sum_i (E_i\,\delta D_i + D_i\,\delta E_i + \delta D_i\,\delta E_i)\,dv - \int\limits_A \sum_i \left(\frac{D_i\,\delta D_i}{\varepsilon_0} + \frac{\delta D_i\,\delta D_i}{2\varepsilon_0}\right) dv.\end{aligned}$$

Since $E_i = (\partial H/\partial D_i)_{T,\,\theta}$ in V and $E_i = D_i/\varepsilon_0$ in A,

$$\begin{aligned}\delta\mathcal{G} = &-\int\limits_{A+V} \sum_i (D_i\,\delta E_i + \delta D_i\,\delta E_i) + \frac{1}{2}\int\limits_V \sum_{i,j}\left(\frac{\partial E_i}{\partial D_j}\right)\delta D_i\,\delta D_j\,dv \\ &+ \frac{1}{2}\int\limits_A \sum_i \frac{\delta D_i\,\delta D_i}{\varepsilon_0}\,dv.\end{aligned}$$

By reversing the argument that enables us to replace eqn. (2.4.8) by eqn. (2.4.7) we can show that the first integral reduces to $(-Q\delta\phi_0 - \delta Q\delta\phi_0)$ and hence vanishes because of the constraint $\delta\phi_0 = 0$; thus

$$\delta \mathcal{G} = \frac{1}{2}\int\limits_V \sum_{i,j}\left(\frac{\partial E_i}{\partial D_j}\right)_{T,\,\theta} \delta D_i\,\delta D_j\,dv + \frac{1}{2}\int\limits_A \sum_i \frac{\delta D_i\,\delta D_i}{\varepsilon_0}\,dv.$$

By inspection $\delta\mathcal{G}$ is a positive definite function of the δD_i if and only if the principal minors of the matrix $[\partial E_i/\partial D_j]$ are positive, at all points in the dielectric. In particular, when this matrix is diagonal the necessary and sufficient condition for stability is

$$\left(\frac{\partial E_i}{\partial D_i}\right)_{T,\,\theta} > 0 \quad (i = 1, 2, 3) \tag{3.2.8}$$

at all points in the dielectric. A similar result obtains for other constraints; the only difference occurs in the subscripts which indicate the independent variables. When $\delta\mathcal{G} > 0$ the state is said to be *locally stable*. It can happen that two or more states satisfy a given set of constraints and *each* of these is locally stable. Under these circumstances the state with the least value of $\mathcal{G}$ is said to be *absolutely stable* and the remainder *metastable*. This latter type of stability is sometimes referred to as *global* stability. Thus, to summarize, global stability is determined by the relative sizes of $\mathcal{G}$ for different states whereas local stability is determined by the extremal properties of $\mathcal{G}$ under the action of small virtual changes in the fields describing the state. In general, for an arbitrary set of constraints these remarks apply to the integral of the thermodynamic potential appropriate to the constraints.

3.2.3 Phase Transitions I

The quantity $\mathcal{G}$ (or whatever function is appropriate for the constraints) is a function of the value of the constraints imposed on the system, i.e. $\mathcal{G} = \mathcal{G}(\theta, \phi_0, \boldsymbol{t}^0)$. If for a given set of constraint values more than one locally stable solution to the field equations exists, then there is a corresponding set of functions $\mathcal{G}^1 < \mathcal{G}^2 < \mathcal{G}^3$, say. The set of states or phase described by $\mathcal{G}^1$ is absolutely stable and the remainder, described by $\mathcal{G}^2$, $\mathcal{G}^3$, metastable (see the previous subsection). It may happen that

as the values of the constraints are altered $\mathcal{G}^2 - \mathcal{G}^1$ changes sign; then $\mathcal{G}^2 < \mathcal{G}^1 < \mathcal{G}^3$ and the phase described by $\mathcal{G}^2$ is now absolutely stable. In these circumstances, if the system passes through only absolutely stab le states, then the system undergoes a *phase transition* (or *phase change*) from a phase described by $\mathcal{G}^1(\mathcal{G}^2 - \mathcal{G}^1 > 0)$ to a phase described by $\mathcal{G}^2(\mathcal{G}^1 - \mathcal{G}^2 > 0)$ when $\mathcal{G}^1 - \mathcal{G}^2 = 0$. The values of the constraints at the transition satisfy the equation $\mathcal{G}^1 - \mathcal{G}^2 = 0$.

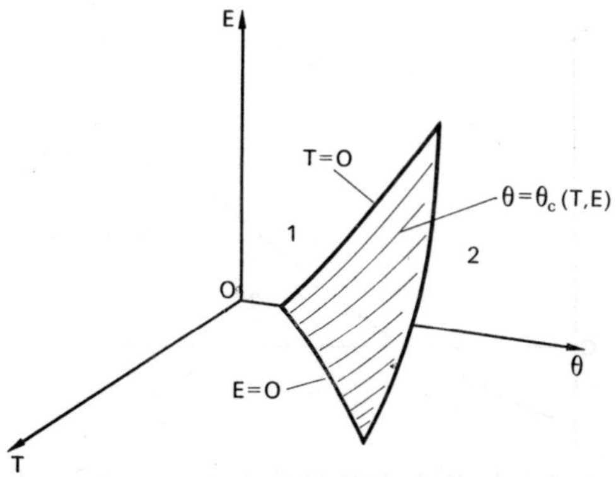

FIG. 3.2.1. A section of a phase boundary $\theta = \theta_c(T, E)$ in the positive octant of the T, E, θ phase space.

Consider a system in which the fields are uniform and the contribution to $\mathcal{G}$ from free space is negligible. Then $\mathcal{G}^1 = V^1 G^1$ and $\mathcal{G}^2 = V^2 G^2$, where G^1, G^2 are the Gibbs functions and V^1, V^2 the volumes of the dielectric in the two phases. For simplicity we shall suppose that the behaviour of the dielectric is described by a single stress tensor component T and a single electric field component E; thus $G^1 = G^1(\theta, T, E)$, $G^2 = G^2(\theta, T, E)$ and the conjugate variables are

$$\sigma^1 = -\left(\frac{\partial G^1}{\partial \theta}\right)_{T,E}, \quad S^1 = -\left(\frac{\partial G^1}{\partial T}\right)_{E,\theta}, \quad D^1 = -\left(\frac{\partial G^1}{\partial E}\right)_{\theta,T} \tag{3.2.9}$$

in one phase and

$$\sigma^2 = -\left(\frac{\partial G^2}{\partial \theta}\right)_{T,E}, \quad S^2 = -\left(\frac{\partial G^2}{\partial T}\right)_{E,\theta}, \quad D^2 = -\left(\frac{\partial G^2}{\partial E}\right)_{\theta,T} \tag{3.2.10}$$

in the other. Now, since we are assuming that the fields are small, $V^1G^1 \cong V^0G^1$ and $V^2G^2 \cong V^0G^2$, where V^0 is the volume of the dielectric in the initial state; hence the condition for a phase transition reduces to $G^1(\theta, T, E) = G^2(\theta, T, E)$. We may regard this as an equation determining the transition temperature $\theta_c = \theta_c(T, E)$.

In a phase space described by the three quantities θ, T, E, the function $\theta = \theta_c(T, E)$ defines a surface (Fig. 3.2.1). The coordinates of points

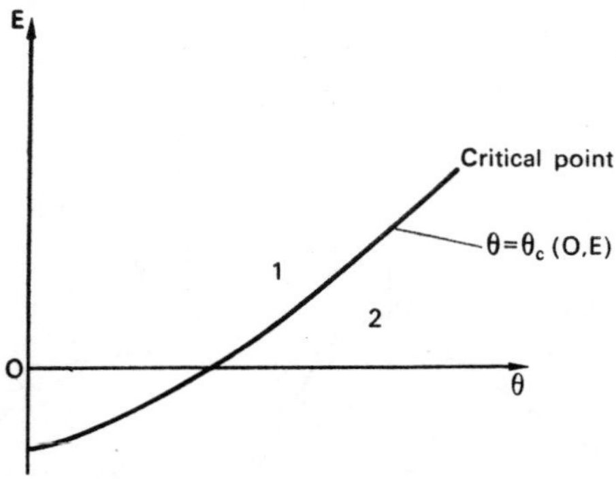

FIG. 3.2.2. The phase diagram generated by the plane $T = 0$ in Fig. 3.2.1.

on one side of this surface represent the absolutely stable states described by the function $\mathcal{G}^2$, and those on the other side the absolutely stable states described by $\mathcal{G}^1$. This surface is referred to as the *phase boundary*. Various *phase diagrams* may be generated by considering the intersection of the phase boundary with different planes (Fig. 3.2.2). If the phase boundary in a phase diagram terminates in a point, this is referred to as a *critical point*.

Consider a point (θ_c^0, T^0, E^0) on the phase boundary. We shall assume that in the neighbourhood of this point: (i) G^1 and G^2 are analytic functions of θ, T, E, and (ii) $(\partial\theta_c/\partial T)_E \neq 0$, $(\partial\theta_c/\partial E)_T \neq 0$, $(\partial T/\partial E)_{\theta_c} \neq 0$.

Consider another point $(\theta_c^0+\mathrm{d}\theta_c,\ T^0+\mathrm{d}T,\ E^0+\mathrm{d}E)$ on the phase boundary. By definition,

$$G^1(\theta_c^0, T^0, E^0) = G^2(\theta_c^0, T^0, E^0)$$

and

$$G^1(\theta_c^0+d\theta_c,\ T^0+dT,\ E^0+dE) = G^2(\theta_c^0+d\theta_c,\ T^0+dT,\ E^0+dE).$$

Since G^1 and G^2 are analytic we may expand in a power series in $d\theta_c$, dT, dE. Thus to terms linear in the expansion variables we get

$$\left(\frac{\partial G^1}{\partial \theta}\right)_{T,E} d\theta_c + \left(\frac{\partial G^1}{\partial T}\right)_{E,\theta} dT + \left(\frac{\partial G^1}{\partial E}\right)_{\theta,T} dE$$

$$= \left(\frac{\partial G^2}{\partial \theta}\right)_{T,E} d\theta_c + \left(\frac{\partial G^2}{\partial T}\right)_{E,\theta} dT + \left(\frac{\partial G^2}{\partial E}\right)_{\theta,T} dE.$$

Using eqns. (3.2.9) and (3.2.10) we can rewrite this equation in the form

$$(\sigma^1 - \sigma^2)^0 \, d\theta_c + (S^1 - S^2)^0 \, dT + (D^1 - D^2)^0 \, dE = 0, \qquad (3.2.11)$$

where the superscript 0 denotes the value of the enclosed function at the point (θ_c^0, T^0, E^0). Hence, provided $\sigma^1 - \sigma^2 \neq 0$ at (θ_c^0, T^0, E^0) we obtain the *Clausius–Clapeyron* relations

$$\left(\frac{\partial \theta_c}{\partial T}\right)_E^0 = -\left(\frac{S^1 - S^2}{\sigma^1 - \sigma^2}\right)^0 = -\theta_c^0 \frac{(S^1 - S^2)^0}{L^0} \qquad (3.2.12)$$

and

$$\left(\frac{\partial \theta_c}{\partial E}\right)_T^0 = -\left(\frac{D^1 - D^2}{\sigma^1 - \sigma^2}\right)^0 = -\theta_c^0 \frac{(D^1 - D^2)^0}{L^0}, \qquad (3.2.13)$$

where $L^0 = \theta_c^0(\sigma^1 - \sigma^2)^0$ is the *latent heat* at the point (θ_c^0, T^0, E^0). $(\partial\theta_c/\partial T)_E^0$ is the rate of change of transition temperature with stress at constant electric field and $(\partial\theta_c/\partial E)_T^0$ has an obvious similar interpretation. By assumption (ii) neither of these partial derivatives vanish, and hence $(\sigma^1 - \sigma^2)^0 = 0$ if and only if $(S^1 - S^2)^0 = 0$ and $(D^1 - D^2)^0 = 0$ [i.e. either the entropy, strain and displacement fields describing the absolutely stable states each suffer a discontinuity at the point (θ_c^0, T^0, E^0), or none of them do].

3.2.4 Phase Transitions II

Let us now consider the case in which the following conditions are satisfied:

$$(\sigma^1 - \sigma^2) = 0, \quad (S^1 - S^2) = 0, \quad (D^1 - D^2) = 0 \qquad (3.2.14)$$

at some point (θ_c, T^0, E^0) on the phase boundary. In these circumstances the quantities $(\partial\theta_c/\partial T)^0_E(\partial\theta_c/\partial E)^0_T$, $(\partial T/\partial E)^0_{\theta_c}$ at this point are determined by the second derivatives of (G^1-G^2); thus, expanding (G^1-G^2) to terms of order two in $d\theta_c = \theta-\theta^0_c$, $dT = T-T^0$, $dE = E-E^0$, we replace eqn. (3.2.11) with

$$\left.\begin{aligned}\frac{1}{2}\left[\left(\frac{\partial\sigma^1}{\partial\theta}\right)^0-\left(\frac{\partial\sigma^2}{\partial\theta}\right)^0\right](d\theta_c)^2+\frac{1}{2}\left[\left(\frac{\partial S^1}{\partial T}\right)^0-\left(\frac{\partial S^2}{\partial T}\right)^0\right](dT)^2\\ +\frac{1}{2}\left[\left(\frac{\partial D^1}{\partial E}\right)^0-\left(\frac{\partial D^2}{\partial E}\right)^0\right](dE)^2+\left[\left(\frac{\partial\sigma^1}{\partial T}\right)^0-\left(\frac{\partial\sigma^2}{\partial T}\right)^0\right]d\theta_c\,dT\\ +\left[\left(\frac{\partial S^1}{\partial E}\right)^0-\left(\frac{\partial S^2}{\partial E}\right)^0\right]dE\,dT+\left[\left(\frac{\partial D^1}{\partial\theta}\right)^0-\left(\frac{\partial D^2}{\partial\theta}\right)^0\right]dE\,d\theta=0.\end{aligned}\right\}\quad(3.2.15)$$

We now introduce the following additional assumption. (iii) At points on the phase boundary in the neighbourhood of (θ^0_c, T^0, E^0), *each* of the three variables θ_c, T, E is a single valued function of the remaining two.

The first point to note is that either all the square brackets in (3.2.15) vanish or none of them do. [It is a simple exercise to verify that this result holds as a consequence of assumptions (ii) and (iii).]

Now let us treat eqn. (3.2.15) as a quadratic equation determining $d\theta_c$ as a function of dT and dE. If $d\theta_c$ is to be a single valued function of these variables, (iii), the quadratic equation must have a double root, i.e. the discriminant of the quadratic must vanish; thus

$$\begin{aligned}&\left\{\left[\left(\frac{\partial\sigma^1}{\partial T}\right)^0-\left(\frac{\partial\sigma^2}{\partial T}\right)^0\right]dT+\left[\left(\frac{\partial D^1}{\partial\theta}\right)^0-\left(\frac{\partial D^2}{\partial\theta}\right)^0\right]dE\right\}^2\\ &-\left[\left(\frac{\partial\sigma^1}{\partial\theta}\right)^0-\left(\frac{\partial\sigma^2}{\partial\theta}\right)^0\right]\left\{\left[\left(\frac{\partial S^1}{\partial T}\right)^0-\left(\frac{\partial S^2}{\partial T}\right)^0\right](dT)^2.\right.\\ &\left.+2\left[\left(\frac{\partial S^1}{\partial E}\right)^0-\left(\frac{\partial S^2}{\partial E}\right)^0\right]dT\,dE+\left[\left(\frac{\partial D^1}{\partial E}\right)^0-\left(\frac{\partial D^2}{\partial E}\right)^0\right](dE)^2\right\}=0.\end{aligned}$$

Since dT, dE are arbitrary the coefficients of $(dT)^2$, $dTdE$ and $(dE)^2$ are zero; hence we obtain the following set of conditions:

$$\left[\left(\frac{\partial\sigma^1}{\partial\theta}\right)^0-\left(\frac{\partial\sigma^2}{\partial\theta}\right)^0\right]=\frac{\left[\left(\frac{\partial\sigma^1}{\partial T}\right)^0-\left(\frac{\partial\sigma^2}{\partial T}\right)^0\right]\left[\left(\frac{\partial D^1}{\partial\theta}\right)^0-\left(\frac{\partial D^2}{\partial\theta}\right)^0\right]}{\left[\left(\frac{\partial S^1}{\partial E}\right)^0-\left(\frac{\partial S^2}{\partial E}\right)^0\right]}, \quad (3.2.16)$$

$$\left[\left(\frac{\partial S^1}{\partial T}\right)^0-\left(\frac{\partial S^2}{\partial T}\right)^0\right]=\frac{\left[\left(\frac{\partial S^1}{\partial E}\right)^0-\left(\frac{\partial S^2}{\partial E}\right)^0\right]\left[\left(\frac{\partial\sigma^1}{\partial T}\right)^0-\left(\frac{\partial\sigma^2}{\partial T}\right)^0\right]}{\left[\left(\frac{\partial D^1}{\partial\theta}\right)^0-\left(\frac{\partial D^2}{\partial\theta}\right)^0\right]}, \quad (3.2.17)$$

$$\left[\left(\frac{\partial D^1}{\partial E}\right)^0-\left(\frac{\partial D^2}{\partial E}\right)^0\right]=\frac{\left[\left(\frac{\partial D^1}{\partial\theta}\right)^0-\left(\frac{\partial D^2}{\partial\theta}\right)^0\right]\left[\left(\frac{\partial S^1}{\partial E}\right)^0-\left(\frac{\partial S^2}{\partial E}\right)^0\right]}{\left[\left(\frac{\partial\sigma^1}{\partial T}\right)^0-\left(\frac{\partial\sigma^2}{\partial T}\right)^0\right]}. \quad (3.2.18)$$

It is clear from the symmetry of these results that these three conditions express the necessary and sufficient condition for eqn. (3.2.15) to satisfy assumption (iii). It follows immediately that the three square brackets appearing on the left-hand side of eqns. (3.2.16), (3.2.17) and (3.2.18) are of the same sign.

From eqns. (3.2.16) and (3.2.17) we deduce

$$\left[\left(\frac{\partial S^1}{\partial T}\right)^0-\left(\frac{\partial S^2}{\partial T}\right)^0\right]\left[\left(\frac{\partial\sigma^1}{\partial\theta}\right)^0-\left(\frac{\partial\sigma^2}{\partial\theta}\right)^0\right]=\left[\left(\frac{\partial\sigma^1}{\partial T}\right)^0-\left(\frac{\partial\sigma^2}{\partial T}\right)^0\right]^2. \quad (3.2.19)$$

Similarly,

$$\left[\left(\frac{\partial\sigma^1}{\partial\theta}\right)^0-\left(\frac{\partial\sigma^2}{\partial\theta}\right)^0\right]\left[\left(\frac{\partial D^1}{\partial E}\right)^0-\left(\frac{\partial D^2}{\partial E}\right)^0\right]=\left[\left(\frac{\partial D^1}{\partial\theta}\right)^0-\left(\frac{\partial D^2}{\partial\theta}\right)^0\right]^2, \quad (3.2.20)$$

$$\left[\left(\frac{\partial D^1}{\partial E}\right)^0-\left(\frac{\partial D^2}{\partial E}\right)^0\right]\left[\left(\frac{\partial S^1}{\partial T}\right)^0-\left(\frac{\partial S^2}{\partial T}\right)^0\right]=\left[\left(\frac{\partial S^1}{\partial E}\right)^0-\left(\frac{\partial S^2}{\partial E}\right)^0\right]^2. \quad (3.2.21)$$

The double root of eqn. (3.2.15) is

$$d\theta_c=-\frac{\left[\left(\frac{\partial\sigma^1}{\partial T}\right)^0-\left(\frac{\partial\sigma^2}{\partial T}\right)^0\right]}{\left[\left(\frac{\partial\sigma^1}{\partial\theta}\right)^0-\left(\frac{\partial\sigma^2}{\partial\theta}\right)^0\right]}dT-\frac{\left[\left(\frac{\partial\sigma^1}{\partial E}\right)^0-\left(\frac{\partial\sigma^2}{\partial E}\right)^0\right]}{\left[\left(\frac{\partial\sigma^1}{\partial\theta}\right)^0-\left(\frac{\partial\sigma^2}{\partial\theta}\right)^0\right]}dE;$$

hence at (θ_c^0, T^0, E^0)

$$\left(\frac{\partial\theta_c}{\partial T}\right)_E^0 = -\frac{\left[\left(\frac{\partial\sigma^1}{\partial T}\right)^0 - \left(\frac{\partial\sigma^2}{\partial T}\right)^0\right]}{\left[\left(\frac{\partial\sigma^1}{\partial\theta}\right)^0 - \left(\frac{\partial\sigma^2}{\partial\theta}\right)^0\right]}, \tag{3.2.22}$$

$$\left(\frac{\partial\theta_c}{\partial E}\right)_T^0 = -\frac{\left[\left(\frac{\partial\sigma^1}{\partial E}\right)^0 - \left(\frac{\partial\sigma^2}{\partial E}\right)^0\right]}{\left[\left(\frac{\partial\sigma^1}{\partial\theta}\right)^0 - \left(\frac{\partial\sigma^2}{\partial\theta}\right)^0\right]}. \tag{3.2.23}$$

From the identity $(\partial T/\partial E)_{\theta c}^0(\partial E/\partial\theta_c)_T^0(\partial\theta_c/\partial T)_E^0 = -1$ and eqns. (3.2.16), (3.2.17), (3.2.18), (3.2.22) and (3.2.23), we deduce

$$\left(\frac{\partial T}{\partial E}\right)_{\theta_c}^0 = -\frac{\left[\left(\frac{\partial S^1}{\partial E}\right)^0 - \left(\frac{\partial S^2}{\partial E}\right)^0\right]}{\left[\left(\frac{\partial S^1}{\partial T}\right)^0 - \left(\frac{\partial S^2}{\partial T}\right)^0\right]}. \tag{3.2.24}$$

Equations (3.2.22), (3.2.23) and (3.2.24) are the *Ehrenfest relations*. Usin g appropriate Maxwell relations and the identities in eqns. (3.2.16), (3.2.17) and (3.2.18), we may write the right-hand sides of these expression s in a variety of ways. For example:

$$\left(\frac{\partial\theta_c}{\partial T}\right)_E^0 = -\frac{\left[\left(\frac{\partial S^1}{\partial E}\right)^0 - \left(\frac{\partial S^2}{\partial E}\right)^0\right]}{\left[\left(\frac{\partial D^1}{\partial\theta}\right)^0 - \left(\frac{\partial D^2}{\partial\theta}\right)^0\right]} = -\frac{\left[\left(\frac{\partial D^1}{\partial T}\right)^0 - \left(\frac{\partial D^2}{\partial T}\right)^0\right]}{\left[\left(\frac{\partial D^1}{\partial\theta}\right)^0 - \left(\frac{\partial D^2}{\partial\theta}\right)^0\right]}.$$

Equations (3.2.16), (3.2.17) and (3.2.18) impose stringent conditions on the behaviour of $G^1 - G^2$ in the neighbourhood of (θ_c^0, T^0, E^0). Consider a point (θ, T, E) close to (θ_c^0, T^0, E^0); using the conditions (3.2.16), (3.2.17) and (3.2.18) we may write

$$\left.\begin{aligned} G^1(\theta, T, E) - G^2(\theta, T, E) &= \left\{1/2\left[\left(\frac{\partial\sigma^1}{\partial\theta}\right)^0 - \left(\frac{\partial\sigma^2}{\partial\theta}\right)^0\right]\right\}\times \\ \times\Bigg\{\left[\left(\frac{\partial\sigma^1}{\partial\theta}\right)^0 - \left(\frac{\partial\sigma^2}{\partial\theta}\right)^0\right](\theta-\theta_c^0) &+ \left[\left(\frac{\partial\sigma^1}{\partial T}\right)^0 - \left(\frac{\partial\sigma^2}{\partial T}\right)^0\right](T-T^0) \\ + \left[\left(\frac{\partial\sigma^1}{\partial E}\right)^0 - \left(\frac{\partial\sigma^2}{\partial E}\right)^0\right](E-E^0)\Bigg\}^2 & \end{aligned}\right\} \tag{3.2.25}$$

correct to second order in the small quantities $(\theta-\theta_c^0)$, $(T-T^0)$, $(E-E^0)$. We see that (G^1-G^2) is positive (negative) definite if the common sign of $[(\partial\sigma^1/\partial\theta)^0-(\partial\sigma^2/\partial\theta)^0]$, $[(\partial S^1/\partial T)^0-(\partial S^2/\partial T)^0]$, $[(\partial D^1/\partial E)^0-(\partial D^2/\partial E)^0]$ is positive (negative). This result means that there is no phase transition in the neighbourhood of (θ_c^0, T^0, E^0) unless one of the following

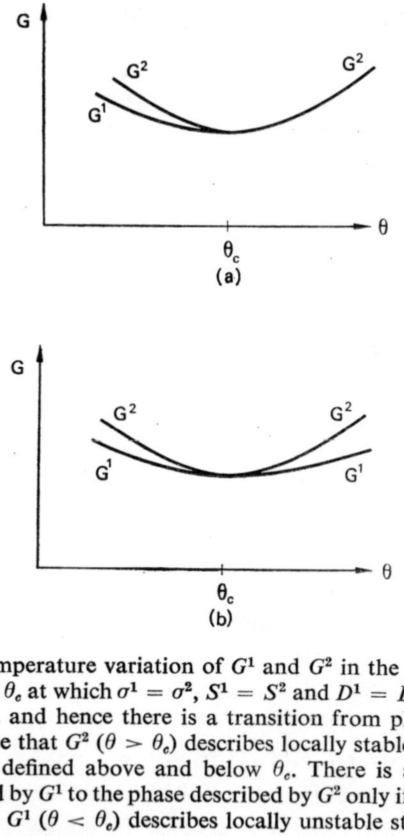

FIG. 3.2.3. The temperature variation of G^1 and G^2 in the vicinity of a transition temperature θ_c at which $\sigma^1 = \sigma^2$, $S^1 = S^2$ and $D^1 = D^2$. In (a) G^1 does not exist above θ_c and hence there is a transition from phase 1 to phase 2 [provided of course that G^2 $(\theta > \theta_c)$ describes locally stable states]. In (b) G^1 is assumed to be defined above and below θ_c. There is a transition from the phase described by G^1 to the phase described by G^2 only if either G^1 $(\theta > \theta_c)$ or G^1 $(\theta < \theta_c)$ describes locally unstable states.

conditions holds. The function G^1, say, which describes the absolutely stable states to one side of the phase boundary either

(a) does not exist† or

† The models described in §§ 5.4.3 and 5.4.4 possess this property.

(b) describes locally unstable states on the other side of the boundary (Fig. 3.2.3). From eqn. (3.2.25) we deduce that

$$\sigma^1(\theta,T,E)-\sigma^2(\theta,T,E)=\sqrt{2\left[\left(\frac{\partial\sigma^1}{\partial\theta}\right)^0-\left(\frac{\partial\sigma^2}{\partial\theta}\right)^0\right][G^1(\theta,T,E)-G^2(\theta,T,E)]},$$

$$S^1(\theta,T,E)-S^2(\theta,T,E)=\sqrt{2\left[\left(\frac{\partial S^1}{\partial T}\right)^0-\left(\frac{\partial S^2}{\partial T}\right)^0\right][G^1(\theta,T,E)-G^2(\theta,T,E)]},$$

$$D^1(\theta,T,E)-D^2(\theta,T,E)=\sqrt{2\left[\left(\frac{\partial D^1}{\partial E}\right)^0-\left(\frac{\partial D^2}{\partial E}\right)^0\right][G^1(\theta,T,E)-G^2(\theta,T,E)]}.$$

In calculating the latter two expressions we have used eqns. (3.2.16), (3.2.17) and (3.2.18). By inspection,

$$\sigma^1(\theta_c,\ T,\ E)-\sigma^2(\theta_c,\ T,\ E)\equiv 0,\qquad S^1(\theta_c,\ T,\ E)-S^2(\theta_c,\ T,\ E)\equiv 0,$$
$$D^1(\theta_c,\ T,E)-D^2(\theta_c,T,E)\equiv 0,$$

i.e. $\sigma^1-\sigma^2$, S^1-S^2, D^1-D^2 vanish at all points on the phase boundary in the neighbourhood of $(\theta_c^0,\ T^0,\ E^0)$. Clearly this result can be extended by analytic continuation to points on the phase boundary beyond the neighbourhood of $(\theta_c^0,\ T^0,\ E^0)$ provided (i), (ii) and (iii) continue to hold.

To summarize; if for some region of the surface $G^1(\theta,\ T,\ E)-G^2(\theta,T,E)=0$, (a) assumptions (i), (ii) and (iii) hold and (b) $\sigma^1-\sigma^2=0$, $S^1-S^2=0$, $D^1-D^2=0$ at some point. Then (A) $\sigma^1-\sigma^2\equiv 0$, $S^1-S^2\equiv 0$, $D^1-D^2\equiv 0$ at all points in this region, (B) the six quantities $[(\partial\sigma^1/\partial\theta)-(\partial\sigma^2/\partial\theta)]$, $[(\partial\sigma^1/\partial T)-(\partial\sigma^2/\partial T)]$, etc., are either all zero or all non-zero, and (C) there is a phase transition provided the functions G^1, G^2 satisfy the conditions envisaged in Fig. 3.2.3.

Let us now consider the special case in which the phase boundary reduces to a curve $\theta_c=\theta_c(T)$ in the $E=0$ plane. This case occurs for one of the models described in Chapter 5. Assume that (a) $\sigma^1-\sigma^2=0$, $S^1-S^2=0$ at some point $(\theta_c^0,\ T^0,\ 0)$ on the boundary, (b) $(d\theta_c/dT)^0$ is finite and non-zero, and (c) θ_c is a single valued function of T and vice versa. Assumption (c) replaces (iii) above.

The derivative $(d\theta_c/dT)^0$ is determined by the equation

$$\left.\begin{aligned}\frac{1}{2}\left[\left(\frac{\partial S^1}{\partial T}\right)^0-\left(\frac{\partial S^2}{\partial T}\right)^0\right](dT)^2+\left[\left(\frac{\partial \sigma^1}{\partial T}\right)^0-\left(\frac{\partial \sigma^2}{\partial T}\right)\right]dT\,d\theta_c\\ +\frac{1}{2}\left[\left(\frac{\partial \sigma^1}{\partial \theta}\right)^0-\left(\frac{\partial \sigma^2}{\partial \theta}\right)^0\right](d\theta_c)^2=0\end{aligned}\right\} \quad (3.2.26)$$

[cf. eqn. (3.2.15)].

Again, the terms in square brackets are either all zero or all non-zero, and the condition that eqn. (3.2.26) yields a double root is

$$\left[\left(\frac{\partial S^1}{\partial T}\right)^0-\left(\frac{\partial S^2}{\partial T}\right)^0\right]\left[\left(\frac{\partial \sigma^1}{\partial \theta}\right)^0-\left(\frac{\partial \sigma^2}{\partial \theta}\right)^0\right]=\left[\left(\frac{\partial \sigma^1}{\partial T}\right)^0-\left(\frac{\partial \sigma^2}{\partial T}\right)^0\right]^2, \quad (3.2.27)$$

in agreement with eqn. (3.2.19).

The rest of the calculation is carried out in a similar fashion to the one above and we find that $(\sigma^1-\sigma^2)$ and (S^1-S^2) vanish at points on the curve $[\theta=\theta_c(T), E=0]$, in the vicinity of $(\theta_c^0, T^0, 0)$.

3.2.5 Ehrenfest Classification

The discussion given in the previous two subsections suggests the following nomenclature for transitions. Consider the partial derivatives of G^1-G^2 (or whichever potentials are appropriate for the constraints) with respect to a set of arguments of G (not necessarily a complete set); at an *n-th order* transition all these partial derivatives of order n are non-zero and all the partial derivatives of lower order are zero. Within the limitations of assumptions (i), (ii) and some variation of (iii), this definition provides a complete classification—the so-called *Ehrenfest classification*—for transitions. Thus in § 3.2.3 the transition is first order while in § 3.2.4 the transitions are both second order.

In his original treatment of second-order transitions, Ehrenfest (1933) *assumed* that the first-order derivatives of (G^1-G^2) vanish *identically* over a region of the phase boundary. The expressions for $(\partial\theta_c/\partial T)_E^0$, etc., given above, follow immediately from this assumption.† Ehren-

† If $\sigma^1(\theta_c, T, E)-\sigma^2(\theta_c, T, E)\equiv 0$, then we may differentiate with respect to T keeping E fixed; thus

$$\left[\left(\frac{\partial \sigma^1}{\partial T}\right)_{\theta, E}-\left(\frac{\partial \sigma^2}{\partial T}\right)_{\theta, E}\right]+\left[\left(\frac{\partial \sigma^1}{\partial \theta}\right)_{T, E}-\left(\frac{\partial \sigma^2}{\partial \theta}\right)_{T, E}\right]\left(\frac{\partial \theta_c}{\partial T}\right)_E=0$$

in agreement with eqn. (3.2.22).

fest's treatment of the second-order phase transition has been criticized on two counts [see discussion in Epstein (1957), p. 128, and Pippard (1957), p. 146]; since (G^1-G^2) is either positive or negative definite at a phase boundary, there can be no phase transition. However, as we have discussed above, this objection does not apply if one of the conditions envisaged in Fig. 3.2.3 holds. Secondly, it would appear that the possibility of four functions (G^1-G^2), $(\sigma^1-\sigma^2)$, (S^1-S^2) and (D^1-D^2) all vanishing in some *region* on the phase boundary is so remote that such an occurrence is too unlikely to be found experimentally. This criticism is refuted by the behaviour of a phenomenological model described in § 5.4.3.

In practice it appears that the Ehrenfest classification as opposed to the Ehrenfest relations has a wider range of applicability than can be expected from the analysis introduced above. For example in the model discussed in § 5.4.3 we find that at the transition the entropy density σ, strain S and displacement D of the absolutely stable states are continuous (i.e. $(\sigma^1-\sigma^2) = 0$, $(S^1-S^2) = 0$, $(D^1-D^2) = 0$) while the six derivatives $(\partial\sigma/\partial\theta)_{T,E}$, $(\partial\sigma/\partial E)_{T,\theta}$, etc. are discontinuous. Hence this is classified as a second-order transition. However, $(\partial D/\partial E)_{T,\theta}$, $(\partial D/\partial\theta)_{T,E}$ and $(\partial D/\partial T)_{E,\theta}$ all suffer *infinite* discontinuities at the transition. Thus, the transition lies within the Ehrenfest classification scheme but the behaviour of the model does not satisfy the assumptions used to deduce this scheme.

3.3 Equations of State

3.3.1 Equations of State

When discussing the forms of the elastic and dielectric equations of state in §§ 2.5.3 and 2.5.4 we ignored the role of temperature and entropy in these equations. With the aid of the thermodynamic potentials we now rectify this omission. Consider the free energy $F = F(\theta, S_\alpha, D_i)$; from F we may generate dielectric, elastic and thermal equations of state of the form

$$T_\alpha = \left(\frac{\partial F}{\partial S_\alpha}\right)_{D,\theta}, \quad E_i = \left(\frac{\partial F}{\partial D_i}\right)_{\theta,S}, \quad \sigma = -\left(\frac{\partial F}{\partial \theta}\right)_{S,D} \tag{3.3.1}$$

[see eqn. (3.1.8)]. The former two equations are analogous to eqns. (2.5.20) and (2.5.21), but here the temperature θ is displayed explicitly. Yet another form of the equations of state is obtained from the Gibbs function $G = G(\theta, T_\alpha, E_i)$:

$$S_\alpha = -\left(\frac{\partial G}{\partial T_\alpha}\right)_{E,\theta}, \quad D_i = -\left(\frac{\partial G}{\partial E_i}\right)_{\theta, T}, \quad \sigma = -\left(\frac{\partial G}{\partial \theta}\right)_{T,E}. \tag{3.3.2}$$

Of course, no additional information is contained in the set (3.3.2) as compared with the set (3.3.1) and vice versa. The choice of set is a matter of convenience. In the former case, eqn. (3.3.1), we choose to treat S_α, E_i, θ as the independent variables, whereas in the latter, eqn. (3.3.2), T_α, E_i, θ are the independent variables.

Experimental data on the equations of state of materials are usually presented in terms of compliances and/or the coefficients describing a polynomial representation of the electric and mechanical equations of state. In the following subsections we introduce and describe these quantities.

3.3.2 Compliances

Consider the differential form of the equations of state in eqn. (3.3.2). We write

$$dS_\alpha = \sum_{\beta=1}^{6}\left(\frac{\partial S_\alpha}{\partial T_\beta}\right)_{E,\theta} dT_\beta + \sum_{j=1}^{3}\left(\frac{\partial S_\alpha}{\partial E_j}\right)_{\theta, T}, \quad dE_j + \left(\frac{\partial S_\alpha}{\partial \theta}\right)_{T,E} d\theta; \tag{3.3.3}$$

$$dD_i = \sum_{\beta=1}^{6}\left(\frac{\partial D_i}{\partial T_\beta}\right)_{E,\theta} dT_\beta + \sum_{j=1}^{3}\left(\frac{\partial D_i}{\partial E_j}\right)_{\theta, T}, \quad dE_j + \left(\frac{\partial D_i}{\partial \theta}\right)_{T,E} d\theta; \tag{3.3.4}$$

$$d\sigma = \sum_{\beta=1}^{6}\left(\frac{\partial \sigma}{\partial T_\beta}\right)_{E,\theta} dT_\beta + \sum_{j=1}^{3}\left(\frac{\partial \sigma}{\partial E_j}\right)_{\theta, T}, \quad dE_j + \left(\frac{\partial \sigma}{d\theta}\right)_{T,E} d\theta. \tag{3.3.5}$$

The coefficients or *compliances* $(\partial S_\alpha/\partial T_\beta)_{E,\theta}$, $(\partial S_\alpha/\partial E_j)_{\theta,T}$, $(\partial S_\alpha/\partial\theta)_{T,E}$, etc., are in general function of T_α, E_i, θ. The *compliance matrix* is the square array of these coefficients (Table 3.3.1). Since the differential forms listed above are based on the Gibbs function, we shall refer to this matrix as the compliance matrix generated by G. The elements

of the matrix obey Maxwell relations of the form

$$\left.\begin{aligned}
\left(\frac{\partial S_\alpha}{\partial T_\beta}\right)_{E,\theta} = \left(\frac{\partial S_\beta}{\partial T_\alpha}\right)_{E,\theta}, &\quad \left(\frac{\partial S_\alpha}{\partial E_i}\right)_{T,\theta} = \left(\frac{\partial D_i}{\partial T_\alpha}\right)_{E,\theta}, \\
&\quad \left(\frac{\partial S_\alpha}{\partial \theta}\right)_{E,T} = \left(\frac{\partial \sigma}{\partial T_\alpha}\right)_{E,\theta}, \\
\left(\frac{\partial D_i}{\partial E_j}\right)_{T,\theta} = \left(\frac{\partial D_j}{\partial E_i}\right)_{T,\theta}, &\quad \left(\frac{\partial D_i}{\partial \theta}\right)_{E,T} = \left(\frac{\partial \sigma}{\partial E_i}\right)_{\theta,T}
\end{aligned}\right\} \quad (3.3.6)$$

[these relations follow immediately from eqn. (3.3.2)]; hence the compliance matrix is symmetric.† Thus of the hundred elements in the array only fifty-five are independent. The following names have been given to these compliances:

$\left(\frac{\partial S_\alpha}{\partial T_\beta}\right)_{E,\theta}$ the *isothermal elastic compliance* at constant electric field.

$\left(\frac{\partial D_i}{\partial T_\beta}\right)_{E,\theta}$ the *isothermal piezoelectric coefficient* at constant electric field.

$\left(\frac{\partial S_\alpha}{\partial \theta}\right)_{T,E}$ the *thermal expansion coefficient* at constant stress and constant electric field.

$\left(\frac{\partial D_i}{\partial E_j}\right)_{T,\theta}$ the *isothermal dielectric permittivity* at constant stress.

$\left(\frac{\partial D_i}{\partial \theta}\right)_{T,E}$ the *pyroelectric coefficient* at constant stress and constant electric field.

The coefficient $(\partial\sigma/\partial\theta)_{T,E}$ has no specific name, but the combination $(\theta/\varrho_0)\,(\partial\sigma/\partial\theta)_{T,E}$ is the specific heat at constant stress and constant electric field.

† Because a minus sign appears in some of the Maxwell relations, not all compliance matrices are symmetric [see, for example, Table (3.3.3)].

TABLE 3.3.1.

The ten-by-ten compliance matrix generated by the Gibbs function G.

$\left(\frac{\partial S_1}{\partial T_1}\right)_{E,\theta}$	$\left(\frac{\partial S_1}{\partial T_2}\right)_{E,\theta}$	$\left(\frac{\partial S_1}{\partial T_3}\right)_{E,\theta}$	$\left(\frac{\partial S_1}{\partial T_4}\right)_{E,\theta}$	$\left(\frac{\partial S_1}{\partial T_5}\right)_{E,\theta}$	$\left(\frac{\partial S_1}{\partial T_6}\right)_{E,\theta}$	$\left(\frac{\partial S_1}{\partial E_1}\right)_{\theta,T}$	$\left(\frac{\partial S_1}{\partial E_2}\right)_{\theta,T}$	$\left(\frac{\partial S_1}{\partial E_3}\right)_{\theta,T}$	$\left(\frac{\partial S_1}{\partial \theta}\right)_{T,E}$
$\left(\frac{\partial S_2}{\partial T_1}\right)_{E,\theta}$	$\left(\frac{\partial S_2}{\partial T_2}\right)_{E,\theta}$	$\left(\frac{\partial S_2}{\partial T_3}\right)_{E,\theta}$	$\left(\frac{\partial S_2}{\partial T_4}\right)_{E,\theta}$	$\left(\frac{\partial S_2}{\partial T_5}\right)_{E,\theta}$	$\left(\frac{\partial S_2}{\partial T_6}\right)_{E,\theta}$	$\left(\frac{\partial S_2}{\partial E_1}\right)_{\theta,T}$	$\left(\frac{\partial S_2}{\partial E_2}\right)_{\theta,T}$	$\left(\frac{\partial S_2}{\partial E_3}\right)_{\theta,T}$	$\left(\frac{\partial S_2}{\partial \theta}\right)_{T,E}$
$\left(\frac{\partial S_3}{\partial T_1}\right)_{E,\theta}$	$\left(\frac{\partial S_3}{\partial T_2}\right)_{E,\theta}$	$\left(\frac{\partial S_3}{\partial T_3}\right)_{E,\theta}$	$\left(\frac{\partial S_3}{\partial T_4}\right)_{E,\theta}$	$\left(\frac{\partial S_3}{\partial T_5}\right)_{E,\theta}$	$\left(\frac{\partial S_3}{\partial T_6}\right)_{E,\theta}$	$\left(\frac{\partial S_3}{\partial E_1}\right)_{\theta,T}$	$\left(\frac{\partial S_3}{\partial E_2}\right)_{\theta,T}$	$\left(\frac{\partial S_3}{\partial E_3}\right)_{\theta,T}$	$\left(\frac{\partial S_3}{\partial \theta}\right)_{T,E}$
$\left(\frac{\partial S_4}{\partial T_1}\right)_{E,\theta}$	$\left(\frac{\partial S_4}{\partial T_2}\right)_{E,\theta}$	$\left(\frac{\partial S_4}{\partial T_3}\right)_{E,\theta}$	$\left(\frac{\partial S_4}{\partial T_4}\right)_{E,\theta}$	$\left(\frac{\partial S_4}{\partial T_5}\right)_{E,\theta}$	$\left(\frac{\partial S_4}{\partial T_6}\right)_{E,\theta}$	$\left(\frac{\partial S_4}{\partial E_1}\right)_{\theta,T}$	$\left(\frac{\partial S_4}{\partial E_2}\right)_{\theta,T}$	$\left(\frac{\partial S_4}{\partial S_3}\right)_{\theta,T}$	$\left(\frac{\partial S_4}{\partial \theta}\right)_{T,E}$
$\left(\frac{\partial S_5}{\partial T_1}\right)_{E,\theta}$	$\left(\frac{\partial S_5}{\partial T_2}\right)_{E,\theta}$	$\left(\frac{\partial S_5}{\partial T_3}\right)_{E,\theta}$	$\left(\frac{\partial S_5}{\partial T_4}\right)_{E,\theta}$	$\left(\frac{\partial S_5}{\partial T_5}\right)_{E,\theta}$	$\left(\frac{\partial S_5}{\partial T_6}\right)_{E,\theta}$	$\left(\frac{\partial S_5}{\partial E_1}\right)_{\theta,T}$	$\left(\frac{\partial S_5}{\partial E_2}\right)_{\theta,T}$	$\left(\frac{\partial S_5}{\partial E_3}\right)_{\theta,T}$	$\left(\frac{\partial S_5}{\partial \theta}\right)_{T,E}$
$\left(\frac{\partial S_6}{\partial T_1}\right)_{E,\theta}$	$\left(\frac{\partial S_6}{\partial T_2}\right)_{E,\theta}$	$\left(\frac{\partial S_6}{\partial T_3}\right)_{E,\theta}$	$\left(\frac{\partial S_6}{\partial T_4}\right)_{E,\theta}$	$\left(\frac{\partial S_6}{\partial T_5}\right)_{E,\theta}$	$\left(\frac{\partial S_6}{\partial T_6}\right)_{E,\theta}$	$\left(\frac{\partial S_6}{\partial E_1}\right)_{\theta,T}$	$\left(\frac{\partial S_6}{\partial E_2}\right)_{\theta,T}$	$\left(\frac{\partial S_6}{\partial E_3}\right)_{\theta,T}$	$\left(\frac{\partial S_6}{\partial \theta}\right)_{T,E}$
$\left(\frac{\partial D_1}{\partial T_1}\right)_{E,\theta}$	$\left(\frac{\partial D_1}{\partial T_2}\right)_{E,\theta}$	$\left(\frac{\partial D_1}{\partial T_3}\right)_{E,\theta}$	$\left(\frac{\partial D_1}{\partial T_4}\right)_{E,\theta}$	$\left(\frac{\partial D_1}{\partial T_5}\right)_{E,\theta}$	$\left(\frac{\partial D_1}{\partial T_6}\right)_{E,\theta}$	$\left(\frac{\partial D_1}{\partial E_1}\right)_{\theta,T}$	$\left(\frac{\partial D_1}{\partial E_2}\right)_{\theta,T}$	$\left(\frac{\partial D_1}{\partial E_3}\right)_{\theta,T}$	$\left(\frac{\partial D_1}{\partial \theta}\right)_{T,E}$
$\left(\frac{\partial D_2}{\partial T_1}\right)_{E,\theta}$	$\left(\frac{\partial D_2}{\partial T_2}\right)_{E,\theta}$	$\left(\frac{\partial D_2}{\partial S_3}\right)_{E,\theta}$	$\left(\frac{\partial D_2}{\partial T_4}\right)_{E,\theta}$	$\left(\frac{\partial D_2}{\partial T_5}\right)_{E,\theta}$	$\left(\frac{\partial D_2}{\partial T_6}\right)_{E,\theta}$	$\left(\frac{\partial D_2}{\partial E_1}\right)_{\theta,T}$	$\left(\frac{\partial D_2}{\partial E_2}\right)_{\theta,T}$	$\left(\frac{\partial D_2}{\partial E_3}\right)_{\theta,T}$	$\left(\frac{\partial D_2}{\partial \theta}\right)_{T,E}$
$\left(\frac{\partial D_3}{\partial T_1}\right)_{E,\theta}$	$\left(\frac{\partial D_3}{\partial T_2}\right)_{E,\theta}$	$\left(\frac{\partial D_3}{\partial T_3}\right)_{E,\theta}$	$\left(\frac{\partial D_3}{\partial T_4}\right)_{E,\theta}$	$\left(\frac{\partial D_3}{\partial T_5}\right)_{E,\theta}$	$\left(\frac{\partial D_3}{\partial T_6}\right)_{E,\theta}$	$\left(\frac{\partial D_3}{\partial E_1}\right)_{\theta,T}$	$\left(\frac{\partial D_3}{\partial E_2}\right)_{\theta,T}$	$\left(\frac{\partial D_3}{\partial E_3}\right)_{\theta,T}$	$\left(\frac{\partial D_3}{\partial \theta}\right)_{T}$
$\left(\frac{\partial \sigma}{\partial T_1}\right)_{E,\theta}$	$\left(\frac{\partial \sigma}{\partial T_2}\right)_{E,\theta}$	$\left(\frac{\partial \sigma}{\partial T_3}\right)_{E,\theta}$	$\left(\frac{\partial \sigma}{\partial T_4}\right)_{E,\theta}$	$\left(\frac{\partial \sigma}{\partial T_5}\right)_{E,\theta}$	$\left(\frac{\partial \sigma}{\partial T_6}\right)_{E,\theta}$	$\left(\frac{\partial \sigma}{\partial E_1}\right)_{\theta,T}$	$\left(\frac{\partial \sigma}{\partial E_2}\right)_{\theta,T}$	$\left(\frac{\partial \sigma}{\partial E_3}\right)_{\theta,T}$	$\left(\frac{\partial \sigma}{\partial \theta}\right)_{T,E}$

TABLE 3.3.2.

Names given to various compliances.

For example, $(\partial D_i/\partial E_j)_{x}, \mu$ is the dielectric permittivity at constant x and μ, where x, μ refer to the appropriate elastic and thermal variables which are held fixed. The entries X indicate that this type of compliance has not been named. We note that $(\partial\sigma/\partial\theta)_{x,I}$ must be multiplied by the factor (θ/ϱ_0) to obtain the specific heat at constant x and I.

	Strain	Stress	Displacement	Electric field	Entropy	Temperature
Strain	—	Elastic compliance	Piezo-electric coefficient	Piezo-electric coefficient	X	Thermal expansion coefficient
Stress	Elastic stiffness	—	Piezo-electric coefficient	Piezo-electric coefficient	X	X
Displacement	Piezo-electric coefficient	Piezo-electric coefficient	—	Dielectric permittivity	X	Pyroelectric coefficient
Electric field	Piezo-electric coefficient	Piezo-electric coefficient	Dielectric permeability	—	X	X
Entropy	X	X	X	X	—	Specific heat
Temperature	X	X	X	X	X	—

Other compliance matrices can be formed by choosing different sets of independent and dependent variables. One can derive various thermodynamic identities which relate the compliances of one compliance

matrix with those of another. Examples are given in eqns. (3.1.19) to (3.1.26) and in the next subsection. (The Maxwell relations provide the only thermodynamic identities between compliances of the same compliance matrix.) Not all compliances have been named. Those that have are displayed in Table 3.3.2.

The compliances provide a measure of the coupling between fields; thus, for example, $(\partial S_\alpha/\partial T_\beta)_{E,\theta}$ measures the rate of change of the strain component S_α with respect to changes in T_β under the constraint of constant electric field and constant temperature [and, of course, the remaining components T_γ $(\gamma \neq \beta)$ constant]. The couplings between the mechanical and electric variables, electric and thermal variables and the mechanical and thermal variables are referred to as follows.

Electro-mechanical coupling

(i) The *direct piezoelectric effect*	$(\partial D_i/\partial T_\alpha)_{E,\theta}$, $(\partial D_i/\partial T_\alpha)_{E,\sigma}$.
(ii) The *converse piezoelectric effect*	$(\partial S_\alpha/\partial E_i)_{T,\theta}$, $(\partial S_\alpha/\partial E_i)_{T,\sigma}$.

Electro-thermal coupling

(i) The *pyroelectric effect*	$(\partial D_i/\partial \theta)_{T,E}$, $(\partial D_i/\partial \theta)_{S,E}$.
(ii) The *electro-caloric effect*	$(\partial \sigma/\partial E_i)_{T,\theta}$, $(\partial \sigma/\partial E_i)_{S,\theta}$.

Thermo-elastic coupling

(i) *Thermal expansion*	$(\partial S_\alpha/\partial \theta)_{T,E}$, $(\partial S_\alpha/\partial \theta)_{T,D}$.
(ii) The *piezo-caloric effect*	$(\partial \sigma/\partial T_\alpha)_{E,\theta}$, $(\partial \sigma/\partial T_\alpha)_{\theta,D}$.

The derivatives on the right provide measures of the coupling described on the left. Thus, for example, $(\partial D_i/\partial T_\alpha)_{E,\theta}$ describes the direct piezoelectric effect under isothermal conditions and $(\partial D_i/\partial T_\alpha)_{E,\sigma}$ the same effect under adiabatic conditions. Referring to the Maxwell relations in eqn. (3.3.6), we see that while the two effects in each type of coupling are given different names the compliances describing the effects are identical; thus, for example,

$$\left(\frac{\partial D_i}{\partial T_\alpha}\right)_{E,\theta} \equiv \left(\frac{\partial S_\alpha}{\partial E_i}\right)_{T,\theta} \quad \text{and} \quad \left(\frac{\partial D_i}{\partial T_\alpha}\right)_{E,\sigma} \equiv \left(\frac{\partial S_\alpha}{\partial E_i}\right)_{T,\sigma}.$$

3.3.3 Identities

In this subsection we show how one may derive identities relating the elements of one compliance matrix to those of another.

Consider the differential forms

$$dT_\alpha = \sum_\beta \left(\frac{\partial T_\alpha}{\partial S_\beta}\right)_{D,\theta} dS_\beta + \sum_j \left(\frac{\partial T_\alpha}{\partial D_j}\right)_{S,\theta} dD_j + \left(\frac{\partial T_\alpha}{\partial \theta}\right)_{S,D} d\theta, \quad (3.3.7)$$

$$dE_i = \sum_\beta \left(\frac{\partial E_i}{\partial S_\beta}\right)_{D,\theta} dS_\beta + \sum_j \left(\frac{\partial E_i}{\partial D_j}\right)_{S,\theta} dD_j + \left(\frac{\partial E_i}{\partial \theta}\right)_{S,D} d\theta, \quad (3.3.8)$$

$$d\sigma = \sum_\beta \left(\frac{\partial \sigma}{\partial S_\beta}\right)_{D,\theta} dS_\beta + \sum_j \left(\frac{\partial \sigma}{\partial D_j}\right)_{S,\theta} dD_j + \left(\frac{\partial \sigma}{\partial \theta}\right)_{S,D} d\theta. \quad (3.3.9)$$

Substituting for dT_α, dE_i from eqns. (3.3.7) and (3.3.8) in eqns. (3.3.3), (3.3.4) and (3.3.5), we get

$$\left.\begin{aligned} dS_\alpha \equiv \sum_\beta \left(\frac{\partial S_\alpha}{\partial T_\beta}\right)_{E,\theta} \Bigg[\sum_\gamma \left(\frac{\partial T_\beta}{\partial S_\gamma}\right)_{D,\theta} dS_\gamma + \sum_j \left(\frac{\partial T_\beta}{\partial D_j}\right)_{S,\theta} dD_j \\ + \left(\frac{\partial T_\beta}{\partial \theta}\right)_{S,D} d\theta\Bigg] + \sum_j \left(\frac{\partial S_\alpha}{\partial E_j}\right)_{T,\theta} \Bigg[\sum_\beta \left(\frac{\partial E_j}{\partial S_\beta}\right)_{D,\theta} dS_\beta \\ + \sum_k \left(\frac{\partial E_j}{\partial D_k}\right)_{S,\theta} dD_k + \left(\frac{\partial E_j}{\partial \theta}\right)_{S,D} d\theta\Bigg] + \left(\frac{\partial S_\alpha}{\partial \theta}\right)_{T,E} d\theta, \end{aligned}\right\} \quad (3.3.10)$$

$$\left.\begin{aligned} dD_i \equiv \sum_\beta \left(\frac{\partial D_i}{\partial T_\beta}\right)_{E,\theta} \Bigg[\sum_\gamma \left(\frac{\partial T_\beta}{\partial S_\gamma}\right)_{D,\theta} dS_\gamma + \sum_j \left(\frac{\partial T_\beta}{\partial D_j}\right)_{S,\theta} dD_j \\ + \left(\frac{\partial T_\beta}{\partial \theta}\right)_{S,D} d\theta\Bigg] + \sum_j \left(\frac{\partial D_i}{\partial E_j}\right)_{\theta,T} \Bigg[\sum_\beta \left(\frac{\partial E_j}{\partial S_\beta}\right)_{D,\theta} dS_\beta \\ + \sum_k \left(\frac{\partial E_j}{\partial D_k}\right)_{S,\theta} dD_k + \left(\frac{\partial E_j}{\partial \theta}\right)_{S,D} d\theta\Bigg] + \left(\frac{\partial D_i}{\partial \theta}\right)_{T,E} d\theta, \end{aligned}\right\} \quad (3.3.11)$$

$$\left.\begin{aligned} d\sigma = \sum_\beta \left(\frac{\partial \sigma}{\partial T_\beta}\right)_{E,\theta} \Bigg[\sum_\gamma \left(\frac{\partial T_\beta}{\partial S_\gamma}\right)_{D,\theta} dS_\gamma + \sum_j \left(\frac{\partial T_\beta}{\partial D_j}\right)_{S,\theta} dD_j \\ + \left(\frac{\partial T_\beta}{\partial \theta}\right)_{D,S} d\theta\Bigg] + \sum_j \left(\frac{\partial \sigma}{\partial E_j}\right)_{S,\theta} \Bigg[\sum_\gamma \left(\frac{\partial E_j}{\partial S_\gamma}\right)_{D,\theta} dS_\gamma \\ + \sum_k \left(\frac{\partial E_j}{\partial D_k}\right)_{S,\theta} dD_k + \left(\frac{\partial E_j}{\partial \theta}\right)_{D,S} d\theta\Bigg] + \left(\frac{\partial \sigma}{\partial \theta}\right)_{E,S} d\theta. \end{aligned}\right\} \quad (3.3.12)$$

Since the differentials dS_α, dD_i, $d\theta$ are independent, their coefficients in eqns. (3.3.10) and (3.3.11) must vanish; hence

$$\left.\begin{aligned}\delta_{\alpha\gamma} = \sum_\beta \left(\frac{\partial S_\alpha}{\partial T_\beta}\right)_{E,\theta} \left(\frac{\partial T_\beta}{\partial S_\gamma}\right)_{D,\theta} + \sum_j \left(\frac{\partial S_\alpha}{\partial E_j}\right)_{T,\theta} \left(\frac{\partial E_j}{\partial S_\gamma}\right)_{D,\theta} \\ (\alpha, \gamma = 1, 2, 3, 4, 5, 6),\end{aligned}\right\} \quad (3.3.13)$$

$$\left.\begin{aligned}0 = \sum_\beta \left(\frac{\partial S_\alpha}{\partial T_\beta}\right)_{E,\theta} \left(\frac{\partial T_\beta}{\partial D_j}\right)_{S,\theta} + \sum_k \left(\frac{\partial S_\alpha}{\partial E_k}\right)_{T,\theta} \left(\frac{\partial E_k}{\partial D_j}\right)_{S,\theta} \\ (\alpha = 1, 2, 3, 4, 5, 6; j = 1, 2, 3),\end{aligned}\right\} \quad (3.3.14)$$

$$\left.\begin{aligned}0 = \sum_\beta \left(\frac{\partial S_\alpha}{\partial T_\beta}\right)_{,\theta} \left(\frac{\partial T_\beta}{\partial \theta}\right)_{S,D} + \sum_j \left(\frac{\partial S_\alpha}{\partial E_j}\right)_{T,\theta} \left(\frac{\partial E_j}{\partial \theta}\right)_{S,D} + \left(\frac{\partial S_\alpha}{\partial \theta}\right)_{T,E} \\ (\alpha = 1, 2, 3, 4, 5, 6),\end{aligned}\right\} \quad (3.3.15)$$

$$\left.\begin{aligned}0 = \sum_\beta \left(\frac{\partial D_i}{\partial T_\beta}\right)_{E,\theta} \left(\frac{\partial T_\beta}{\partial S_\gamma}\right)_{D,\theta} + \sum_j \left(\frac{\partial D_i}{\partial E_j}\right)_{\theta,T} \left(\frac{\partial E_j}{\partial S_\gamma}\right)_{D,\theta} \\ (\gamma = 1, 2, 3, 4, 5, 6; i = 1, 2, 3),\end{aligned}\right\} \quad (3.3.16)$$

$$\delta_{ij} = \sum_\beta \left(\frac{\partial D_i}{\partial T}\right)_{E,\theta} \left(\frac{\partial T_\beta}{\partial D_j}\right)_{S,\theta} + \sum_k \left(\frac{\partial D_i}{\partial E_k}\right)_{\theta,T} \left(\frac{\partial E_k}{\partial D_k}\right)_{S,\theta} \quad (i, j = 1, 2, 3), \quad (3.3.17)$$

$$0 = \sum_\beta \left(\frac{\partial D_i}{\partial T_\beta}\right)_{E,\theta} \left(\frac{\partial T_\beta}{\partial \theta}\right)_{S,D} + \sum_j \left(\frac{\partial D_i}{\partial E_j}\right)_{\theta,T} \left(\frac{\partial E_j}{\partial \theta}\right)_{S,D} + \left(\frac{\partial D_i}{\partial \theta}\right)_{T,E} \quad (i = 1, 2, 3), \quad (3.3.18)$$

where $\delta_{\alpha\gamma}$ is the Kronecker delta.

Comparing eqn. (3.3.9) with eqn. (3.3.12) we get

$$\left(\frac{\partial \sigma}{\partial S_\gamma}\right)_{D,} = \sum_\beta \left(\frac{\partial \sigma}{\partial T_\beta}\right)_{E,\theta} \left(\frac{\partial T_\beta}{\partial S_\gamma}\right)_{D,\theta} + \sum_j \left(\frac{\partial \sigma}{\partial E_j}\right)_{S,\theta} \left(\frac{\partial E_j}{\partial S_\gamma}\right)_{D,\theta} \quad (\gamma = 1, 2, 3, 4, 5, 6), \quad (3.3.19)$$

$$\left(\frac{\partial \sigma}{\partial D_j}\right)_{S,\theta} = \sum_\beta \left(\frac{\partial \sigma}{\partial T_\beta}\right)_{E,\theta} \left(\frac{\partial T_\beta}{\partial D_j}\right)_{S,\theta} + \sum_k \left(\frac{\partial \sigma}{\partial E_k}\right)_{S,\theta} \left(\frac{\partial E_k}{\partial D_j}\right)_{S,\theta} \quad (j = 1, 2, 3), \quad (3.3.20)$$

$$\left(\frac{\partial \sigma}{\partial \theta}\right)_{S,D} = \sum_\beta \left(\frac{\partial \sigma}{\partial T_\beta}\right)_{E,\theta} \left(\frac{\partial T_\beta}{\partial \theta}\right)_{D,S} + \sum_j \left(\frac{\partial \sigma}{\partial E_j}\right)_{S,\theta} \left(\frac{\partial E_j}{\partial \theta}\right)_{D,S} + \left(\frac{\partial \sigma}{\partial \theta}\right)_{S,E}. \quad (3.3.21)$$

Equations (3.3.13) to (3.3.21) provide the complete set of relationships between the elements of the compliance matrix of G and those of F. Given the values for the elements of one matrix we may calculate the elements of the other by means of these equations. A similar procedure yields the complete set of identities relating the elements of any two compliance matrices. In particular the identities in eqns. (3.1.19) to (3.1.24) can be obtained by this approach.

3.3.4 LINEAR EQUATIONS OF STATE I

Let us assume that we may represent the free energy F by a polynomial in the arguments S_α, D_i [cf. § 2.5.4]. Consider the following quadratic representation:

$$\left.\begin{aligned} F(S_\alpha, D_i, \theta) = (F)^0 &+ \sum_{\alpha=1}^{6}\left(\frac{\partial F}{\partial S_\alpha}\right)^0 S_\alpha + \sum_{i=1}^{3}\left(\frac{\partial F}{\partial D_i}\right)^0 D_i \\ &+\frac{1}{2}\sum_{\alpha,\beta}\left(\frac{\partial^2 F}{\partial S_\alpha\,\partial S_\beta}\right)^0 S_\alpha S_\beta + \sum_{\alpha,i}\left(\frac{\partial^2 F}{\partial S_\alpha\,\partial D_i}\right)^0 S_\alpha D_i \\ &+\frac{1}{2}\sum_{i,j}\left(\frac{\partial^2 F}{\partial D_i\,\partial D_j}\right)^0 D_i D_j, \end{aligned}\right\} \tag{3.3.22}$$

where the superscript 0 represents the enclosed function with S_α, D_i set equal to zero. Thus the quantities $(F)^0$, $(\partial F/\partial S_\alpha)^0$, etc., are functions only of θ. Because of the nine identities $(\partial^2 F/\partial S_\alpha \partial S_\beta)^0 \equiv (\partial^2 F/\partial S_\beta \partial S_\alpha)^0$ and $(\partial^2 F/\partial D_i \partial D_j)^0 \equiv (\partial^2 F/\partial D_j \partial D_i)^0$, there are at most fifty-five† different temperature dependent functions appearing in this representation.

Substituting F from eqn. (3.3.22) we obtain the following *linear isothermal equations of state* for the components T_α, E_i:

$$T_\alpha = \left(\frac{\partial F}{\partial S_\alpha}\right)_{D,\theta} = T_\alpha^0 + \sum_{\beta=1}^{6} c_{\alpha\beta}^{D,\theta} S_\beta - \sum_{j=1}^{3} h_{j\alpha}^{\theta} D_j, \tag{3.3.23}$$

$$E_i = \left(\frac{\partial F}{\partial D_i}\right)_{S,\theta} = E_i^0 - \sum_{\beta=1}^{6} h_{i\beta}^{\theta} S_\beta + \sum_{j=1}^{3} \beta_{ij}^{S,\theta} D_j, \tag{3.3.24}$$

† Explicitly these are composed of one function $(F)^0$, six functions $(\partial F/\partial S_\alpha)^0$, three functions $(\partial F/\partial D_i)^0$, twenty-one functions $(\partial^2 F/\partial S_\alpha\,\partial S_\beta)^0$, eighteen functions $(\partial^2 F/\partial S_\alpha\,\partial D_i)^0$ and six functions $(\partial^2 F/\partial D_i\,\partial D_j)^0$.

where we have set

$$\left(\frac{\partial F}{\partial S_\alpha}\right)^0_{D,\theta} = T^0_\alpha \qquad \left(\frac{\partial F}{\partial D_i}\right)^0_{S,\theta} = E^0_i, \tag{3.3.25}$$

$$\left.\begin{aligned} \left(\frac{\partial^2 F}{\partial S_\alpha\,\partial S_\beta}\right)^0_{D,\theta} &= \left(\frac{\partial T_\alpha}{\partial S_\beta}\right)^0_{D,\theta} = c^{D,\theta}_{\alpha\beta}, \\ \left(\frac{\partial^2 F}{\partial S_\alpha\,\partial D_i}\right)^0_{\theta} &= \left(\frac{\partial T_\alpha}{\partial D_i}\right)^0_{\theta,S} = -h^\theta_{i\alpha}, \\ \left(\frac{\partial^2 F}{\partial D_i\partial D_j}\right)^0_{S,\theta} &= \left(\frac{\partial E_i}{\partial D_j}\right)^0_{S,\theta} = \beta^{S,\theta}_{ij}. \end{aligned}\right\} \tag{3.3.26}$$

$T^0_\alpha(\theta)$ is the *thermal stress* and $E^0_i(\theta)$ the *thermal electric field.* These *functions* represent the values of the fields T_α, E_i which must be applied to achieve the condition $S_\alpha = 0$, $D_i = 0$ for arbitrary θ. In an obvious extension of the notation and nomenclature of § 2.5.4, $c^{D,\theta}_{\alpha\beta}$ is the *linear isothermal elastic stiffness at constant displacement*, $h^\theta_{i\alpha}$ the *linear isothermal piezoelectric coefficient* and $\beta^{S,\theta}_{ij}$ the *linear isothermal permeability at constant strain.* Equations (3.3.23) and (3.3.24) represent a generalization of the linear equations of state, (2.5.28) and (2.5.29), to the case in which thermal effects play a role.† In this notation the quadratic representation for F takes the form

$$\left.\begin{aligned} F(S_\alpha, D_i, \theta) = (F)^0 + \sum_{\alpha=1}^{6} T^0_\alpha S_\alpha + \sum_{i=1}^{3} E^0_i D_i + \frac{1}{2}\sum_{\alpha,\beta} c^{D,\theta}_{\alpha\beta} S_\alpha S_\beta \\ -\sum_{i,\alpha} h^\theta_{i\alpha} S_\alpha D_i + \frac{1}{2}\sum_{i,j} \beta^{S,\theta}_{ij} D_i D_j \end{aligned}\right\} \tag{3.3.27}$$

† In a strictly linear theory we must replace each expression by its leading non-vanishing term. Thus in the theory described above we should treat T^0_α, E^0_i as infinitesimals vanishing at $\theta = \theta_i$ (if T^0_α, E^0_i are not infinitesimals then a strictly linear theory predicts that $T_\alpha = T^0_\alpha$, $E_i = E^0_i$, i.e. functions only of θ) and neglect the temperature variation of the coefficients $c^{D,\theta}_{\alpha\beta}$, $h^\theta_{i\alpha}$ and $\beta^{S,\theta}_{ij}$. In the case of the ferroelectrics T^0_α are small over a wide range of temperatures about θ_i (see footnote on p. 113) and $E^0_i \equiv 0$. It also appears that the temperature dependence of the linear elastic and dielectric equations of state of these crystals is satisfactorily described by treating $c^{D,\theta}_{\alpha\beta}$, $h^\theta_{i\alpha}$ and $\beta^{S,\theta}_{ij}$ in eqns. (3.3.23) and (3.3.24) as functions of θ. It is this theory—quasi-linear in the temperature dependence—which is developed in §§ 3.3.4 and 3.3.5. To our knowledge there have been no experimental or theoretical investigations of the range of validity of this theory.

and the corresponding entropy density σ is

$$\left.\begin{aligned}\sigma = \sigma^0 - \sum_{\alpha=1}^{6}\left(\frac{dT_\alpha^0}{d\theta}\right)S_\alpha - \sum_{i=1}^{3}\left(\frac{dE_i^0}{d\theta}\right)D_i - \frac{1}{2}\sum_{\alpha,\beta}\left(\frac{dc_{\alpha\beta}^{D,\theta}}{d\theta}\right)S_\alpha S_\beta \\ +\sum_{i,\alpha}\left(\frac{dh_{i\alpha}^{\theta}}{d\theta}\right)S_\alpha D_i - \frac{1}{2}\sum_{i,j}\left(\frac{d\beta_{ij}^{S,\theta}}{d\theta}\right)D_i D_j,\end{aligned}\right\} \quad (3.3.28)$$

where $\sigma^0 = -(\partial F/\partial\theta)^0$.

The elements of the compliance matrix of F in the linear theory are

$$\left(\frac{\partial T_\alpha}{\partial S_\beta}\right)_{D,\theta} = c_{\alpha\beta}^{D,\theta}, \quad \left(\frac{\partial T_\alpha}{\partial D_i}\right)_{S,\theta} = -h_{i\alpha}^{\theta}, \quad \left(\frac{\partial E_i}{\partial D_j}\right)_{S,\theta} = \beta_{ij}^{S,\theta}, \quad (3.3.29)$$

$$\left(\frac{\partial \sigma}{\partial S_\alpha}\right)_{D,\theta} = -\left(\frac{dT_\alpha^0}{d\theta}\right) - \sum_{\beta=1}^{6}\left(\frac{dc_{\alpha\beta}^{D,\theta}}{d\theta}\right)S_\beta + \sum_{j=1}^{3}\left(\frac{dh_{j\alpha}^{\theta}}{d\theta}\right)D_j, \quad (3.3.30)$$

$$\left(\frac{\partial \sigma}{\partial D_i}\right)_{S,\theta} = -\left(\frac{dE_i^0}{d\theta}\right) + \sum_{\beta=1}^{6}\left(\frac{dh_{i\beta}^{\theta}}{d\theta}\right)S_\beta - \sum_{j=1}^{3}\left(\frac{d\beta_{ij}^{S,\theta}}{d\theta}\right)D_j, \quad (3.3.31)$$

$$\left.\begin{aligned}\left(\frac{\partial \sigma}{\partial \theta}\right)_{S,D} = \left(\frac{d\sigma^0}{d\theta}\right) - \sum_{\alpha=1}^{6}\left(\frac{d^2T_\alpha^0}{d\theta^2}\right)S_\alpha - \sum_{i=1}^{3}\left(\frac{d^2E_i}{d\theta^2}\right)D_i \\ -\frac{1}{2}\sum_{\alpha,\beta}\left(\frac{d^2c_{\alpha\beta}^{D,\theta}}{d\theta^2}\right)S_\alpha S_\beta + \sum_{i,\alpha}\left(\frac{d^2h_{i\alpha}^{\theta}}{d\theta^2}\right)S_\alpha D_i \\ -\frac{1}{2}\sum_{i,j}\left(\frac{d^2\beta_{ij}^{S,\theta}}{d\theta^2}\right)D_i D_j.\end{aligned}\right\} \quad (3.3.32)$$

When $S_\alpha = 0$, $D_i = 0$ the compliance matrix of F has the form displayed in Table 3.3.3. This is easily verified from eqns. (3.3.29) to (3.3.32).

When $T_\alpha \equiv 0$, $E_i \equiv 0$ we set $S_\alpha = \gamma_\alpha(\theta)$ the *thermal expansion* in zero electric field and $D_i = P_i^0(\theta)$ the *spontaneous polarization* at zero stress. Crystals with $P_i^0(\theta) \not\equiv 0$ are referred to as *pyroelectric*. In the initial state $T_\alpha = 0$, $E_i = 0$, $S_\alpha = 0$, $D_i = 0$, and $\theta = \theta_i$; hence

$$\gamma_\alpha(\theta_i) = 0 \quad (\alpha = 1, 2, 3, 4, 5, 6); \quad P_j^0(\theta_i) = 0 \quad (j = 1, 2, 3) \quad (3.3.33)$$

Aside from satisfying these conditions, the functions $\gamma_\alpha(\theta)$, $P_i^0(\theta)$ are in

TABLE 3.3.3.

The zero displacement, zero strain compliance matrix generated by the free energy F in the notation of eqns. (3.3.23) and (3.3.24).

$c_{11}^{D,\theta}$	$c_{12}^{D,}$	$c_{13}^{D,\theta}$	$c_{14}^{D,\theta}$	$c_{15}^{D,\theta}$	$c_{16}^{D,\theta}$	$-h_{11}^{\theta}$	$-h_{21}^{\theta}$	$-h_{31}^{\theta}$	$-\frac{dT_1^0}{d\theta}$
$c_{21}^{D,\theta}$	$c_{22}^{D,\theta}$	$c_{23}^{D,\theta}$	$c_{24}^{D,\theta}$	$c_{25}^{D,\theta}$	$c_{26}^{D,\theta}$	$-h_{12}^{\theta}$	$-h_{22}^{\theta}$	$-h_{32}^{\theta}$	$-\frac{dT_2^0}{d\theta}$
$c_{31}^{D,\theta}$	$c_{32}^{D,\theta}$	$c_{33}^{D,\theta}$	$c_{34}^{D,\theta}$	$c_{35}^{D,\theta}$	$c_{36}^{D,\theta}$	$-h_{13}^{\theta}$	$-h_{23}^{\theta}$	$-h_{33}^{\theta}$	$-\frac{dT_3^0}{d\theta}$
$c_{41}^{D,\theta}$	$c_{42}^{D,\theta}$	$c_{43}^{D,\theta}$	$c_{44}^{D,\theta}$	$c_{45}^{D,\theta}$	$c_{46}^{D,\theta}$	$-h_{14}^{\theta}$	$-h_{24}^{\theta}$	$-h_{34}^{\theta}$	$-\frac{dT_4^0}{d\theta}$
$c_{51}^{D,\theta}$	$c_{52}^{D,\theta}$	$c_{53}^{D,\theta}$	$c_{54}^{D,\theta}$	$c_{55}^{D,\theta}$	$c_{56}^{D,\theta}$	$-h_{15}^{\theta}$	$-h_{25}^{\theta}$	$-h_{35}^{\theta}$	$-\frac{dT_5^0}{d\theta}$
$c_{61}^{D,}$	$c_{62}^{D,\theta}$	$c_{63}^{D,\theta}$	$c_{64}^{D,\theta}$	$c_{65}^{D,\theta}$	$c_{66}^{D,\theta}$	$-h_{16}^{\theta}$	$-h_{26}^{\theta}$	$-h_{36}^{\theta}$	$-\frac{dT_6}{d\theta}$
$-h_{11}^{\theta}$	$-h_{12}^{\theta}$	$-h_{13}^{\theta}$	$-h_{14}^{\theta}$	$-h_{15}^{\theta}$	$-h_{16}^{\theta}$	$\beta_{11}^{S,\theta}$	$\beta_{12}^{S,\theta}$	$\beta_{13}^{S,\theta}$	$-\frac{dE_1^0}{d\theta}$
$-h_{21}^{\theta}$	$-h_{22}^{\theta}$	$-h_{23}^{\theta}$	$-h_{24}^{\theta}$	$-h_{25}^{\theta}$	$-h_{26}^{\theta}$	$\beta_{21}^{S,\theta}$	$\beta_{22}^{S,\theta}$	$\beta_{23}^{S,\theta}$	$-\frac{dE_2^0}{d\theta}$
$-h_{31}^{\theta}$	$-h_{32}^{\theta}$	$-h_{33}^{\theta}$	$-h_{34}^{\theta}$	$-h_{35}^{\theta}$	$-h_{36}^{\theta}$	$\beta_{31}^{S,\theta}$	$\beta_{32}^{S,\theta}$	$\beta_{33}^{S,\theta}$	$-\frac{dE_3^0}{d\theta}$
$\frac{dT_1^0}{d\theta}$	$\frac{dT_2^0}{d\theta}$	$\frac{dT_3^0}{d\theta}$	$\frac{dT_4^0}{d\theta}$	$\frac{dT_5^0}{d\theta}$	$\frac{dT_6^0}{d\theta}$	$\frac{dE_1^0}{d\theta}$	$\frac{dE_2^0}{d\theta}$	$\frac{dE_3^0}{d\theta}$	$\frac{d\sigma^0}{d\theta}$

general completely arbitrary. The quantities $(d\gamma_\alpha/d\theta)$, $(dP_i^0/d\theta)$ are respectively the *thermal expansion coefficients and the pyroelectric coefficients at zero stress and electric field. In the linear theory* the functions $\gamma_\alpha(\theta)$, $P_i^0(\theta)$ are determined by eqns. (3.3.23) and (3.3.24) with the left sides put equal to zero; thus

$$0 = T_\alpha^0 + \sum_{\beta=1}^{6} c_{\alpha,\beta}^{D,\theta}\gamma_\beta - \sum_{j=1}^{3} h_{j\alpha}P_j^0, \tag{3.3.34}$$

$$0 = E_i^0 - \sum_{\beta=1}^{6} h_{i\beta}^{\theta}\gamma_\beta + \sum_{j=1}^{3} \beta_{i,j}^{S,\theta}P_j^0. \tag{3.3.35}$$

Subtracting eqn. (3.3.34) from eqn. (3.3.23) and eqn. (3.3.35) from eqn. (3.3.24) we get the following alternative form for the linear equa-

tions of state:

$$T_\alpha = \sum_{\beta=1}^{6} c_{\alpha,\beta}^{D,\theta}(S_\beta - \gamma_\beta) - \sum_{j=1}^{3} h_{j\alpha}^{\theta}(D_j - P_j^0), \tag{3.3.36}$$

$$E_i = -\sum_{\beta=1}^{6} h_{i\beta}^{\theta}(S_\beta - \gamma_\beta) + \sum_{j=1}^{3} \beta_{ij}^{S,\theta}(D_j - P_j^0). \tag{3.3.37}$$

Comparing eqns. (3.3.35), (3.3.36) and (2.5.28), (2.5.29), we see that the former set of equations reduce to the latter if

$$|S_\alpha| \gg |\gamma_\alpha| \quad (\alpha = 1, 2, 3, 4, 5, 6), \tag{3.3.38}$$

$$|D_i| \gg |P_i^0| \quad (i = 1, 2, 3). \tag{3.3.39}$$

3.3.5 Linear Equations of State II

Other forms of the linear equations of state may be more convenient in particular circumstances. Thus, for example, we may solve eqns. (3.3.36) and (3.3.37) for $(S_\alpha - \gamma_\alpha)$, $(D_i - P_i^0)$ as functions of T_α, E_i. We write the solutions in the form [compare eqns. (2.5.35) and (2.5.36)]

$$S_\alpha = \gamma_\alpha + \sum_{\beta=1}^{6} s_{\alpha\beta}^{E,\theta} T_\beta + \sum_{j=1}^{3} d_{j\alpha}^{\theta} E_j, \tag{3.3.40}$$

$$D_i = P_i^0 + \sum_{\beta=1}^{6} d_{i\beta}^{\theta} T_\beta + \sum_{j=1}^{3} \varepsilon_{ij}^{T,\theta} E_j, \tag{3.3.41}$$

where the coefficients $s_{\alpha\beta}^{E,\theta}$, $d_{j\alpha}^{\theta}$, $E_{ij}^{T,\theta}$ are related to $c_{\alpha\beta}^{D,\theta}$, $h_{i\alpha}^{\theta}$, $\beta_{ij}^{S,\theta}$ in the fashion described in § 2.5.4.

We note that eqns. (3.3.40) and (3.3.41) are generated by the following quadratic representation for G:

$$\left.\begin{aligned} G(T_\alpha, E_i, \theta) = (G)^0 - \sum_{\alpha=1}^{6} \gamma_\alpha T_\alpha - \sum_{i=1}^{3} P_i^0 E_i - \tfrac{1}{2}\sum_{\alpha,\beta} s_{\alpha\beta}^{E,\theta} T_\alpha T_\beta \\ - \sum_{i,\alpha} d_{i\alpha}^{\theta} T_\alpha E_i - \tfrac{1}{2}\sum_{i,j} \varepsilon_{ij}^{T,\theta} E_i E_j, \end{aligned}\right\} \tag{3.3.42}$$

hence

$$\left.\begin{aligned} s_{\alpha\beta}^{E,\theta} = \left(\frac{\partial S_\alpha}{\partial T_\beta}\right)_{E,\theta}^0 = -\left(\frac{\partial^2 G}{\partial T_\alpha\,\partial T_\beta}\right)_{E,\theta}^0; \quad d_{i\alpha}^{\theta} = \left(\frac{\partial S_\alpha}{\partial E_i}\right)_{T,\theta}^0 = -\left(\frac{\partial^2 G}{\partial E_i\,\partial T_\alpha}\right)_{\theta}^0; \\ \varepsilon_{ij}^{T,\theta} = +\left(\frac{\partial D_i}{\partial E_j}\right)_{T,\theta}^0 = -\left(\frac{\partial^2 G}{\partial E_i\,\partial E_j}\right)_{T,\theta}^0. \end{aligned}\right\} \tag{3.3.43}$$

The symbols with the superscript 0 in eqns. (3.3.42) and (3.3.43) represent the enclosed functions with the independent variables T_α, E_i set equal to zero. These quantities are thus functions only of θ. The following nomenclature is commonly used in the literature of ferroelectricity:

$s_{\alpha\beta}^{E,\theta}$ the *linear isothermal elastic compliance at constant electric field.*

$d_{i\alpha}^{\theta}$ the *linear isothermal piezoelectric coefficient.*

$\varepsilon_{ij}^{T,\theta}$ the *linear isothermal dielectric permittivity at constant stress.*

The entropy density σ derived from eqns. (3.3.42) is

$$\left.\begin{aligned}\sigma = \sigma^0 + \sum_{\alpha=1}^{6}\left(\frac{d\gamma_\alpha}{d\theta}\right)T_\alpha + \sum_{i=1}^{3}\left(\frac{dP_i^0}{d\theta}\right)E_i + \frac{1}{2}\sum_{\alpha,\beta}\left(\frac{ds_{\alpha\beta}^{E,\theta}}{d\theta}\right)T_\alpha T_\beta \\ + \sum_{i,\alpha}\left(\frac{dd_{i\alpha}^{\theta}}{d\theta}\right)T_\alpha E_i + \frac{1}{2}\sum_{i,j}\left(\frac{d\varepsilon_{ij}^{T,\theta}}{d\theta}\right)E_i E_j \,.\end{aligned}\right\} \quad (3.3.44)$$

Using eqns. (3.3.40) to (3.3.44) it is a simple exercise to show that in terms of the notation P_i^0, γ_α, $s_{\alpha\beta}^{E,\theta}$, $d_{i\alpha}^{\theta}$, $\varepsilon_{ij}^{T,\theta}$, the compliance matrix of G in zero stress and electric field takes the form displayed in Table 3.3.4.

It is evident from the discussion given in this and the previous subsection that this type of analysis can be applied to the quadratic representation of any thermodynamic potential. It is also evident how one can generate a notation and nomenclature for the quadratic coefficients appearing in these representations—we merely add a symbol indicating the appropriate thermal constraint to the notation described in Table 2.5.1 and an appropriate adjective (adiabatic or isothermal) to the nomenclature.† Thus, for example,

$c_{\alpha\beta}^{D,\sigma}$ is the *linear adiabatic elastic stiffness at constant displacement*;

$h_{i\alpha}^{\sigma}$ is the *linear adiabatic piezoelectric coefficient*; and

$\beta_{ij}^{S,\sigma}$ is the *linear adiabatic dielectric permeability at constant strain.*

This last set of coefficients describes the linear adiabatic variation of stress and electric field with changes in strain and displacement.

† When there can be no ambiguity it is customary to omit superscripts.

TABLE 3.3.4.

The zero electric field, zero stress compliance matrix generated by the Gibbs function G in the notation of eqns. (3.3.40) and (3.3.41).

$s_{11}^{E,\theta}$	$s_{12}^{E,\theta}$	$s_{13}^{E,\theta}$	$s_{14}^{E,\theta}$	$s_{15}^{E,\theta}$	$s_{16}^{E,\theta}$	d_{11}^{θ}	d_{21}^{θ}	d_{31}^{θ}	$(d\gamma_1/d\theta)$
$s_{21}^{E,\theta}$	$s_{22}^{E,\theta}$	$s_{23}^{E,\theta}$	$s_{24}^{E,\theta}$	$s_{25}^{E,\theta}$	$s_{26}^{E,\theta}$	d_{12}^{θ}	d_{22}^{θ}	d_{32}^{θ}	$(d\gamma_2/d\theta)$
$s_{31}^{E,\theta}$	$s_{32}^{E,\theta}$	$s_{33}^{E,\theta}$	$s_{34}^{E,\theta}$	$s_{35}^{E,\theta}$	$s_{36}^{E,\theta}$	d_{13}^{θ}	d_{23}^{θ}	d_{33}^{θ}	$(d\gamma_3/d\theta)$
$s_{41}^{E,\theta}$	$s_{42}^{E,\theta}$	$s_{43}^{E,\theta}$	$s_{44}^{E,\theta}$	$s_{45}^{E,\theta}$	$s_{46}^{E,\theta}$	d_{14}^{θ}	d_{24}^{θ}	d_{34}^{θ}	$(d\gamma_4/d\theta)$
$s_{51}^{E,\theta}$	$s_{52}^{E,\theta}$	$s_{53}^{E,\theta}$	$s_{54}^{E,\theta}$	$s_{55}^{E,\theta}$	$s_{56}^{E,\theta}$	d_{15}^{θ}	d_{25}^{θ}	d_{35}^{θ}	$(d\gamma_5/d\theta)$
$s_{61}^{E,\theta}$	$s_{62}^{E,\theta}$	$s_{63}^{E,\theta}$	$s_{64}^{E,\theta}$	$s_{65}^{E,\theta}$	$s_{66}^{E,\theta}$	d_{16}^{θ}	d_{26}^{θ}	d_{36}^{θ}	$(d\gamma_6/d\theta)$
d_{11}^{θ}	d_{12}^{θ}	d_{13}^{θ}	d_{14}^{θ}	d_{15}^{θ}	d_{16}^{θ}	$\varepsilon_{11}^{T,\theta}$	$\varepsilon_{12}^{T,\theta}$	$\varepsilon_{13}^{T,\theta}$	$(dP_1^0/d\theta)$
d_{21}^{θ}	d_{22}^{θ}	d_{23}^{θ}	d_{24}^{θ}	d_{25}^{θ}	d_{26}^{θ}	$\varepsilon_{21}^{T,\theta}$	$\varepsilon_{22}^{T,\theta}$	$\varepsilon_{23}^{T,\theta}$	$(dP_2^0/d\theta)$
d_{31}^{θ}	d_{32}^{θ}	d_{33}^{θ}	d_{34}^{θ}	d_{35}^{θ}	d_{36}^{θ}	$\varepsilon_{31}^{T,\theta}$	$\varepsilon_{32}^{T,}$	$\varepsilon_{33}^{T,\theta}$	$(dP_3^0/d\theta)$
$\dfrac{d\gamma_1}{d\theta}$	$\dfrac{d\gamma_2}{d\theta}$	$\dfrac{d\gamma_3}{d\theta}$	$\dfrac{d\gamma_4}{d\theta}$	$\dfrac{d\gamma_5}{d\theta}$	$\dfrac{d\gamma_6}{d\theta}$	$\dfrac{dP_1^0}{d\theta}$	$\dfrac{dP_2^0}{d\theta}$	$\dfrac{dP_3^0}{d\theta}$	$\left(\dfrac{d\sigma^0}{d\theta}\right)$

We have already emphasized, § 3.3.1, that the information contained in any given set of equations of state is fully equivalent to the information contained in any other set. This means in particular that the sets of functions of θ [namely $(c_{\alpha\beta}^{D,\theta}, h_{i\alpha}^{\theta}, \beta_{ij}^{S,\theta})$, $(s_{\alpha\beta}^{E,\sigma}, h_{i\alpha}^{\sigma}, E_{ij}^{T,\sigma})$, etc.] are interdependent. We close this subsection by indicating how the various identities describing these connections may be determined.

(i) $(c_{\alpha\beta}^{E,\theta}, e_{j\alpha}^{\theta}, \varepsilon_{ij}^{S,\theta}) \rightarrow (c_{\alpha\beta}^{D,\theta}, h_{j\alpha}^{\theta}, \beta_{ij}^{S,\theta})$

From eqns. (2.5.40) to (2.5.42)

$$c_{\alpha\beta}^{E,\theta} = c_{\alpha\beta}^{D,\theta} - \sum_{j,k} h_{j\alpha}^{\theta}(\beta_{jk}^{S,\theta})^{-1} h_{k\beta}^{\theta}, \tag{3.3.45}$$

$$e_{j\alpha}^{\theta} = \sum_{k} h_{k\alpha}^{\theta}(\beta_{jk}^{S,\theta})^{-1}, \tag{3.3.46}$$

$$\varepsilon_{ij}^{S,\theta} = (\beta_{ij}^{S,\theta})^{-1}. \tag{3.3.47}$$

(ii) $(s_{\alpha\beta}^{E,\theta}, d_{j\alpha}^{\theta}, \varepsilon_{ij}^{T,\theta}) \rightarrow (c_{\alpha\beta}^{E,\theta}, e_{j\alpha}^{\theta}, \varepsilon_{ij}^{S,\theta})$

Using the technique described in § (2.5.4), we find that

$$s_{\alpha\beta}^{E,\theta} = (c_{\alpha\beta}^{E,\theta})^{-1}, \tag{3.3.48}$$

$$d_{j\alpha}^{\theta} = \sum_{\gamma} (c_{\alpha\gamma}^{E,\theta})^{-1} e_{j\gamma}, \tag{3.3.49}$$

$$\varepsilon_{ij}^{T,\theta} = \varepsilon_{ij}^{S,\theta} + \sum_{\beta,\gamma} e_{i\beta}^{\theta} (c_{\gamma\beta}^{E,\theta})^{-1} e_{j\gamma}, \tag{3.3.50}$$

where the $(c_{\alpha\beta}^{E,\theta})^{-1}$ are the elements of the matrix reciprocal to $[c_{\alpha\beta}^{E,\theta}]$.

(iii) $(s_{\alpha\beta}^{E,\sigma}, d_{i\alpha}^{\sigma}, \varepsilon_{ij}^{T,\sigma}) \rightarrow (s_{\alpha\beta}^{E,\theta}, d_{i\alpha}^{\theta}, \varepsilon_{ij}^{T,\theta})$

From eqns. (3.1.19) to (3.1.21) we find:

$$s_{\alpha\beta}^{E,\sigma} = s_{\alpha\beta}^{E,\theta} - \theta \left(\frac{d\gamma_\alpha}{d\theta}\right) \left(\frac{d\gamma_\beta}{d\theta}\right) \Big/ \varrho_0 C_{E,T}^0, \tag{3.3.51}$$

$$d_{i\alpha}^{\sigma} = d_{i\alpha}^{\theta} - \theta \left(\frac{dP_i^0}{d\theta}\right) \left(\frac{d\gamma_\alpha}{d\theta}\right) \Big/ \varrho_0 C_{E,T}^0, \tag{3.3.52}$$

$$\varepsilon_{ij}^{T,\sigma} = \varepsilon_{ij}^{T,\theta} - \theta \left(\frac{dP_i^0}{d\theta}\right) \left(\frac{dP_j^0}{d\theta}\right) \Big/ \varrho_0 C_{E,T}^0, \tag{3.3.53}$$

where $C_{E,T}^0$ represents the specific heat in zero electric field and at zero stress.

3.3.6 Transformation of Axes

In this subsection we describe the effects of a transformation of the coordinate axes on the fields and coefficients appearing above. It is convenient for this discussion (a) to drop the Voigt notation, and (b) to use the infinitesimal strain tensor S_{ij} in place of the strain S_α. The components of the fields X_i, u_i,† S_{ij}, T_{ij}, D_i, E_i are measured in some rectangular coordinate system F. Consider a second rectangular coordinate system F'. Let these two systems have a common origin. We write the coordinate transformation relating these axes in the form

$$X_i' = \sum_{=1}^{3} R_{ij} X_j,$$

† See § 2.2.1.

where $[R_{ij}]$ is a matrix with the following properties:

$$\sum_{i=1}^{3} R_{ij}R_{ik} = \delta_{jk}, \quad \sum_{i=1}^{3} R_{ji}R_{ki} = \delta_{jk}, \quad \det R_{ij} = \pm 1.$$

We may regard this matrix as representing the operation R (rotation and/or inversion) which carries the unprimed axes into coincidence with the primed axes. R belongs to the Full Orthogonal group.†

We denote components measured in F' by primes. Since $\boldsymbol{u}, \boldsymbol{D}, \boldsymbol{E}$ are vectors, their components transform as follows:

$$u_i' = \sum_{j=1}^{3} R_{ij}u_i, \quad D_i' = \sum_{j=1}^{3} R_{ij}D_j, \quad E_i' = \sum_{j=1}^{3} R_{ij}E_j.$$

By definition, § 2.2.1, $S_{ij} = \frac{1}{2}(\partial u_i/\partial X_j + \partial u_l/\partial X_i)$; hence

$$S_{ij}' = \tfrac{1}{2}(\partial u_i'/\partial X_j' + \partial u_j'/\partial X_i') = \tfrac{1}{2}\sum_{k,l} R_{ik}R_{jl}(\partial u_k/\partial X_l + \partial u_l/\partial X_k)$$

or

$$S_{ij}' = \sum_{k,l} R_{ik}R_{jl}S_{kl}. \tag{3.3.54}$$

Thus the infinitesimal strain S_{ij} is a second-rank tensor. It follows immediately that the strain S_α is not a tensor.

From eqn. (2.3.8)

$$t_i = \sum_j T_{ji}n_j.$$

Since t_i, n_i are components of vectors, T_{ij} is a second-rank tensor, i.e.

$$T_{ij}' = \sum_{k,l} R_{ik}R_{jl}T_{kl}. \tag{3.3.55}$$

To determine the transformation properties of the coefficients T_{ij}^0, E_i^0, $c_{\alpha\beta}^{D,\theta}$, $h_{i\alpha}^{\theta}$, $\beta_{ij}^{S,\theta}$ we first refer to eqn. (3.3.27) and write:

$$\left.\begin{aligned} F(S_{ij}, D_i, \theta) = (F)^0 &+ \sum_{i,j} T_{ij}^0 S_{ij} + \sum_i E_i^0 D_i \\ &+ \tfrac{1}{2}\sum_{i,j} c_{ijkl}^{D,\theta} S_{ij}S_{kl} + \sum_{i,j,k} h_{ijk}^{\theta} S_{jk}D_i \\ &+ \tfrac{1}{2}\sum_{i,j} \beta_{ij}^{S,\theta} D_iD_j. \end{aligned}\right\} \tag{3.3.56}$$

† The set of all three-dimensional rotations plus inversion.

It is a simple exercise to show that

$$T_{ij}^{0} = T_{\alpha}^{0}, \quad c_{ijkl}^{D,\theta} = c_{\alpha\beta}^{D,\theta}, \; h_{kij}^{\theta} = h_{k\alpha}^{\theta}, \tag{3.3.57}$$

with $ij \to \alpha$, $kl \to \beta$ in the Voigt notation.

Since the free energy F is a scalar and D_i, S_{ij} are tensors of rank one and two respectively, it follows that E_i^0, T_{ij}, $\beta_{ij}^{S,\theta}$, h_{ijk}^{θ}, and $c_{ijkl}^{D,\theta}$ are tensors of rank one, two, two, three and four respectively, i.e. their transformation properties are as follows:

$$E_i^{0\prime} = \sum_j R_{ij} E_j^0, \quad T_{ij}^{0\prime} = \sum_{k,l} R_{ik} R_{jl} T_{kl}^0, \quad \beta_{ij}^{D,\theta\prime} = \sum_{k,l} R_{ik} R_{jl} \beta_{kl}^{D,\theta}, \tag{3.3.58}$$

$$h_{ijk}^{\theta\prime} = \sum_{l,m,n} R_{il} R_{jm} R_{kn} h_{lmn}^{\theta}, \quad c_{ijkl}^{D,\theta\prime} = \sum_{m,n,o,p} R_{im} R_{jn} R_{ko} R_{lp} c_{mnop}^{D,\theta}; \tag{3.3.59}$$

Physically these results mean that in general the temperature dependent functions describing a representation for F in one set of axes are linear combinations of the analogous functions for another set of axes.

Consider now the potential G. We write the polynomial representation for $G(T_{ij}, E_i, \theta)$ analogous to eqn. (3.3.42) in the form

$$\left. \begin{aligned} G = (G)^0 - \sum_{i,j} \gamma_{ij} T_{ij} - \sum_i P_i^0 E_i - \tfrac{1}{2} \sum_{i,j,k,l} s_{ijkl}^{E,\theta} T_{ij} T_{kl} \\ - \sum_{i,j,k} d_{ijk}^{\theta} E_i T_{jk} - \tfrac{1}{2} \sum_{i,j} \varepsilon_{ij}^{T,\theta} E_i E_j. \end{aligned} \right\} \tag{3.3.60}$$

In contrast to the results given in eqn. (3.3.57) we find that:

$$\gamma_\alpha = \gamma_{ij} \quad (i = j), \qquad \gamma_\alpha = 2\gamma_{ij} \quad (i \neq j), \tag{3.3.61}$$

$$d_{i\alpha}^{\theta} = d_{ijk}^{\theta} \quad (j = k), \qquad d_{i\alpha}^{\theta} = 2 d_{ijk}^{\theta} \quad (j \neq k), \tag{3.3.62}$$

$$\left. \begin{aligned} s_{\alpha\beta}^{E,\theta} &= s_{ijkl}^{E,\theta} \quad (i = j, \quad k = l), \\ s_{\alpha\beta}^{E,\theta} &= 2 s_{ijkl}^{E,\theta} \quad (i = j, \quad k \neq l \quad \text{or} \quad i \neq j, k = l) \\ s_{\alpha\beta}^{E,\theta} &= 4 s_{ijkl}^{E,\theta} \quad (i \neq j, \quad k \neq l). \end{aligned} \right\} \tag{3.3.63}$$

Thus, because of the appearance of the factors 2 and 4 the coefficients γ_α, $d_{i\alpha}^{\theta}$ and $s_{\alpha\beta}^{E,\theta}$ unlike E_i^0, T_{ij}^0, etc., are not elements of tensors. It is evident from these results that some care is required when a transformation of coordinates is carried out on a set of equations of state expressed in the Voigt notation.

3.3.7 Material Symmetry

In addition to symmetry associated with interchanges of subscripts, the coefficients of a polynomial representation of a thermodynamic potential display *material symmetry*. It is found experimentally that for any given material the coefficients are invariants (in the sense described below) under the operations of a sub-group of the full orthogonal group. This sub-group is the material symmetry group characteristic of the substance. In the case of a crystalline dielectric the sub-group is one of the crystallographic point groups;† for an *isotropic material* the sub-group is the full orthogonal group.

For definiteness consider the case of the thermodynamic potential F. Let R be an operator of the material symmetry group and let it have a matrix representation R_{ij}‡ in the coordinate axes X_1, X_2, X_3; since R is an operator of the full orthogonal group,

$$\sum_j R_{ij}R_{kj} = \delta_{ik}, \quad \sum_i R_{ij}R_{ik} = \delta_{jk}, \quad \det R_{ij} = \pm 1. \qquad (3.3.64)$$

Consider the coefficients $c_{ijkl}^{D,\theta}$, $\beta_{ij}^{S,\theta}$. These tensors are invariants under the operations of the sub-group if and only if

$$c_{ijkl}^{D,\theta} \equiv \sum_{m,n,o,p} R_{im}R_{jn}R_{ko}R_{lp}c_{mnop}^{D,\theta}, \qquad (3.3.65)$$

$$\beta_{ij}^{S,\theta} \equiv \sum_{k,l} R_{ik}R_{jl}\beta_{kl}^{S,\theta}, \qquad (3.3.66)$$

for all operators R of the sub-group. Similar expressions apply in the case of other coefficients. Comparing eqns. (3.3.58) and (3.3.59) with (3.3.65) and (3.3.67) we see that if two coordinate frames are connected by an operation of the symmetry group then the set of temperature dependent functions ($c_{ijkl}^{D,\theta}$, etc.) describing F in one coordinate frame is identical to the analogous set for the other coordinate frame.

† Clearly this material symmetry arises because of the underlying lattice structure of the solid. Further discussion of this connection lies outside the scope of the present treatment.

‡ The operations R are defined relative to certain characteristic directions and/or planes in the material and the form of the representation will depend on the orientation of these directions and planes relative to the coordinate axes X_1, X_2, X_3.

TABLE 3.3.5.

Two examples of the patterns of symmetry imposed by the crystallographic point groups on the compliance matrices.

(a) refers to the orthorhombic symmetry 222 and (b) to the cubic symmetry *m*3*m*. The arrays are modelled on the arrangements shown in Tables 3.3.1, 3.3.3 and 3.3.4. Thus the leading 6×6 block on the main diagonal refers to the compliances $\partial T_\alpha/\partial S_\beta$ or $\partial S_\alpha/\partial T_\beta$, etc. Each full dot represents a non-vanishing element and each small dot a vanishing element. The lines connect elements which are identical. The relations (across the main diagonal) imposed by the Maxwell relations are not shown. The elements in these arrays are measured in the standard crystallographic axes of the point group of interest.

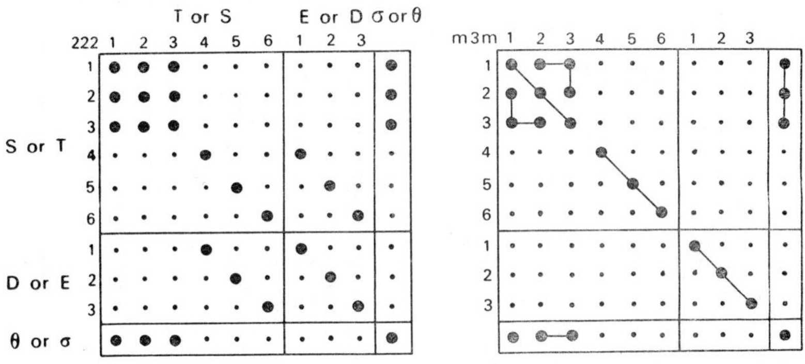

(a) (b)

Th.e demands of material symmetry as expressed for example by eqns. (3.3.65) and (3.3.66) may impose conditions on the elements of the tensor. Thus some elements may be required to vanish identically[†] and otthers be interrelated. The full pattern of symmetries imposed on the coefficients E_i^0, T_{ij}^0, $\beta_{ij}^{S,\theta}$, h_{ijk}^θ and $c_{ijkl}^{D,\theta}$ by each crystallographic point group have been worked out (Nye, 1964). The results are normally displayed as a matrix array using the Voigt subscript notation. Two

† We recall that the elements or coefficients are functions of θ.

TABLE 3.3.6.

The thirty-two crystallographic point groups arranged by crystal systems. A check mark indicates that the point group is polar or piezoelectric as the case may be.

Crystal system	Symbol		Polar	Piezoelectric
	International	Schoenflies		
Triclinic	1	C_1	✓	✓
	$\bar{1}$	C_i		
Tetragonal	4	C_4	✓	✓
	$\bar{4}$	S_4		✓
	$4/m$	C_{4h}		
	422	D_4		✓
	$4mm$	C_{4v}	✓	✓
	$\bar{4}2m$	D_{2d}		✓
	$4/mmm$	D_{4h}		
Hexagonal	6	C_6	✓	✓
	$\bar{6}$	C_{3h}		✓
	$6/m$	C_{6h}		
	622	D_6		✓
	$6mm$	C_{6v}	✓	✓
	$\bar{6}m2$	D_{3h}		✓
	$6/mmm$	D_{6h}		
Monoclinic	2	C_2	✓	✓
	m	C_s	✓	✓
	$2/m$	C_{2h}		
Orthorhombic	222	D_2		✓
	$mm2$	C_{2v}	✓	✓
	mmm	D_{2h}		
Trigonal	3	C_3	✓	✓
	$\bar{3}$	S_6		
	32	D_3		✓
	$3m$	C_{3v}	✓	✓
	$\bar{3}m$	D_{3d}		
Cubic	23	T		✓
	$m3$	T_h		
	432	O		
	$\bar{4}3m$	T_d		✓
	$m3m$	O_h		

examples are given in Table 3.3.5. In the case of the potential F these arrays represent the compliance matrix of F in zero strain and zero displacement (Table 3.3.3). By inspection we see that for the material symmetry 222,

$$c_{14}^{D,\theta} \equiv 0, \quad c_{11}^{D,\theta} \not\equiv 0, \quad h_{11}^{\theta} \equiv 0, \quad h_{14}^{\theta} \not\equiv 0, \quad \text{etc.},$$

and for material symmetry $m3m$,

$$c_{11}^{D,\theta} \equiv c_{22}^{D,\theta} \equiv c_{33}^{D,\theta} \not\equiv 0, \quad h_{i\alpha}^{\theta} \equiv 0 \quad \text{all} \quad i, \alpha, \quad \text{etc.}$$

A similar discussion applies to any other thermodynamic potential. For a given crystallographic point group the sets of coefficients describing the polynomial representations of the different potentials display the same pattern of symmetries.† Thus, for example, in the case of material symmetry $m3m$, the quadratic coefficients describing G obey the following relations:

$$s_{11}^{E,\,\theta} \equiv s_{22}^{E,\,\theta} \equiv s_{33}^{E,\,\theta} \not\equiv 0, \quad d_{i\alpha}^{\theta} \equiv 0 \quad \text{all} \quad i, \alpha, \quad \text{etc.}$$

The following two subclassifications of the point groups are of particular interest in the physics of crystals:

(a) *Polar groups.* These are the point groups which permit the first rank tensor invariant E_i^0 to have at least one non-vanishing component. There are ten polar groups (Table 3.3.6). Only these groups permit a spontaneous polarization within the linear approximation.

(b) *Piezoelectric groups.* These are the point groups which permit the third-rank tensor invariant h_{ijk}^{θ} ($\equiv h_{ikj}^{\theta}$) to have at least one non-vanishing component. There are twenty piezoelectric groups (Table 3.3.6).

We note that no point group requires all the components of the second rank tensor invariant γ_{ij} to vanish identically. However, the γ_{ij} vanish by definition at the temperature θ_i of the initial state.

3.3.8 Quasi-linear Theory

While the thermodynamic theory developed in § 3.1 is strictly valid only for small elastic and electric transformations from the initial state, § 3.3.1, we can, in the spirit of the quasi-linear theory of § 2.5.5, attempt to use this thermodynamic theory to describe non-linear effects. For the most part this is the approach that has been used in the literature to analyse the non-linear behaviour of the ferroelectrics (see remarks at the end of § 2.5.6). We shall confine our discussion to those points of direct interest in the field of ferroelectricity.

† This occurs because these sets of coefficients are interrelated (see, for example, the items at the end of § 3.3.5).

The free energy representation in the quasi-linear approximation is

$$\left.\begin{aligned}F(S_\alpha, D_i, \theta) = {} & (F)^0+\sum_\alpha T^0_\alpha S_\alpha+\sum_j E^0_j D_j+\tfrac{1}{2}\sum_{\alpha,\beta} c^{D,\theta}_{\alpha\beta} S_\alpha S_\beta \\ & -\sum_{i,\alpha} h^\theta_{i\alpha} S_\alpha D_i+\tfrac{1}{2}\sum_{i,j}\beta^{S,\theta}_{ij} D_i D_j+\sum_{i,j,\alpha} q^\theta_{ij\alpha} S_\alpha D_i D_j \\ & +\tfrac{1}{3}\sum_{i,j,k}\eta^{S,\theta}_{ijk} D_i D_j D_k+\tfrac{1}{4}\sum_{i,j,k,l}\xi^{S,\theta}_{ijkl} D_i D_j D_k D_l \\ & +\tfrac{1}{5}\sum_{i,j,k,l,m}\mu^{S,\theta}_{ijklm} D_i D_j D_k D_l D_m \\ & +\tfrac{1}{6}\sum_{i,j,k,l,m,n}\zeta^{S,\theta}_{ijklmn} D_i D_j D_k D_l D_m D_n .\end{aligned}\right\} \quad (3.3.67)$$

This expression should be compared with that given in eqn. (2.5.43). The differences are (a) the appearance of the terms $(F)^0$, $\sum_\alpha, T^0_\alpha S_\alpha$ and $\sum_i E^0_i D_i$, and (b) the explicit acknowledgement that in general all the coefficients may be functions of temperature. It is a simple exercise to show that $\mu^{S,\theta}_{ijk}$, $q^\theta_{ij\alpha}$, $\xi^{S,\theta}_{ijkl}$, $\mu^{S,\theta}_{ijklm}$ and $\zeta^{S,\theta}_{ijklmn}$ are tensors of rank three, four, four, five and six respectively (compare treatment of linear coefficients in § 3.3.6). The symmetries of these coefficients under subscript interchanges are listed in § 2.5.5. For a given specimen these coefficients are also tensor invariants under the operations of the appropriate material symmetry group.

The equations of state for the stress and electric field components generated by this representation are

$$T_\alpha = T^0_\alpha+\sum_\beta c^{D,\theta}_{\alpha\beta} S_\beta-\sum_j h^\theta_{j\alpha} D_j+\sum_{i,j} q^\theta_{ij\alpha} D_i D_j \quad (\alpha = 1, 2, 3, 4, 5, 6), \quad (3.3.68)$$

$$\left.\begin{aligned}E_i = {} & E^0_i-\sum_\beta h^\theta_{i\beta} S_\beta+\sum_j \beta^{S,\theta}_{ij} D_j+2\sum_{j,\beta} q^\theta_{ij\beta} D_j S_\beta \\ & +\sum_{j,k}\eta^{S,\theta}_{ijk} D_j D_k+\sum_{j,k,l}\xi^{S,\theta}_{ijkl} D_j D_k D_l \\ & +\sum_{j,k,l,m}\mu^{S,\theta}_{ijklm} D_j D_k D_l D_m \\ & +\sum_{j,k,l,m,n}\zeta^{S,\theta}_{ijklmn} D_j D_k D_l D_m D_n \quad (i = 1, 2, 3).\end{aligned}\right\} \quad (3.3.69)$$

When $T_\alpha = 0$, $E_i = 0$, we set $S_\alpha = \gamma_\alpha$ the thermal expansion and $D_i = P_i^0$ the spontaneous polarization. γ_α, P_i^0 satisfy the following nine equations:

$$0 = T_\alpha^0 + \sum_\beta c_{\alpha\beta}^{D,\theta}\,\gamma_\beta - \sum_j h_{j\alpha}^\theta P_j^0 + \sum_{i,j} q_{ij\alpha}^\theta P_i^0 P_j^0, \tag{3.3.70}$$

$$\left.\begin{aligned} 0 = E_i^0 &- \sum_\beta h_{i\beta}^\theta \gamma_\beta + \sum_j \beta_{ij}^{S,\theta} P_j^0 + 2\sum_{j,\beta} q_{ij\beta}^\theta P_j^0 \gamma_\beta + \sum_{j,k} \eta_{ijk}^{S,\theta} P_j^0 P_k^0 \\ &+ \sum_{j,k,l} \xi_{ijkl}^{S,\theta} P_j^0 P_k^0 P_l^0 + \sum_{j,k,l,m} \mu_{ijklm}^{S,\theta} P_j^0 P_k^0 P_l^0 P_m^0 \\ &+ \sum_{j,k,l,m,n} \zeta_{ijklmn}^{S,\theta} P_j^0 P_k^0 P_l^0 P_m^0 P_n^0. \end{aligned}\right\} \tag{3.3.71}$$

Introducing the matrix $[s_{\alpha\beta}^{D,\theta}]$ reciprocal to $[c_{\alpha\beta}^{D,\theta}]$, see § 3.3.5, we may solve eqn. (3.3.70) and obtain γ_β in the form

$$\gamma_\alpha = \gamma_\alpha^0 + \sum_{j,\beta} h_{j\beta}^\theta s_{\alpha\beta}^{D,\theta} P_j^0 - \sum_{i,j,\beta} q_{ij\beta}^\theta s_{\alpha\beta}^{D,\theta} P_i^0 P_j^0, \tag{3.3.72}$$

where

$$\gamma_\alpha^0 = -\sum_\beta s_{\alpha\beta}^{D,\theta} T_\beta^0 \tag{3.3.73}$$

is the *normal thermal expansion*. The remaining two terms on the right of eqn. (3.3.72) together represent the *spontaneous strain* S_α^0 "induced" by the spontaneous polarization. In the case of the ferroelectrics γ_α^0 is a slowly varying function of θ,† hence the adjective "normal", whereas S_α^0 or its temperature derivative exhibits a discontinuity at the ferroelectric phase transition.

Equations (3.3.70) and (3.3.71) provide a non-linear generalization of eqns. (3.3.34) and (3.3.35).

Combining eqns. (3.3.68) and (3.3.73) we get the following alternative form of the equation of state for T_α:

$$T_\alpha = \sum_\beta c_{\alpha\beta}^{D,\theta}(S_\beta - \gamma_\beta^0) - \sum_j h_{j\alpha}^\theta D_j + \sum_{i,j} q_{ij\alpha}^\theta D_i D_j. \tag{3.3.74}$$

Eliminating γ_α between eqn. (3.3.71) and (3.3.72) we obtain three

† Typically in the ferroelectrics $\gamma_\beta^0 = \alpha_\beta(\theta - \theta_i)$, where $\alpha_\beta \sim 10^{-5}/°K$. Thus within 100 degrees of θ_i, the thermal stress T_α^0 induces strains $\gtrsim 10^{-3}$ (i.e. infinitesimal strains).

coupled quintic equations for the components P_i^0; thus

$$0 = E_i^0 - \sum_\beta h_{i\beta}^\theta \gamma_\beta^0 + \sum_j \beta_{ij}^{T,\theta} P_j^0 + \sum_{j,k} \eta_{ijk}^{T,\theta} P_j^0 P_k^0 + \sum_{j,k,l} \xi_{ijkl}^{T,\theta} P_j^0 P_k^0 P_l^0$$
$$+ \sum_{j,k,l,m} \mu_{ijklm}^{T,\theta} P_j^0 P_k^0 P_l^0 P_m^0 + \sum_{i,j,k,l,m,n} \zeta_{ijklmn}^{T,\theta} P_j^0 P_k^0 P_l^0 P_m^0 \quad (i = 1, 2, 3), \tag{3.3.75}$$

where

$$\beta_{ij}^{T,\theta} = \beta_{ij}^{S,\theta} - \sum_{\beta,\gamma} h_{i\beta}^\theta h_{j\gamma}^\theta s_{\beta\gamma}^{D,\theta} + 2\sum_\beta q_{ij\beta}^\theta \gamma_\beta^0, \tag{3.3.76}$$

$$\eta_{ijk}^{T,\theta} = \eta_{ijk}^{S,\theta} + \sum_{\beta,\gamma} (h_{i\beta}^\theta q_{jk\gamma}^\theta s_{\beta\gamma}^{D,\theta} + 2h_{k\gamma}^\theta q_{ij\beta}^\theta S_{\beta\gamma}^{D,\theta}), \tag{3.3.77}$$

$$\xi_{ijkl}^{T,\theta} = \xi_{ijkl}^{S,\theta} - 2\sum_{\beta,\gamma} q_{ij\beta}^\theta q_{kl\gamma}^\theta s_{\beta\gamma}^{D,\theta}, \tag{3.3.78}$$

$$\mu_{ijklm}^{T,\theta} = \mu_{ijklm}^{S,\theta}, \tag{3.3.79}$$

$$\zeta_{ijklmn}^{T,\theta} = \zeta_{ijklmn}^{S,\theta}. \tag{3.3.80}$$

[Cf. eqns. (2.5.52) to (2.5.57).]

The real roots of eqn. (3.3.75) are the components of a polarization vector or vectors in the electric field and stress-free state. When two or more such roots $\boldsymbol{P}^0$ exist, stability criteria, § 3.2, determine which root correctly describes the response of the system. It is clear that the state can be ferroelectric only if these equations possess at least two roots of the form $\boldsymbol{P}^0$ and $-\boldsymbol{P}^0$. This is a necessary condition for ferroelectricity; we doubt that in general it is sufficient. The question of sufficiency has not been discussed in the literature. Since E_i^0 and $\sum_\beta h_{i\beta}^\theta \gamma_\beta$ are tensors of rank one, eqns. (3.3.75) have a solution $\boldsymbol{P}^0 \equiv 0$ if and only if the material symmetry of F is non-polar. This last condition appears to be satisfied by the ferroelectrics discovered so far; hence at any given temperature these crystals always possess an initial state ($E_i = 0$, $T_\alpha = 0$, $D_i = 0$) as envisaged in §§ 2.5 and 4.1. We emphasize that this non-polar condition is not a consequence of the definition of ferroelectricity.

When the three equations $E_i = 0$, (3.3.69), have two or more real roots in the components D_i, then the D_i, regarded as functions of the E_i, are multivalued functions of these components. This means that,

if we write the D_i as a power series in the E_i [i.e. invert the power series expansion of eqn. (3.3.69)], we obtain different power series expansions, each one appropriate to some range of D_i values. Consider, for example, the expansion in the region of the root $D_i = P_i^{0''}$,† where $E_i(P_i^{0''}, S_\alpha = 0, \theta) = 0$ $(i, j = 1, 2, 3)$. We set

$$D_i = P_i^{0''} + \sum_\beta e_{i\beta}^\theta S_\beta + \sum_j \varepsilon_{ij}^{S,\theta} E_j + \ldots.$$

The point we wish to make in this paragraph can be made without introducing the additional higher order terms explicitly. The thermodynamic potential generating this equation of state takes the form

$$I(S_\alpha, E_i, \theta) = -\sum_i P_i^{0''} E_i - \sum_{i,\beta} e_{i\beta}^\theta E_i S_\beta - \tfrac{1}{2} \sum_{i,j} \varepsilon_{ij}^{S,\theta} E_i E_j + . \quad (3.3.81)$$

Inserting this expression for D_i into the right-hand side of eqn. (3.3.69) we obtain relations between the coefficients $P_i^{0''}$, $e_{i\beta}^\theta$, $\varepsilon_{ij}^{S,\theta}$ and E_i^0, $h_{i\beta}^\theta$), $\beta_{ij}^{S,\theta}$, $q_{ij\beta}^\theta$, $\eta_{ijk}^{S,\theta}$, $\xi_{ijkl}^{S,\theta}$, $\mu_{ijklm}^{S,\theta}$, $\zeta_{ijklmn}^{S,\theta}$. Thus

$$\left.\begin{aligned} 0 = E_i^0 &+ \sum_j \beta_{ij}^{S,\theta} P_j^{0''} + \sum_{j,k} \eta_{ijk}^{S,\theta} P_j^{0''} P_k^{0''} + \sum_{j,k,l} \xi_{ijkl}^{S,\theta} P_j^{0''} P_k^{0''} P_l^{0''} \\ &+ \sum_{j,k,l,m} \mu_{ijklm}^{S,\theta} P_j^{0''} P_k^{0''} P_l^{0''} P_m^{0''} \\ &+ \sum_{i,j,k,l,m,n} \zeta_{ijklmn}^{S,\theta} P_j^{0''} P_k^{0''} P_l^{0''} P_m^{0''} P_n^{0''} \end{aligned}\right\} \quad (3.3.82)$$

$$\left.\begin{aligned} \delta_{ik} = \sum_j \varepsilon_{jk}^{S,\theta} \Big(\beta_{ij}^{S,\theta} &+ 2 \sum_l \eta_{ijl}^{S,\theta} P_l^{0''} + 3 \sum_{l,m} \xi_{ijlm}^{S,\theta} P_l^{0''} P_m^{0''} \\ &+ 4 \sum_{l,m,n} \mu_{ijlmn}^{S,\theta} P_l^{0''} P_m^{0''} P_n^{0''} + 5 \sum_{l,m,n,o} \zeta_{ijlmno}^{S,\theta} P_l^{0''} P_m^{0''} P_n^{0''} P_o^{0''}\Big), \end{aligned}\right\} \quad (3.3.83)$$

$$\left.\begin{aligned} 0 = -h_{i\beta}^\theta &+ 2 \sum_j q_{ij\beta}^\theta P_j^{0''} + \sum_j e_{j\beta}^\theta \Big(\beta_{ij}^{S,\theta} + 2 \sum_k \eta_{ijk}^{S,\theta} P_k^{0''} \\ &+ 3 \sum_{k,l} \xi_{ijkl}^{S,\theta} P_k^{0''} P_l^{0''} + 4 \sum_{k,l,m} \mu_{ijklm}^{S,\theta} P_k^{0''} P_l^{0''} P_m^{0''} \\ &+ 5 \sum_{k,l,m,n} \zeta_{ijklmn}^{S,\theta} P_k^{0''} P_l^{0''} P_m^{0''} P_n^{0''}\Big). \end{aligned}\right\} \quad (3.3.84)$$

† These are the components of the spontaneous polarization vector of the clamped crystal. In general they differ from the P_i^0, the components for the free crystal. Strictly, we should write $P_i^{0'}$ for the latter (see end of § 2.5.4), but as the spontaneous polarization of physical interest is almost invariably the free value, it is customary to drop the single prime.

Equations (3.3.83) and (3.3.84) can be regarded as equations determining $\varepsilon_{ij}^{S,\theta}$, $e_{j\beta}^{\theta}$ respectively in terms of the coefficients describing F. Now $P_i^{0''}$ are components of a polar vector and hence in general the symmetry displayed by $e_{j\beta}^{\theta}$, $\varepsilon_{ij}^{S,\theta}$, etc., is not the same as the symmetry displayed by $h_{j\beta}^{\theta}$, $\beta_{ij}^{S,\theta}$, etc. In other words *the material symmetry displayed by I need not coincide with the material symmetry of F* despite the fact those two potentials generate equivalent equations of state. Thus in general the material symmetry of a crystal is a function of the choice of thermodynamic potential used to describe its properties.

Now let us consider the following elements of the compliance matrix of F. From eqns. (3.3.68) and (3.3.69)

$$\left(\frac{\partial T_\alpha}{\partial S_\beta}\right)_{D,\theta} = c_{\alpha\beta}^{D,\theta}, \quad \left(\frac{\partial T_\alpha}{\partial D_i}\right)_{S,\theta} = -h_{i\alpha}^{\theta} + 2\sum_j q_{ij\alpha}^{\theta} D_j, \tag{3.3.85}$$

$$\left.\begin{aligned} \left(\frac{\partial E_i}{\partial D_j}\right)_{\theta,S} &= \beta_{ij}^{S,\theta} + 2\sum_\beta q_{ij\beta}^{\theta} S_\beta + 2\sum_k \eta_{ijk}^{S,\theta} D_k + 3\sum_{k,l} \xi_{ijkl}^{S,\theta} D_k D_l \\ &\quad + 4\sum_{k,l,m} \mu_{ijklm}^{S,\theta} D_k D_l D_m + 5\sum_{k,l,m,n} \zeta_{ijklmn}^{S,\theta} D_k D_l D_m D_n . \end{aligned}\right\} \tag{3.3.86}$$

We may talk of the symmetry properties of a compliance matrix in the sense that in particular instances some of the elements of the matrix may satisfy identities in addition to those imposed by the Maxwell relations. It is clear from the expressions in eqns. (3.3.85) and (3.3.86) that the symmetry displayed by the compliance matrix of F is a function of both the material symmetry of F and the values of S_α and D_i imposed, i.e. generally speaking the compliance matrix will not display the material symmetry of the representation of the associated thermodynamic potential.† It is clear that this result hinges on the multi-valued nature of the equations of state. In the case of the ferroelectrics it appears that the components of $\boldsymbol{E}$ are *always* single valued functions of the components of $\boldsymbol{D}$ but not vice versa. Thus for these crystals the material symmetry of polynomial representations of potentials with arguments E_i is, in general, lower than that of the potentials with arguments D_i. In contrast to the results described above, for a given

† This statement is true even in the linear theory, see eqns. (3.3.29) and (3.3.30).

state of the system the compliance matrices of the set of all possible thermodynamic potentials display the same symmetry properties. The proof follows immediately from the existence of thermodynamic identities of the type discussed in § 3.3.3. Thus, while the material symmetries of the polynomial representations of the various thermodynamic potentials which describe the same state may not coincide, the symmetries of the associated compliance matrices do.

It is customary to refer to the *symmetry of a phase of a crystal*; by this one means the symmetry of any compliance matrix along the path $E_i = 0$ $(i = 1, 2, 3)$, $T_\alpha = 0$ $(\alpha = 1, 2, 3, 4, 5, 6)$. In a pyroelectric phase $P_i^0 \not\equiv 0$; thus $(\partial P_i^0/\partial\theta) \not\equiv 0$ and hence the symmetry of a pyroelectric phase is polar.

One further point based on eqns. (3.3.85) and (3.3.86) is worth making. If, as appears to be the case for the ferroelectrics, the same free energy representation describes both the paraelectric phase, $P_i^0 \equiv 0$, and the ferroelectric phase, $P_i^0 \not\equiv 0$, then the lower symmetry of the latter phase compared with the former arises solely from the presence of the spontaneous polarization. For example, consider the expression given for the piezoelectric coefficient in eqn. (3.3.85).

For $P_i^0 \equiv 0$,

$$\left(\frac{\partial T_\alpha}{\partial D_i}\right)_{S,\theta}\Bigg|_{\substack{E=0\\T=0}} = -h_{i\alpha}^\theta, \tag{3.3.87}$$

whereas for $P_i^0 \not\equiv 0$,

$$\left(\frac{\partial T_\alpha}{\partial D_i}\right)_{S,\theta}\Bigg|_{\substack{E=0\\T=0}} = -h_{i\alpha}^\theta + 2\sum_j q_{ij\alpha}^\theta P_j^0. \tag{3.3.88}$$

The difference in symmetry, if any, between these expressions stems from the electrostrictive term in eqn. (3.3.88). If it should happen for a particular pair of values i, α, that $h_{i\alpha} \equiv 0$ and $\sum_j q_{ij\alpha}^\theta P_\alpha^0 \not\equiv 0$, then the spontaneous polarization is said to "induce" an extra non-zero element in the zero electric field zero stress compliance matrix. If we take as an example KDP, we find that for $\bar{4}2m$ material symmetry the only non-vanishing electrostrictive coefficients are q_{111}^θ, q_{222}^θ, q_{112}^θ, q_{221}^θ, q_{333}^θ, q_{113}^θ, q_{223}^θ, q_{331}^θ, q_{332}^θ, q_{234}^θ, q_{135}^θ, q_{126}^θ, and these are subject to the following

identities:

$$q^{\theta}_{111} \equiv q^{\theta}_{222} \quad q^{\theta}_{112} \equiv q^{\theta}_{221} \quad q^{\theta}_{113} \equiv q^{\theta}_{223} \quad q^{\theta}_{331} \equiv q^{\theta}_{332} \quad q^{\theta}_{234} \equiv q^{\theta}_{135}.^{\dagger}$$

With the spontaneous polarization parallel to the X_3 axis, the only induced piezoelectric coefficients are then

$$(h^{\theta}_{31})^{\text{ind}} \equiv (h^{\theta}_{32})^{\text{ind}} = -2q_{331}P^0_3,$$
$$(h^{\theta}_{33})^{\text{ind}} = -2q_{333}P^0_3,$$
$$(h^{\theta}_{24})^{\text{ind}} \equiv (h^{\theta}_{15})^{\text{ind}} = -2q_{234}P^0_3.$$

Combining these coefficients with the linear piezoelectric coefficients of symmetry $\bar{4}2m$, we get the array shown in Table 3.3.7. This pattern

TABLE 3.3.7.

Array of piezoelectric compliances

The crossed full dots represent elements induced by the spontaneous polarization and the remaining dots the linear piezoelectric coefficients permitted by the point group symmetry $\bar{4}2m$.

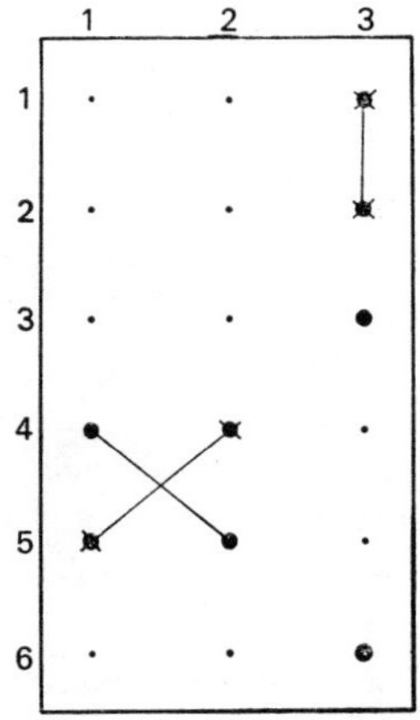

† Nye (1964) gives tables showing the symmetry properties of the so-called photoelastic coefficients for the thirty-two crystallographic point groups. These coefficients possess the same symmetry properties as the electrostrictive coefficients and hence Nye's tables apply immediately to the latter.

coincides with the pattern displayed by the linear piezoelectric coefficients with symmetry *mm*2 (see Table 1.2.1), the symmetry of the ferroelectric phase of KDP. By extending this calculation to the other elements we may then show that the extra elements which appear in the zero stress and zero field compliance matrix because of the lowering of the symmetry of KDP as it is cooled through the critical point, can be accounted for, at least qualitatively, in terms of the spontaneous polarization and the electrostrictive coefficients.

For future reference we write out the quasi-linear polynomial representation for the free energy $F(S_\alpha, D_i, \theta)$ when the material symmetry is $m3m$.

$$\left.\begin{aligned}F(S_\alpha, D_i, \theta) = {} & (F)^0 - 3(\gamma^0)^2 \tfrac{1}{2}(c_{11}^{D,\theta} + c_{12}^{D,\theta}) \\ & + \tfrac{1}{2} c_{11}^{D,\theta}[(S_1-\gamma^0)^2 + (S_2-\gamma^0)^2 + (S_3^2-\gamma^0)^2] \\ & + c_{12}^{D,\theta}[(S_1-\gamma^0)(S_2-\gamma^0) + (S_2-\gamma^0)(S_3-\gamma^0) \\ & + (S_3-\gamma^0)(S_1-\gamma^0)] + \tfrac{1}{4} c_{44}^{D,\theta}(S_4^2+S_5^2+S_6^2) \\ & + \tfrac{1}{2}\beta_{11}^{S,\theta}(D_1^2+D_2^2+D_3^2) + q_{11}^\theta(S_1D_1^2+S_2D_2^2+S_3D_3^2) \\ & + q_{12}^\theta[S_1(D_2^2+D_3^2)+S_2(D_3^2+D_1^2)+S_3(D_1^2+D_2^2)] \\ & + q_{44}^\theta(S_4D_2D_3+S_5D_3D_1+S_6D_1D_2) \\ & + \tfrac{1}{4}\xi_{11}^{S,\theta}(D_1^4+D_2^4+D_3^4) \\ & + \tfrac{3}{2}\xi_{12}^{S,\theta}(D_1^2D_2^2+D_2^2D_3^2+D_3^2D_1^2) \\ & + \tfrac{1}{6}\zeta_{111}^{S,\theta}(D_1^6+D_2^6+D_3^6) \\ & + \tfrac{5}{2}\zeta_{112}^{S,\theta}[D_1^4(D_2^2+D_3^2)+D_2^4(D_3^2+D_1^2) \\ & + D_3^4(D_1^2+D_2^2)] + 15\zeta_{123}^{S,\theta}D_1^2D_2^2D_3^2,\end{aligned}\right\} \quad (3.3.89)$$

where we have introduced the Voigt subscript notation for the pairs of Roman subscripts appearing in $\xi_{ijkl}^{S,\theta}$ and $\zeta_{ijklmn}^{S,\theta}$. The coefficients $\eta_{ijk}^{S,\theta}$ and $\mu_{ijklm}^{S,\theta}$ all vanish because of the demands of the inversion operator —one of the operations of the group $m3m$. The equations of state generated by this representation are of the form:

$$T_1 = c_{11}^{D,\theta}(S_1-\gamma^0) + c_{12}^{D,\theta}[(S_2-\gamma^0)+(S_3-\gamma^0)] + q_{11}^\theta D_1^2 + q_{12}^\theta[D_2^2+D_3^2], \quad (3.3.90)$$

$$T_4 = c_{44}^{D,\theta}S_4 + q_{44}^\theta D_2D_3, \quad (3.3.91)$$

$$\left.\begin{aligned}E_1 = {} & \beta_{11}^{S,\theta}D_1 + 2q_{11}^\theta S_1D_1 + 2q_{12}^\theta(S_2+S_3)D_1 + q_{44}^\theta(S_5D_3+S_6D_2) \\ & + \xi_{11}^{S,\theta}D_1^3 + 3\xi_{12}^{S,\theta}D_1(D_2^2+D_3^2) + \zeta_{111}^{S,\theta}D_1^5 + 5\zeta_{112}^{S,\theta}[2D_1^3(D_2^2+D_3^2) \\ & + D_1(D_2^4+D_3^4)] + 30\zeta_{123}^{S,\theta}D_1D_2^2D_3^2.\end{aligned}\right\} \quad (3.3.92)$$

3.3.9 *D* or *P*?

Hitherto we have chosen to describe the dielectric response of an insulator in terms of the two fields $\boldsymbol{D}$ and $\boldsymbol{E}$. In particular instances it may prove more convenient to manipulate $\boldsymbol{P}$, the polarization, explicitly rather than $\boldsymbol{D}$ the displacement field. The equations of state relating $\boldsymbol{P}$ and $\boldsymbol{E}$ are clearly derivable from those relating $\boldsymbol{D}$ and $\boldsymbol{E}$. The latter are subject to the restrictions of the Maxwell relations and hence the former are also affected by these restrictions. It is the purpose of this subsection to show that the restrictions take the same form in each case.

Consider the potential $G(\theta, T_\alpha, E_i)$; from eqn. (3.1.10)

$$D_i = -(\partial G/\partial E_i)_{\theta, T}$$

and hence, since $D_i = \varepsilon_0 E_i + P_i$,

$$P_i = -\left(\frac{\partial G}{\partial E_i}\right)_{\theta, T} - \varepsilon_0 E_i = -\left(\frac{\partial \tilde{G}}{\partial E_i}\right)_{\theta, T},$$

where $\tilde{G} = G(\theta, T_\alpha, E_i) + \sum_i \varepsilon_0 E_i^2/2$.

The differential form of $\tilde{G}$ is

$$d\tilde{G} = -\sigma d\theta - \sum_\alpha S_\alpha \, dT_\alpha - \sum_i P_i \, dE_i.$$

Thus the function $\tilde{G}(\theta, T_\alpha, E_i)$ generates the equations of state

$$\sigma = -\left(\frac{\partial \tilde{G}}{\partial \theta}\right)_{T, E}, \quad S_\alpha = -\left(\frac{\partial \tilde{G}}{\partial T_\alpha}\right)_{E, \theta}, \quad P_i = -\left(\frac{\partial \tilde{G}}{\partial E_i}\right)_{\theta, T},$$

which are subject to the restrictions

$$\left(\frac{\partial \sigma}{\partial T_\alpha}\right)_{E, \theta} = \left(\frac{\partial S_\alpha}{\partial \theta}\right)_{E, T}, \quad \left(\frac{\partial S_\alpha}{\partial E_i}\right)_{\theta, T} = \left(\frac{\partial P_i}{\partial T_\alpha}\right)_{\theta, E}, \quad \left(\frac{\partial P_i}{\partial \theta}\right)_{T, E} = \left(\frac{\partial \sigma}{\partial E_i}\right)_{T, \theta}.$$

These are analogous to the Maxwell relations described in § 3.1.2. Clearly by applying various Legendre transformations to $\tilde{G}$ we may generate a whole range of relations between E_i, P_i, T_α, S_α, θ, σ. The existence of these relations means that in all the identities we have derived in the previous subsections we may replace D_i by P_i to obtain a new set of identities for the variables P_i, E_i. Thus as far as equations of state are concerned there is no intrinsic advantage in using $\boldsymbol{D}$ rather than $\boldsymbol{P}$. However, this is not the case for questions of stability because,

as far as we are aware, the functions such as $\tilde{G}$ which generate the equations of state for P_i do not possess, in general, extremal properties (cf. § 3.2). Hence for problems in stability $\boldsymbol{D}$ rather than $\boldsymbol{P}$ is a more convenient variable to use.

3.4 Neighbourhood Temperature

The Clausius–Clapeyron equations and the Ehrenfest relations only apply in situations where the derivatives of the thermodynamic potential describing the system undergo finite discontinuties. We now describe another type of identity which holds even when the derivatives are singular. In the case of the ferroelectrics the identities are useful for categorizing the singularities in the compliance matrix.

3.4.1 Neighbourhood Temperature

Consider again the system described by the six thermodynamic coordinates T, S, E, D, σ and θ (see § 3.2.3). Let the system undergo a phase transition at points on a phase boundary $\theta = \theta_c(T, E)$. The *neighbourhood temperature* τ is defined

$$\tau(\theta, T, E) = \theta_c - \theta(T, E). \tag{3.4.1}$$

Thus $\tau \equiv 0$ on the phase boundary. From this definition,

$$\left(\frac{\partial\tau}{\partial\theta}\right)_{T, E} = 1, \quad \left(\frac{\partial\tau}{\partial T}\right)_{E, \theta} = -\left(\frac{\partial\theta_c}{\partial T}\right)_E, \quad \left(\frac{\partial\tau}{\partial E}\right)_{T, \theta} = -\left(\frac{\partial\theta_c}{\partial E}\right)_T \tag{3.4.2}$$

and

$$\left(\frac{\partial\theta}{\partial T}\right)_{\tau, E} = \left(\frac{\partial\theta_c}{\partial T}\right)_E, \quad \left(\frac{\partial\theta}{\partial E}\right)_{\tau, T} = \left(\frac{\partial\theta_c}{\partial E}\right)_T. \tag{3.4.3}$$

Consider the identity

$$\left(\frac{\partial\sigma}{\partial T}\right)_{\theta, E} = \frac{\partial(\sigma, \theta, E)}{\partial(T, \theta, E)} = \frac{\partial(\sigma, \theta, E)}{\partial(T, \tau, E)} \Big/ \frac{\partial(T, \theta, E)}{\partial(T, \tau, E)}$$

$$= \left(\frac{\partial\sigma}{\partial T}\right)_{\tau, E} - \left(\frac{\partial\sigma}{\partial\tau}\right)_{T, E}\left(\frac{\partial\theta}{\partial T}\right)_{\tau, E}\left(\frac{\partial\tau}{\partial\theta}\right)_{T, E}.$$

Using the chain rule $\left(\frac{\partial\sigma}{\partial\tau}\right)_{T, E} = \left(\frac{\partial\sigma}{\partial\theta}\right)_{T, E}\left(\frac{\partial\theta}{\partial\tau}\right)_{T, E}$ and eqns. (3.4.2) and

(3.4.3) we obtain

$$\left(\frac{\partial\sigma}{\partial T}\right)_{\theta, E} = \left(\frac{\partial\sigma}{\partial T}\right)_{\tau, E} - \left(\frac{\partial\sigma}{\partial\theta}\right)_{T, E}\left(\frac{\partial\theta_c}{\partial T}\right)_E. \tag{3.4.4}$$

Similarly,

$$\left(\frac{\partial\sigma}{\partial T}\right)_{\theta, E} = \left(\frac{\partial S}{\partial\theta}\right)_{T, E} = \left(\frac{\partial S}{\partial\theta}\right)_{\tau, E} - \left(\frac{\partial S}{\partial T}\right)_{\theta, E}\left(\frac{\partial\theta_c}{\partial T}\right)_E^{-1}. \tag{3.4.5}$$

We note that eqns. (3.4.4) and (3.4.5) are exact and that by definition $(\partial\theta_c/\partial T)_E$ is independent of θ.

Let us suppose that in some region of phase space, $(\partial S/\partial\theta)_{\tau, E}$ is a slowly varying function of θ in comparison with $(\partial S/\partial\theta)_{T, E}$. It follows immediately that the temperature variations of $(\partial S/\partial\theta)_{T, E}$ and $(\partial S/\partial T)_{\theta, E}$ differ only by the temperature independent scaling factor $-(\partial\theta_c/\partial T)_E^{-1}$. Thus a plot of $(\partial S/\partial\theta)_{T, E}$ versus $(\partial S/\partial T)_{\theta, E}$ for constant T, E and different θ should yield an approximate straight line with slope $(\partial\theta_c/\partial T)_E^{-1}$. Clearly the deviations from the straight line will depend on the relative sizes of the temperature variations of $(\partial S/\partial\theta)_{T, E}$ and $(\partial S/\partial\theta)_{\tau, E}$. The circumstances envisaged here occur in the region of the ferroelectric phase boundary (see, for example, § 5.4.5) and the identities analogous to eqns. (3.4.4) and (3.4.5) in the case when all the components of stress and electric field are introduced explicitly provide a useful means of correlating various "anomalous" properties of the ferroelectrics (see § 3.4.5). It should be emphasized that it is a matter of experimental observation that in a real crystal the quantities such as $(\partial S/\partial\theta)_{\tau, E}$ are indeed slowly varying functions of temperature. There is no physical principle (as far as we are aware) which would lead one to expect this behaviour *a priori*.

3.4.2 Ehrenfest Relations

The identities discussed above have an interesting connection with the Ehrenfest relations. Consider the identity (3.4.4) at a point on the phase boundary, thus

$$\left(\frac{\partial\sigma^1}{\partial T}\right)_{E, \theta} = \left(\frac{\partial\sigma^1}{\partial T}\right)_{\tau, E} - \left(\frac{\partial\sigma^1}{\partial\theta}\right)_{T, E}\left(\frac{\partial\theta_c}{\partial T}\right)_E, \tag{3.4.6}$$

$$\left(\frac{\partial\sigma^2}{\partial T}\right)_{E, \theta} = \left(\frac{\partial\sigma^2}{\partial T}\right)_{\tau, E} - \left(\frac{\partial\sigma^2}{\partial\theta}\right)_{T, E}\left(\frac{\partial\theta_c}{\partial T}\right)_E, \tag{3.4.7}$$

where the superscripts refer to the two phases. If $(\partial\sigma/\partial T)_{\tau, E}$ is continuous across the phase boundary we can subtract eqn. (3.4.7) from eqn. (3.4.6) and solve for $(\partial\theta_c/\partial T)_E$ to get

$$\left(\frac{\partial\theta_c}{\partial T}\right)_E = -\left[\left(\frac{\partial\sigma^1}{\partial T}\right)_{E,\theta} - \left(\frac{\partial\sigma^2}{\partial T}\right)_{E,\theta}\right] \Big/ \left[\left(\frac{\partial\sigma^1}{\partial\theta}\right)_{T,E} - \left(\frac{\partial\sigma^2}{\partial\theta}\right)_{T,E}\right]$$

i.e. one of the Ehrenfest relations, eqn. (3.2.23). Similarly, if $(\partial S/\partial\theta)_{\tau, E}$ is continuous across the phase boundary we obtain, from eqn. (3.4.5),

$$\left(\frac{\partial\theta_c}{\partial T}\right)_E = -\left[\left(\frac{\partial S^1}{\partial T}\right)_{\theta,E} - \left(\frac{\partial S^2}{\partial T}\right)_{\theta,E}\right] \Big/ \left[\left(\frac{\partial S^1}{\partial\theta}\right)_{T,E} - \left(\frac{\partial S^2}{\partial\theta}\right)_{T,E}\right].$$

By eliminating $(\partial\theta_c/\partial T)_E$ from these two equations we get the relation given in eqn. (3.2.19). Conversely, if the Ehrenfest relations hold at a point on the phase boundary, $(\partial S/\partial\theta)_{\tau, E}$ and $(\partial\sigma/\partial T)_{\tau, E}$ are continuous across the boundary.

3.4.3 The Cylindrical Approximation

Let us set $E = 0$ and let us suppose that

$$\left(\frac{\partial\sigma}{\partial T}\right)_\tau \cong \text{constant}, \tag{3.4.8a}$$

$$\left(\frac{\partial\theta}{\partial T}\right)_\tau \cong \text{constant}. \tag{3.4.8b}$$

(These conditions are satisfied in some of the ferroelectric phase transitions discussed in Chapter 5.) Equation (3.4.8a) can be integrated to give

$$\sigma(\theta, T) = \left(\frac{\partial\sigma}{\partial T}\right)_\tau T + g(\tau)$$

or

$$\sigma(\theta, T) = \left(\frac{\partial\sigma}{\partial T}\right)_\tau T + g[\theta - \theta_c(T)].$$

For $T = 0$

$$\sigma(\theta, 0) = g[\theta - \theta_c(0)].$$

Thus g is the entropy at zero stress.

Hence

$$\sigma(\theta, T) = \left(\frac{\partial\sigma}{\partial T}\right)_\tau T + \sigma[\theta - \theta_c(T) + \theta_c(0), 0]$$

or

$$\sigma(\theta, T) = \left(\frac{\partial\sigma}{\partial T}\right)_\tau T + \sigma\left[\theta - \left(\frac{\partial\theta_c}{\partial T}\right) T, 0\right] \tag{3.4.9}$$

by eqn. (3.4.8b). In this picture the strong temperature dependence of σ, if any, is described by the function $\sigma(\theta, 0)$. The expression for $\sigma(\theta, T)$ in eqn. (3.4.9) describes a cylindrical surface defined by the intersection of the family of straight lines

$$[\sigma(\theta, T) - \sigma(\theta_0, 0)] \Big/ \left(\frac{\partial\sigma}{\partial T}\right)_\tau = [\theta - \theta_0] \Big/ \left(\frac{\partial\theta_c}{\partial T}\right) = T \tag{3.4.10}$$

with the curve $\sigma = \sigma(\theta_0, 0)$ in the $T = 0$ plane. As a result, the approximation embodied in eqn. (3.4.8) is referred to as the *cylindrical approximation.*

3.4.4 General Case†

We now develop the complete set of identities based on the neighbour temperature for the general case of a dielectric whose behaviour is described by the fields T_α, S_α, E_i, D_i, σ and θ. The neighbourhood temperature $\tau = \theta - \theta_c(T_\alpha, E_i)$ is a function of the nine quantitites T_α, E_i. In this case the phase boundary $\tau \equiv 0$ is an eight-dimensional surface in a nine-dimensional phase space. Consider the identity

$$\left(\frac{\partial\sigma}{\partial T_\alpha}\right)_{\theta, E} = \frac{\partial(\sigma, \theta, T', E)}{\partial(T_\alpha, \theta, T', E)}, \tag{3.4.11}$$

where T' represents the set of all components T_μ except T_α. The transformation leading to eqn. (3.4.4) applies to the present identity and yields

$$\left(\frac{\partial S_\alpha}{\partial\theta}\right)_{T, E} = \left(\frac{\partial\sigma}{\partial T_\alpha}\right)_{\theta, E} = \left(\frac{\partial\sigma}{\partial T_\alpha}\right)_{\tau, E, T'} - \left(\frac{\partial\sigma}{\partial\theta}\right)_{T, E}\left[\left(\frac{\partial\theta_c}{\partial T_\alpha}\right)_E\right]. \tag{3.4.12}$$

† In this and the next subsection we shall follow for the most part the treatment of Janovec (1966).

Similarly,

$$\left(\frac{\partial S_\alpha}{\partial T_\beta}\right)_{\theta, E} = \left(\frac{\partial S_\alpha}{\partial T_\beta}\right)_{\tau, E, T'} - \left(\frac{\partial S_\alpha}{\partial \theta}\right)_{T, E} \left[\left(\frac{\partial \theta_c}{\partial T_\beta}\right)_E\right], \qquad (3.4.13)$$

$$\left(\frac{\partial S_\alpha}{\partial T_\beta}\right)_{\theta, E} = \left(\frac{\partial S_\alpha}{\partial T_\beta}\right)_{\tau, E, T'} + \left(\frac{\partial S_\alpha}{\partial T_\gamma}\right)_{E, \theta} \left[\left(\frac{\partial \theta_c}{\partial T_\beta}\right)_E \Big/ \left(\frac{\partial \theta_c}{\partial T_\gamma}\right)_E\right], \qquad (3.4.14)$$

$$\left(\frac{\partial S_\alpha}{\partial E_i}\right)_{\theta, T} = \left(\frac{\partial S_\alpha}{\partial E_i}\right)_{\tau, E', T} + \left(\frac{\partial S_\alpha}{\partial T_\beta}\right)_{E, \theta} \left[\left(\frac{\partial \theta_c}{\partial E_i}\right)_T \Big/ \left(\frac{\partial \theta_c}{\partial T_\beta}\right)_E\right], \qquad (3.4.15)$$

$$\left(\frac{\partial D_i}{\partial T_\alpha}\right)_{\theta, E} = \left(\frac{\partial D_i}{\partial T_\alpha}\right)_{\tau, E, T'} + \left(\frac{\partial D_i}{\partial E_j}\right)_{T, \theta} \left[\left(\frac{\partial \theta_c}{\partial T_\alpha}\right)_E \Big/ \left(\frac{\partial \theta_c}{\partial E_j}\right)_T\right], \qquad (3.4.16)$$

$$\left(\frac{\partial D_i}{\partial E_j}\right)_{\theta, T} = \left(\frac{\partial D_i}{\partial E_j}\right)_{\tau, E', T} - \left(\frac{\partial D_i}{\partial \theta}\right)_{T, E} \left[\left(\frac{\partial \theta_c}{\partial E_j}\right)_T\right], \qquad (3.4.17)$$

$$\left(\frac{\partial D_i}{\partial \theta}\right)_{E, T} = \left(\frac{\partial \sigma}{\partial E_i}\right)_{\theta, T} = \left(\frac{\partial \sigma}{\partial E_i}\right)_{\tau, E', T} - \left(\frac{\partial \sigma}{\partial \theta}\right)_{T, E} \left[\left(\frac{\partial \theta_c}{\partial E_i}\right)_T\right], \qquad (3.4.18)$$

$$\left(\frac{\partial D_i}{\partial E_j}\right)_{\theta, T} = \left(\frac{\partial D_i}{\partial E_j}\right)_{\tau, E', T} + \left(\frac{\partial D_i}{\partial E_k}\right)_{T, \theta} \left[\left(\frac{\partial \theta_c}{\partial E_j}\right)_T \Big/ \left(\frac{\partial \theta_c}{\partial E_k}\right)_T\right], \qquad (3.4.19)$$

$$\left(\frac{\partial S_\alpha}{\partial E_i}\right)_{\theta, T} = \left(\frac{\partial S_\alpha}{\partial E_i}\right)_{\tau, E', T} - \left(\frac{\partial S_\alpha}{\partial \theta}\right)_{T, E} \left[\left(\frac{\partial \theta_c}{\partial E_i}\right)_T\right], \qquad (3.4.20)$$

$$\left(\frac{\partial D_i}{\partial T_\alpha}\right)_{\theta, E} = \left(\frac{\partial D_i}{\partial T_\alpha}\right)_{\tau, E, T'} - \left(\frac{\partial D_i}{\partial \theta}\right)_{T, E} \left[\left(\frac{\partial \theta_c}{\partial T_\alpha}\right)_E\right]. \qquad (3.4.21)$$

The prime on the subscript $T(E)$ indicates that all except one of the components of the stress tensor (electric field) are to be held fixed during the differentiation. For example all the components of the stress tensor except T_α are held fixed in the first differential coefficient on the right of eqn. (3.4.12). The connections between the various derivatives or compliances given in these eight equations are drawn

schematically below

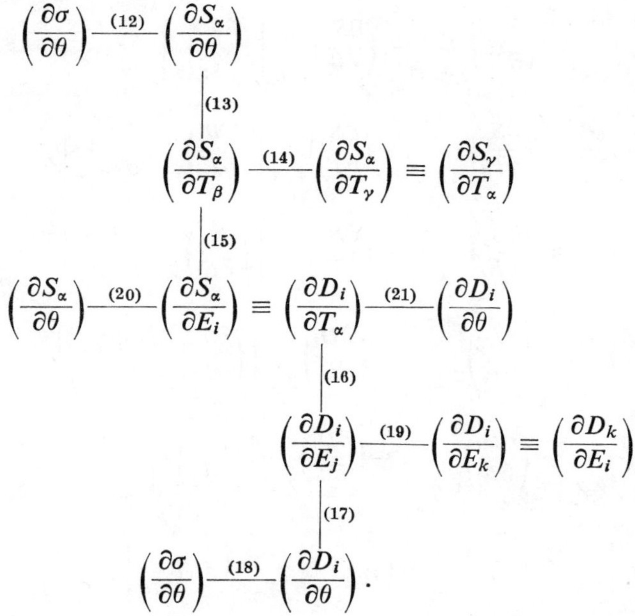

The numbers appearing beside the "bonds" refer to the equations listed above. It is immediately clear from this scheme that *all* the second-order derivatives of the Gibbs function can be linked one to another by identities based on the neighbourhood temperature τ. The precise form of the link between any two derivatives is easily obtained using the set of equations (3.4.12) to (3.4.21).

It may happen in particular cases that some of the derivatives $(\partial\theta_c/\partial T_\alpha)_E$, $(\partial\theta_c/\partial E_i)_T$ vanish; in these circumstances the associated links are broken.

3.4.5 TGS

We now apply the analysis described above to the case of the ferroelectric crystal tri-glycine sulphate (or TGS in abbreviated form).

In the paraelectric phase, TGS displays monoclinic symmetry $2/m$ and in the ferroelectric phase monoclinic symmetry 2. The sponta-

neous polarization lies along the monoclinic b- or X_2-axis. TGS is an example of model B described in Chapter 5, with $\xi^T > 0$. In this case (i) the derivative $(\partial\theta_c/\partial E_i)$ does not exist for $i = 2$ and vanishes for $i = 1, 3$; (ii) $\theta_c = \theta_0$ the Curie–Weiss temperature.

Thus there are no links involving $(\partial\theta_c/\partial E_i)$ $(i = 1, 2, 3)$. An extension of the result given in eqn. (5.4.20) to the case of TGS gives, as a first-order approximation for small stresses,

$$\theta_0 = \theta_0^T - (2/A^T) \sum_{\alpha=1}^{6} Q_{2\alpha} T_\alpha, \tag{3.4.22}$$

where θ_0^T is the Curie–Weiss temperature at zero stress, A^T the inverse Curie–Weiss constant and $Q_{\alpha\beta}$ the electrostrictive coefficients. Thus

$$\left(\frac{\partial\theta_c}{\partial T_\alpha}\right) = \left(\frac{\partial\theta_0}{\partial T_\alpha}\right) = -\frac{2Q_{2\alpha}}{A}. \tag{3.4.23}$$

TGS has material symmetry $2/m$ and hence, using Nye's tables (Nye, 1964),† we find that for this ferroelectric Q_{21}, Q_{22}, Q_{23} and Q_{25} are all non-zero and distinct while $Q_{24} = Q_{26} = 0$. Therefore for small stresses the only neighbourhood temperature links between the compliances are those formed by the four independent quantities $(\partial\theta_c/\partial T_\alpha)$ $(\alpha = 1, 2, 3, 5)$ and the numerical values of these derivatives are directly related by the electrostrictive coefficients $Q_{2\alpha}$ $(\alpha = 1, 2, 3, 5)$.

In Table 3.4.1 we display the forms of the compliance matrix with symmetry $2/m$ and with symmetry 2 (these are the symmetries of the zero electric field and zero stress compliance matrices of TGS in, respectively, the paraelectric phase and ferroelectric phase) and all the links resulting from the four quantities listed above. These links are described by the following three equations:

$$\left(\frac{\partial S_\alpha}{\partial\theta}\right)_{T, E} = \left(\frac{\partial\sigma}{\partial T_\alpha}\right)_{\tau, E, T'} - \left(\frac{\partial\sigma}{\partial\theta}\right)_{T, E} \left[\left(\frac{\partial\theta_c}{\partial T_\alpha}\right)_E\right] \quad (\alpha = 1, 2, 3, 5), \tag{3.4.24}$$

$$\left(\frac{\partial S_\alpha}{\partial T_\beta}\right)_{\theta, E} = \left(\frac{\partial S_\alpha}{\partial T_\beta}\right)_{\tau, E, T'} - \left(\frac{\partial S_\alpha}{\partial\theta}\right)_{T, E} \left[\left(\frac{\partial\theta_c}{\partial T_\alpha}\right)_E\right] \quad (\alpha, \beta = 1, 2, 3, 5), \tag{3.4.25}$$

$$\left(\frac{\partial D_2}{\partial T_\alpha}\right)_{\theta, E} = \left(\frac{\partial D_2}{\partial T_\alpha}\right)_{\tau, E, T'} - \left(\frac{\partial D_2}{\partial\theta}\right)_{T, E} \left[\left(\frac{\partial\theta_c}{\partial T_\alpha}\right)_E\right] \quad (\alpha = 1, 2, 3, 5). \tag{3.4.26}$$

† See footnote on p. 118.

TABLE 3.4.1.

Arrays representing the symmetry of the compliance matrix generated by G for the point groups (a) 2 *and* (b) $2/m$.

The lines show the connections described by eqns. (3.4.24). (3.4.25) and (3.4.26).

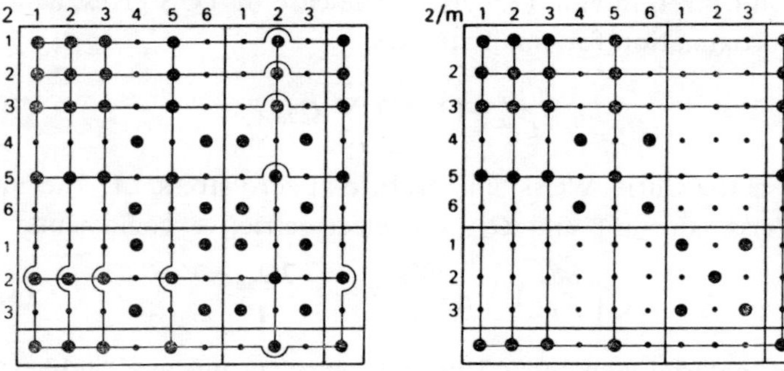

(a) (b)

Using the data measured in the ferroelectric phase, Janovec (1966) has plotted various anomalous compliance elements of TGS one against another as suggested by these three equations. It appears from his results that the cylindrical approximation holds for TGS in the temperature range $\theta_c - 16°C < \theta < \theta_c - 1°C$.

Bibliography

Forsbergh (1956), Mason (1950, 1966) and Nye (1964) discuss various thermodynamic identities for the elastic dielectric. The coordinate transformation properties of the fields and coefficients appearing in the polynomial representations of equations of state are described in detail in these references.

Landau and Lifshitz (1960) give a clear account of electrostatic work and stability in rigid dielectrics.

The neighbourhood temperature was introduced by Buckingham and Fairbanks (1961) to analyse the properties of liquid helium near the λ-line. These authors also proved that the asymptotic form of a neighbourhood temperature identity is equivalent to the cylindrical approximation which was originally discussed by Pippard (1957).

CHAPTER 4

The Non-linear Elastic Dielectric

IN THIS chapter we present a theory of the static non-linear behaviour of an elastic dielectric,† § 4.1. The field equations, boundary conditions and equations of state reduce in the small field approximation to precisely those of the linear theory, described in Chapter 2. However, the quasi-linear theory of that chapter appears as a special case of the general theory only under certain conditions, and these conditions cannot be deduced on the basis of the linear theory alone, § 4.2. Thus the general theory provides us with an indication of the range of validity of the commonly used quasi-linear theory and an alternative approximate non-linear theory outside this range. The discussion in § 4.1 should serve as an introduction to some of the problems and the techniques for their solution which one meets in the area of non-linear field theories.

Because the mathematical formalism of the non-linear theory is of necessity complicated, we have described the non-linear behaviour of a simple system—a dielectric slab—to illustrate the source and physical nature of some of the terms peculiar to the non-linear theory, § 4.3.

4.1 The Non-linear Elastic Dielectric

4.1.1 FINITE DEFORMATION

The deformation of a dielectric is described by a mapping $\boldsymbol{x} = \boldsymbol{x}(\boldsymbol{X})$ (see § 2.2.1), where $\boldsymbol{X}$ denotes the position vectors of material points

† References to the theory of the time-dependent and thermal non-linear behaviour are given in the bibliography at the end of the chapter.

when the dielectric is in some reference configuration and $\boldsymbol{x}$ the corresponding position vectors in the configuration of interest. We shall refer these two vector fields $\boldsymbol{x}$, $\boldsymbol{X}$ to different reference frames; thus x_i ($i = 1, 2, 3$) denotes the components of $\boldsymbol{x}$ measured in the set of rectangular coordinate axes x_1, x_2, x_3 fixed in the laboratory and X_I ($I = 1, 2, 3$) denotes the components of $\boldsymbol{X}$ in the set of rectangular coordinate axes X_1, X_2, X_3, again fixed in the laboratory. Small Roman letters are reserved for subscripts of components measured in the first frame and capital Roman letter for those measured in the second frame. The introduction of two reference frames proves convenient when we discuss the invariance properties of the energy density. We shall use the semicolon notation for partial derivatives with respect to position coordinates; thus, for example, $x_{i;K} \equiv \partial x_i/\partial X_K$ denotes the deformation gradient.

To obtain a strain measure in the case of finite deformations we again consider the behaviour of an arbitrary line element $d\boldsymbol{X}$ at $\boldsymbol{X}$ (cf. § 2.2.1). Let us suppose that as a result of the deformation the line element $d\boldsymbol{X}$ is translated, rotated and altered in length to become the line element $d\boldsymbol{x}$ at $\boldsymbol{x}$. Consider the difference $|d\boldsymbol{x}|^2 - |d\boldsymbol{X}|^2$, where $|d\boldsymbol{x}|$ denotes the magnitude of $d\boldsymbol{x}$, etc. Expanding the components dx_i in terms of the dX_J we get

$$|d\boldsymbol{x}|^2 - |d\boldsymbol{X}|^2 = dx_i\,dx_i - dX_J\,dX_J = (x_{i;J}\,x_{i;K} - \delta_{JK})\,dX_J\,dX_K,$$

where δ_{JK} is the Kronecker delta. In these, as in all later expressions in this section, the Einstein summation convention is employed. The *finite Lagrangian strain tensor* η_{KJ} is defined

$$\eta_{KJ} = \tfrac{1}{2}(x_{i;K}\,x_{i;J} - \delta_{KJ}) = \eta_{JK},$$

and hence

$$|d\boldsymbol{x}|^2 - |d\boldsymbol{X}|^2 = 2\eta_{KJ}\,dX_K\,dX_J.$$

Thus η_{KJ} provides a measure of the finite local deformation at the point x in the dielectric. If we let the two coordinate frames coincide and introduce the displacement vector $\boldsymbol{u} = \boldsymbol{x} - \boldsymbol{X}$, then it is a simple exercise to show that η_{KJ} reduces to the infinitesimal strain tensor S_{KJ} (§ 2.2.1) when the deformations are small, $|\partial u_i/\partial X_J| \ll 1$.

4.1.2 Variational Principle

Consider an isolated system consisting of a conductor and a homogeneous dielectric, both of finite extent. Let us suppose that the system possesses an initial state in which there is no charge distribution on the conductor and no dipole moment distribution in the dielectric. The deformations of the dielectric will be measured relative to its configuration in this initial state. Let charges be brought from infinity and placed on the conductor and let mechanical surface tractions be applied to both the conductor and dielectric. If this process is carried out quasi-statically then the total electric and mechanical work done on the system is stored as potential energy $\mathcal{U}$, say. Let us refer the positions of *all* points in space, when the system is in this final state, to the rectangular coordinate axes x_1, x_2, x_3. The deformation suffered by the dielectric as a result of the introduction of charges and surface tractions is described by the three functions $x_i = x_i(X_J)$ introduced in the previous subsection.

We write the total energy $\mathcal{U}$ in the form

$$\mathcal{U} = \int_{V+A+C} \Psi \, dv,$$

where the subscripts V, A, C refer to the regions occupied in the final state by the dielectric, free space and conductor respectively. We choose an energy density Ψ of the form

$$\left.\begin{aligned} \Psi &= \psi(x_i, x_{i;K}, D_i)\,\varrho && \text{in } V, \\ \Psi &= D_i D_i/2\varepsilon_0 && \text{in } A, \\ \Psi &\equiv 0, && \text{in } C, \end{aligned}\right\} \tag{4.1.1}$$

where ϱ is the mass density and ψ is the energy density in the dielectric; $x_{i;K}$ are the deformation gradients and D_i are the components of some vector field $\boldsymbol{D}$ which spans all space outside the conductor. Thus if a mass of material $dm = \varrho\, dv$ is shifted from a point $\boldsymbol{X}$ to a point $\boldsymbol{x}$ as a result of the deformation, energy $\psi(x_i, x_{i;K}, D_i)\,\varrho\, dv$ is stored in dv. Because the dielectric is assumed to be homogeneous in the initial state ψ is not an explicit function of $\boldsymbol{X}$. We show below that the demands of the conservation or balance of linear momentum remove $\boldsymbol{x}$ as an

explicit variable in ψ. We shall assume that the vector $\boldsymbol{D}$ satisfies the field equation

$$D_{i;\,i} = 0 \quad \text{in } A \text{ and } V \tag{4.1.2}$$

and the boundary condition

$$\oint_C D_i n_i \, dS = Q, \tag{4.1.3}$$

where n_i are the direction cosines of the outward normal to the conductor surface and Q the net charge on the conductor. The field equation (4.1.2) implies that the normal component of $\boldsymbol{D}$ is continuous across any dielectric free space interface (see § 2.4.2). Later we shall identify $\boldsymbol{D}$ with the electric displacement field, but for the present we regard it as some field whose variations are subject to eqns. (4.1.2) and (4.1.3) and the variational principle, introduced below.

Let the position vector of each material point in the dielectric undergo a small variation $x_i \to x_i + \delta x_i$ and let the vector field $\boldsymbol{D}$ undergo a small variation $\boldsymbol{D} \to \boldsymbol{D}'$ at all points in space. The variation in $\boldsymbol{D}$ is assumed to be small in the sense that at all points in the system, which are either outside or inside the dielectric throughout the deformation, $\boldsymbol{D}' - \boldsymbol{D}$ is an infinitesimal. For these points we put $D_i' - D_i = \delta D_i$. The conductor boundaries are to be held fixed during this variation. We take as our variational principle, that $\mathcal{U}$ be an extremum with respect to the variations δx_i, δD_i subject to the following constraints:

(a) $(\delta D_i)_{;\,i} = 0$ in V and A; from eqn. (4.1.2);
(b) $\delta x_i = 0$ on V;
(c) $\delta Q = 0$ on C.

Aside from these constraints the fields δx_i, δD_i are treated as completely arbitrary functions of position. Constraint (a) is imposed by the field equation (4.1.2), while the choice of (b) and (c) reflects the physical limitations on the constraints one can apply in practice to such a system. That is to say, while we are free to constrain the net charge on the conductor to any value and the boundary of the dielectric to any shape we please, once this choice is made the corresponding charge distribution on the conductor (described by $\boldsymbol{D}$) and the mass distribution in the

dielectric (described by $x_{i;K}$), for example, are determined by the physical laws governing the system. It is these physical laws in the form of field equations and boundary conditions that we seek from the variational principle.

To simplify the following analysis, we shall assume that for all the surface elements the variation δx_i is such that $\delta x_i n_i < 0$, where the n_i are the direction cosines of the outward normals to the dielectric surface. Let the new region occupied by the dielectric after the variation be

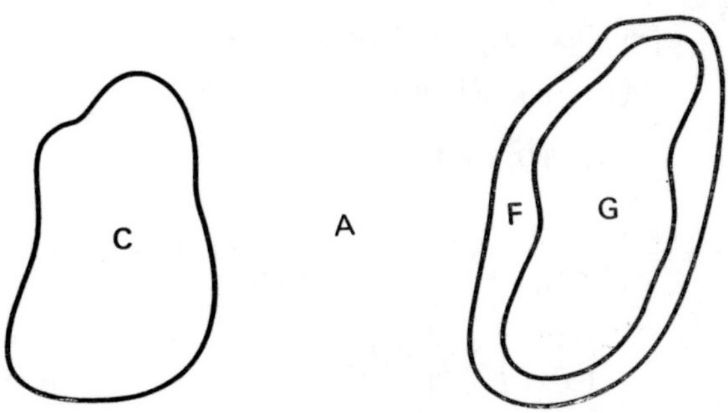

FIG. 4.1.1. Diagram of the regions referred to in the text. C, the region occupied by the conductor. A, the region outside the conductor and dielectric before and after the virtual deformation of the dielectric. $G+F$, the region occupied by the dielectric before the deformation. G, the region occupied by the dielectric after the deformation.

denoted by G (Fig. 4.1.1) and the region now free of dielectric by F. Thus $V = F+G$. As a result of this choice of surface displacements there are no regions free of dielectric before and occupied by dielectric after the variation. This restriction, while simplifying our discussion, in no way impairs the generality of the results [see eqn. (4.1.9) below].

Introducing Lagrangian multipliers ϕ, t_i^0 and ϕ_0 for the constraints (a), (b) and (c) above respectively, we write the condition for an extremum in the form

$$0 = \delta\mathcal{U} - \int_{A+G} \phi(\delta D_i)_{;i}\, dv - \phi_0 \oint_C \delta D_i n_i\, dS - \oint_V t_i^0\, \delta x_i\, dS. \qquad (4.1.4)$$

The variations δx_i, δD_i may now be treated as completely arbitrary fields.

4.1.3 Field Equations and Boundary Conditions

To determine the variation $\delta\mathcal{U}$ consider first the energy stored in the dielectric. An element of mass $dm = \varrho\, dv$ at a position x_i is moved to a position $x_i+\delta x_i$. If the field D_i at x_i is varied from D_i to $D_i+\delta D_i$,† then at the position $x_i+\delta x_i$ the new value is

$$D_i+\delta D_i+(D_{i;\,k}+\delta D_{i;\,k})\,\delta x_k \cong D_i+\delta D_i+D_{i;\,k}\,\delta x_k.$$

Thus the change in energy associated with this volume element is

$$\delta(d\mathcal{U}) = \psi[x_i+\delta x_i,\, x_{i;\,K}+(\delta x_i)_{;\,K},\, D_i+\delta D_i+D_{i;\,k}\,\delta x_k]\,\varrho\, dv$$
$$-\psi(x_i,\, x_{i;\,K},\, D_i)\,\varrho\, dv.$$

To terms linear in δx_i, δD_i we have

$$\delta(d\mathcal{U}) = \left[\left(\frac{\partial\psi}{\partial x_{i;\,K}}\right)(\delta x_i)_{;\,K}+\left(\frac{\partial\psi}{\partial D_i}\right)(\delta D_i+D_{i;\,k}\,\delta x_k)\right.$$
$$\left.+\left(\frac{\partial\psi}{\partial x_i}\right)\delta x_i\right]\varrho\, dv. \qquad (4.1.5)‡$$

The change in the energy stored in the vacuum is

$$\delta\mathcal{U}_{VAC} = \int\limits_{A+F}\frac{D_i'D_i'}{2\varepsilon_0}\,dv-\int\limits_{A}\frac{D_iD_i}{2\varepsilon_0}\,dv.$$

To terms linear in δx_i, δD_i,

$$\delta\mathcal{U}_{VAC} = -\oint\limits_{V}\frac{D_i^A D_i^A}{2\varepsilon_0}\,\delta x_j n_j\, dS+\int\limits_{A}\frac{D_i\,\delta D_i}{\varepsilon_0}\,dv, \qquad (4.1.6)$$

where the superscripts A are used to emphasize that the D_i^A are the components of the field D_i at points in the vacuum; n_i is the unit vector

† If the point x_i lies outside the dielectric after the deformation the quantity $D_i+\delta D_i$ is not the displacement field D_i' at x_i but rather the *function* which gives the field D_i' at points within the dielectric.

‡ Consider the term $(\partial\psi/\partial D_i)\,[\delta D_i+D_{i;\,k}\,\delta x_k]$. If we divide through by the time interval δt and let $\delta t \to 0$, we get $(\partial\psi/\partial D_i)\,[(\partial D_i/\partial t)+D_{i;\,k}(\partial x_k/\partial t)]$. The quantity $(\partial\psi/\partial D_i)D_{i;\,k}(\partial x_k/\partial t)$ is the convective contribution to ψ arising from the functional dependence of ψ on $\mathbf{D}$.

parallel to the outward normal to the dielectric surface. The variation in $\mathcal{U}$ correct to terms linear in δx_i, δD_i is obtained by integrating the expression given by eqn. (4.1.5) over the region V and adding to this $\delta\mathcal{U}_{VAC}$; hence

$$\delta\mathcal{U} = \int_V \left[\frac{\partial\psi}{\partial x_{i;K}}(\delta x_i)_{;K} + \frac{\partial\psi}{\partial D_i}(\delta D_i + D_{i;j}\,\delta x_j) + \frac{\partial\psi}{\partial x_i}\delta x_i\right]\varrho\,dv$$
$$-\oint_V \frac{D_j^A D_j^A}{2\varepsilon_0}\delta x_i n_i\,dS + \int_A \frac{D_i\,\delta D_i}{\varepsilon_0}\,dv.$$

Using the identity $(\delta x_i)_{;K} = (\delta x_i)_{;j}(x_j)_{;K}$ and Gauss' theorem we obtain the expression

$$\left.\begin{aligned}\delta\mathcal{U} = &\int_V \left\{\left[\varrho\left(\frac{\partial\psi}{\partial D_i}\right)D_{i;j} - \left(\varrho\frac{\partial\psi}{\partial x_{j;K}}x_{i;K}\right)_{;i} + \varrho\left(\frac{\partial\psi}{\partial x_i}\right)\right]\delta x_j\right.\\ &\left.+\varrho\left(\frac{\partial\psi}{\partial D_i}\right)\delta D_i\right\}dv + \oint_V \left[\varrho\left(\frac{\partial\psi}{\partial x_{i;K}}\right)x_{j;K}n_j\right.\\ &\left.-\frac{D_j^A D_j^A}{2\varepsilon_0}n_i\right]\delta x_i\,dS + \int_A \frac{D_j\,\delta D_j}{\varepsilon_0}\,dv.\end{aligned}\right\} \tag{4.1.7}$$

Consider now the second term in eqn. (4.1.4). This can be written in the form

$$\int_{A+G}\phi(\delta D_i)_{;i}\,dv = \int_{A+G}(\phi\,\delta D_i)_{;i}\,dv - \int_{A+G}\phi_{;i}\,\delta D_i\,dv$$

or, from Gauss' theorem,

$$\int_{A+G}\phi(\delta D_i)_{;i}\,dv = -\oint_C \phi\,\delta D_i n_i\,dS - \oint_{A\cap F}\phi^A\,\delta D_i^A n_i\,dS$$
$$+\oint_G \phi^V\,\delta D_i^V n_i\,dS - \int_{A+G}\phi_{;i}\,\delta D_i\,dv, \tag{4.1.8}$$

where $A \cap F$ refers to the surface common to regions A and F and the n_i are the direction cosines of the outward normals to the surfaces of the conductor and dielectric. We have assumed that $\phi \to 0$ and $\delta D_i \to 0$ at infinity sufficiently quickly that the surface integral $\oint \phi\,\delta D_i n_i\,dS$ at infinity vanishes.

In terms of the primed notation, we have

$$-\oint_{A\cap F} \phi^A\,\delta D_i^A n_i\,dS+\oint_G \phi^V\,\delta D_i^V n_i\,dS$$
$$=-\oint_{A\cap F} \phi^A(D_i^{A'}-D_i^A)\,n_i\,dS+\oint_G \phi^V(D_i^{V'}-D_i^V)\,n_i\,dS.$$

Consider the *functions* ϕ^A and D_i^A, originally introduced for points within A, at points in F. Adding and subtracting a term involving these functions, we get

$$-\oint_{A\cap F} \phi^A\,\delta D_i^A n_i\,dS+\oint_G \phi^V\delta D_i^V n_i\,dS = \oint_{A\cap F} (\phi^V-\phi^A)(D_i^{A'}-D_i^A)\,n_i\,dS$$
$$-\oint_{A\cap F} \phi^V(D_i^{A'}-D_i^A)\,n_i\,dS+\oint_G \phi^V(D_i^{V'}-D_i^V)\,n_i\,dS$$
$$= \oint_{A\cap F} (\phi^V-\phi^A)(D_i^{A'}-D_i^A)\,n_i\,dS-\int_F \phi^V_{;i}(D_i^{A'}-D_i^V)\,dv.$$

In this last step, we have used Gauss' theorem and the boundary conditions $D_i^{A'}n_i = D_i^{V'}n_i$ on G and $D_i^A n_i = D_i^V n_i$ on V.

Combining this last result with eqn. (4.1.8), we get

$$\int_{A+G} \phi(\delta D_i)_{;i}\,dv =-\int_{A+G} \phi_{;i}\,\delta D_i\,dv-\oint_C \phi\,\delta D_i n_i\,dS$$
$$+\int_{F\cap A} (\phi^V-\phi^A)\,\delta D_i^A n_i\,dS-\int_F \phi^V_{;i}(D_i^{A'}-D_i^V)\,dv.$$

In the linear approximation this equation reads

$$\int_{A+G} \phi(\delta D_i)_{;i}\,dv =-\int_{A+V} \phi_{;i}\,\delta D_i\,dv-\oint_C \phi\,\delta D_i n_i\,dS$$
$$+\oint_V (\phi^V-\phi^A)\delta D_i^A n_i\,dS+\oint_V \phi^V_{;i}(D_i^A-D_i^V)\delta x_j n_j\,dS.$$

Since the normal component of D_i and the tangential component of grad ϕ† are continuous across a dielectric free-space interface, the inte-

† The proof of this statement for the case of grad ϕ follows immediately from the identity curl grad $\phi \equiv 0$ (see § 2.4.2).

grand in the last integral on the right may be replaced by an alternative expression; thus

$$\left.\begin{aligned}\int_{A+G} \phi(\delta D_i)_{;\,i}\,dv = -\int_{A+V} \phi_{;\,i}\delta D_i\,dv - \oint_C \phi\,\delta D_i n_i\,dS \\ +\oint_V (\phi^V - \phi^A)\,\delta D_i^A n_i\,dS \\ +\tfrac{1}{2}\oint_V (\phi^A_{;\,i} + \phi^V_{;\,i})(D_i^A - D_i^V)\,\delta x_j n_j\,dS.\end{aligned}\right\} \quad (4.1.9)^\dagger$$

Hence the condition for an extremum is, from eqns. (4.1.4), (4.1.7) and (4.1.9),

$$\int_V \left\{\left[\varrho\left(\frac{\partial\psi}{\partial D_i}\right) D_{i;\,j} - \left(\varrho\frac{\partial\psi}{\partial x_{j;\,K}} x_{i;\,K}\right)_{;\,i} + \varrho\left(\frac{\partial\psi}{\partial x_j}\right)\right]\delta x_j + \left(\varrho\frac{\partial\psi}{\partial D_i} + \phi_{;\,i}\right)\delta D_i\right\} dv$$

$$+\oint_V \left\{(\phi^A - \phi^V)\,\delta D_i^A n_i + \left[\left(\varrho\frac{\partial\psi}{\partial x_{i;\,K}} x_{j;\,K}\right) n_j - \frac{1}{2}(\phi^A_{;\,j} + \phi^A_{;\,j})(D_j^A - D_j^V)\,n_i - \frac{D_j^A D_j^A}{2\varepsilon_0} n_i - t_i^0\right]\delta x_i\right\} dS$$

$$+\int_A \left(\frac{D_j}{\varepsilon_0} + \phi_{;\,j}\right)\delta D_j\,dv + \oint_C (\phi - \phi_0)\,\delta D_i n_i\,dS = 0.$$

† If $\delta x_i n_i > 0$ for all the surface elements on V, then it is evident that

$$\begin{aligned}\int_{A+G} \phi(\delta D_i)_{;\,i}\,dv &= -\int_{A+V} \phi_{;\,i}\,\delta D_i\,dv - \oint_C \phi\,\delta D_i n_i\,dS + \oint_V (\phi^V - \phi^A)\,\delta D_i^A n_i\,dS \\ &\quad + \oint_V \phi^A_{;\,i}(D_i^A - D_i^V)\,\delta x_j n_j\,dS = -\int_{A+V} \phi_{;\,i}\,\delta D_i\,dv \\ &\quad - \oint_C \phi\,\delta D_i n_i\,dS + \oint_V (\phi^V - \phi^A)\,\delta D_i^A n_i\,dS \\ &\quad + \tfrac{1}{2}\oint_V (\phi^V_{;\,i} + \phi^A_{;\,i})(D_i^A - D_i^V)\,\delta x_j n_j\,ds\end{aligned}$$

using the continuity of (a) the normal component of D_i and (b) the tangential components of $\phi_{;\,i}$ across any interface. This final expression agrees with eqn. (4.1.9) derived in the assumption that $\delta x_i n_i < 0$ on V. Thus it is clear that eqn. (4.1.9) holds in the general case in which $\delta x_i n_i \gtrless 0$ on V.

Equating the coefficients of the arbitrary changes δx_i and δD_i to zero, we obtain the following field equations and boundary conditions.

In the dielectric,

$$\left(\varrho \frac{\partial \psi}{\partial x_{i;K}} x_{j;K}\right)_{;j} - \varrho\left(\frac{\partial \psi}{\partial x_i}\right) - \varrho\left(\frac{\partial \psi}{\partial D_j}\right) D_{j;i} = 0, \tag{4.1.10}$$

$$\varrho\left(\frac{\partial \psi}{\partial D_i}\right) + \phi_{;i} = 0. \tag{4.1.11}$$

In free space,

$$\frac{D_i}{\varepsilon_0} + \phi_{;i} = 0. \tag{4.1.12}$$

At the surface of the dielectric,

$$\phi^V - \phi^A = 0, \tag{4.1.13}$$

$$\varrho \frac{\partial \psi}{\partial x_{i;K}} x_{j;K} n_j - \left[\frac{D_j^A D_j^A}{2\varepsilon_0} - \frac{1}{2}(\phi^V_{;j} + \phi^A_{;j})(D_j^V - D_j^A)\right] n_i - t_i^0 = 0. \tag{4.1.14}$$

At the surface of the conductor,

$$\phi - \phi_0 = 0. \tag{4.1.15}$$

We may now give the Lagrangian multipliers ϕ, ϕ_0, t_i^0 and the field D_i a physical interpretation. ϕ is the electrostatic potential, continuous across a dielectric-vacuum boundary [eqn. (4.1.13)]; ϕ_0 is the potential of the conductor [eqn. (4.1.15)]. We set $E_i = -\phi_{;i}$ and hence $\boldsymbol{E}$ is the electric field and $\boldsymbol{D}$ the electric displacement field [eqns. (4.1.11) and (4.1.12)]. Finally, we show that t_i^0 is the stress field due to the applied mechanical tractions. Consider two solutions to the above field equations and boundary conditions, D_i, x_i and $D_i + \delta D_i$, $x_i + \delta x_i$, where δD_i and δx_i are infinitesimals. In particular we have $(D_i + \delta D_i)_{;i} = 0$ and hence, from eqn. (4.1.4),

$$\delta \mathcal{U} = \phi_0 \oint_C \delta D_i n_i \, dS + \oint_V t_i^0 \, \delta x_i \, dS. \tag{4.1.16}$$

The first term on the right-hand side is the electric work so that the second term is the mechanical work; t_i^0 is therefore the applied stress.

Replacing $\phi_{;i}$ by $-E_i$ we write eqn. (4.1.14) in the form

$$t_i^0 = \varrho \frac{\partial \psi}{\partial x_{i;K}} x_{j;K} n_j - \left[\frac{D_j^A D_j^A}{2\varepsilon_0} + \frac{1}{2}(E_j^V + E_j^A)(D_j^V - D_j^A)\right] n_i. \quad (4.1.17)$$

We shall refer to the quantity in square brackets as the Maxwell stress in analogy with the Maxwell stress acting on the conductor.

4.1.4 Work and Energy

It is instructive to examine the relationship between the energy change $\delta\mathcal{U}$ and the electrical and mechanical work $\delta\mathcal{W}$ in more detail. From eqn. (4.1.7) and the field equation (4.1.10)

$$\delta\mathcal{U} = \int_V \varrho\left(\frac{\partial \psi}{\partial D_i}\right) \delta D_i \, dv + \oint_V \left[\varrho \frac{\partial \psi}{\partial x_{i;K}} x_{j;K} n_j - \left(\frac{D_j^A D_j^A}{2\varepsilon_0}\right) n_i\right] \delta x_i \, dS$$
$$+ \int_A \frac{D_i \, \delta D_i}{\varepsilon_0} dv = \int_{A+V} E_i \, \delta D_i \, dv$$
$$+ \oint_V \left[\varrho \frac{\partial \psi}{\partial x_{i;K}} x_{j;K} n_j - \frac{D_j^A D_j^A}{2\varepsilon_0} n_i\right] \delta x_i \, dS.$$

Consider now the case in which the boundaries of the dielectric are held fixed, i.e. $\delta x_i \equiv 0$ on V. An infinitesimal charge δQ added to the conductor causes an infinitesimal change δD_i^Q in D_i; the associated energy change is then

$$\delta\mathcal{U} = \int_{A+V} E_i \, \delta D_i^Q \, dv.$$

The corresponding work is

$$\delta\mathcal{W} = \phi_0 \, \delta Q = \phi_0 \oint_C \delta D_i^Q n_i \, dS,$$

and hence the conservation of energy $\delta\mathcal{U} = \delta\mathcal{W}$ gives

$$\int_{A+V} E_i \, \delta D_i^Q \, dv = \phi_0 \oint_C \delta D_i^Q n_i \, dS. \quad (4.1.18)$$

This result has been derived previously in § 2.4.3.

If, on the other hand, the dielectric boundaries are deformed while the net charge on the conductor remains fixed, i.e. $\delta x_i \not\equiv 0$ on V and $\delta Q \equiv 0$, there results a change δD_i^x in the field D_i such that

$$\oint_C \delta D_i^x n_i \, dS = 0.$$

The associated change in energy is

$$\delta \mathcal{U} = \int_{V+A} E_i \, \delta D_i^x \, dv + \oint_V \left[\varrho \frac{\partial \psi}{\partial x_{i;K}} x_{j;K} n_i - \left(\frac{D_j^A D_j^A}{2\varepsilon_0} \right) n_i \right] \delta x_i \, dS.$$

and the mechanical work done

$$\delta \mathcal{W} = \oint_V t_i^0 \, \delta x_i \, dS.$$

From the conservation of energy we have

$$\int_{V+A} E_i \, \delta D_i^x \, dv + \oint_V \left[\varrho \frac{\partial \psi}{\partial x_{i;K}} x_{j;K} n_j - \left(\frac{D_j^A D_j^A}{2\varepsilon_0} \right) n_i \right] \delta x_i \, dS = \oint_V t_i^0 \, \delta x_i \, dS$$

and on substituting for t_i^0 from eqn. (4.1.17) we get

$$\int_{V+A} E_i \, \delta D_i^x \, dv = -\tfrac{1}{2} \oint_V (E_j^V + E_j^A)(D_j^V - D_j^A) \, \delta x_i n_i \, dS. \qquad (4.1.19)$$

This result contrasts with that given by eqn. (4.1.18).

For the general case $\delta x_i \neq 0$ on V and $\delta Q \neq 0$ there results a change δD_i in D_i, and from eqns. (4.1.18) and (4.1.19) we have that this change satisfies the identity

$$\int_{V+A} E_i \, \delta D_i \, dv = \phi_0 \oint_C \delta D_i n_i \, dS - \tfrac{1}{2} \oint_V (E_j^V + E_j^A)(D_j^V - D_j^A) \, \delta x_i n_i \, dS. \qquad (4.1.20)$$

The integrand of the surface integral over V appearing in eqns. (4.1.19) and (4.1.20) is a non-linear term and therefore can be neglected in the linear case. Thus, when in the quasi-linear theory we replace the exact

expression for the electric and mechanical work,

$$\delta \mathcal{W} = \oint_V t_i^0 \, \delta x_i \, dS + \phi_0 \oint_C \delta D_i n_i \, dS,$$

by the approximate expression

$$\delta \mathcal{W} = \oint_V t_i^0 \, \delta x_i \, dS + \int_{V+A} E_i \, \delta D_i \, dv$$

we are neglecting the second integral on the right-hand side of eqn. (4.1.20).

4.1.5 Conservation of Momentum

The demands of the conservation of momentum, both linear and angular, put restrictions on the physically allowable functional forms of the energy density. This point has been discussed previously for the linear theory (§ 2.5.3). In this section we investigate the corresponding restrictions imposed on the function $\psi(x_i, x_{i;K}, D_i)$.

When a system is in static equilibrium, linear momentum is conserved if the sum of the forces applied to the system vanishes; angular momentum is conserved if the sum of the moments of these forces about any point vanishes. In the system of dielectric and conductor discussed above the dielectric is held in equilibrium by two types of forces—one the electrostatic interaction between the dielectric and conductor and the other the mechanical surface tractions described by the field t_i^0; similarly, the conductor is held in static equilibrium by electrical and mechanical forces, the former due to the presence of the dielectric, and the latter we shall assume to be surface tractions described by some fields t_i^c, defined over the conductor surface. Now the electrostatic force acting on the dielectric is equal in magnitude but opposite in direction to the electrostatic force acting on the conductor and hence the net force applied to the dielectric and conductor is of the form

$$\oint_C t_i^c \, dS + \oint_V t_i^0 \, dS \qquad (i = 1, 2, 3);$$

and the torque of these forces about the origin of coordinates, the

x_i, is

$$e_{ijk}\left[\oint_C x_j t_k^c \, dS + \oint_V x_j t_k^0 \, dS\right] \qquad (i = 1, 2, 3).$$

Thus the conditions for the conservation of linear and angular momentum are

$$\oint_V t_i^0 \, dS + \oint_C t_i^c \, dS \equiv 0 \qquad (i = 1, 2, 3), \tag{4.1.21}$$

$$e_{ijk}\left[\oint_V x_j t_k^0 \, dS + \oint_C x_j t_k^c \, dS\right] \equiv 0 \qquad (i = 1, 2, 3). \tag{4.1.22}$$

We now derive explicit expressions for each of these integrals in terms of the fields ψ, $x_{i;\,K}$, D_i and E_i. Consider first the applied surface tractions t_i^0; from eqn. (4.1.17)

$$\oint_V t_i^0 \, dS = \oint_V \left[\varrho\left(\frac{\partial\psi}{\partial x_{i;\,K}} x_{j;\,K}\right) n_j - \left\{\frac{D_j^A D_j^A}{2\varepsilon_0} + \frac{1}{2}(E_j^A + E_j^V)(D_j^V - D_j^A)\right\} n_i\right] dS.$$

Using Gauss' theorem and the field equation we can transform the right-hand side to get

$$\left.\begin{aligned}\oint_V t_i^0 \, dS &= \int_V \left[E_j D_{j;\,i} + \varrho\frac{\partial\psi}{\partial x_i}\right] dv \\ &\quad - \oint_V \left[\frac{D_j^A D_j^A}{2\varepsilon_0} + \frac{1}{2}(E_j^A + E_j^V)(D_j^V - D_j^A)\right] n_i \, dS \\ &= \int_V \varrho\left(\frac{\partial\psi}{\partial x_i}\right) dv + \oint_V \left\{\left[E_j^A D_j^V - \frac{D_j^A D_j^A}{2\varepsilon_0}\right.\right. \\ &\quad \left.\left. - \frac{1}{2}(E_j^A + E_j^V)(D_j^V - D_j^A)\right] n_i - E_i^V D_j^V n_j\right\} dS.\end{aligned}\right\} \tag{4.1.23}$$

In this last step we have used the identity $E_j D_{j;\,i} = (E_j D_j)_{;\,i} - (E_i D_j)_{;\,j}$. Since we are not concerned with the details of the behaviour of the conductor, we assume it to be rigid. Thus a mechanical stress field

$t_i^c = -\omega E_i/2$, where ω is the surface charge density, will hold the conductor in static equilibrium. From the properties of the electric field at the surface of the conductor, $(\omega E_i/2)$ has the same value as $[(D_i D_j/\varepsilon_0) - \delta_{ij}(D_k D_k/2\varepsilon_0)]\, n_j$. Thus

$$\oint_C t_i^c \, dS = -\oint_C \left[\frac{D_i D_j}{\varepsilon_0} - \delta_{ij}\frac{D_k D_k}{2\varepsilon_0}\right] n_j \, dS.$$

Using Gauss' theorem and the field equation $D_{i;\,i} = 0$, we get

$$\oint_C t_i^c \, dS = \oint_V \left[\frac{D_i^A D_j^A}{\varepsilon_0} - \delta_{ij}\frac{D_k^A D_k^A}{2\varepsilon_0}\right] n_j \, dS. \qquad (4.1.24)$$

From eqns. (4.1.23) and (4.1.24) the net mechanical force applied to the conductor and dielectric is

$$\left.\begin{aligned} \oint_V t_i^0 \, dS + \oint_C t_i^c \, dS = \int_V \varrho\left(\frac{\partial\psi}{\partial x_i}\right) dv + \oint_V [(D_j^A E_i^A - D_j^V E_i^V)\, n_j \\ + \tfrac{1}{2}(E_k^V - E_k^A)(D_k^A + D_k^V)\, n_i]\, dS. \end{aligned}\right\} \qquad (4.1.25)$$

The surface integral on the right-hand side is identically zero; to prove this we note that at the dielectric free-space boundary (a) $E_i^V - E_i^A$ is parallel to n_i and (b) $D_k^A n_k = D_k^V n_k$. Put $E_i^V - E_i^A = \lambda n_i$, then the integrand in question takes the form $-\lambda n_i D_j^A n_j + \frac{1}{2}\lambda n_k (D_k^A + D_k^V) n_i \equiv 0$.

Thus the net mechanical force applied to the system vanishes identically (i.e. linear momentum is conserved) if and only if

$$\int_V \varrho\left(\frac{\partial\psi}{\partial x_i}\right) dv = 0.$$

If ψ is an intrinsic property of the material making up the dielectric, then this identity is satisfied by ψ for any arbitrary shape and size of dielectric. Under these circumstances the identity holds if and only if

$$\left(\frac{\partial\psi}{\partial x_i}\right) \equiv 0 \quad (i = 1, 2, 3) \qquad (4.1.26)$$

at each point in the dielectric. Thus the energy density ψ cannot be an explicit function of the x_i.

Turning to the determination of the net torque, we have, from eqn. (4.1.17),

$$e_{ijk}\oint_V x_j t_k^0\, dS = e_{ijk}\oint_V x_j\left[\varrho\left(\frac{\partial\psi}{\partial x_{k;K}}\right)x_{r;K}n_r - \left\{\frac{D_r^A D_r^A}{2\varepsilon_0}+\frac{1}{2}(E_r^V+E_r^A)(D_r^V-D_r^A)\right\}n_k\right]dS$$

$$= e_{ijk}\left[\iint_V\left\{\delta_{rj}\varrho\left(\frac{\partial\psi}{\partial x_{k;K}}\right)x_{r;K}+x_j E_r D_{r;k}\right\}dv - \oint_V x_j\left\{\frac{D_r^A D_r^A}{2\varepsilon_0}+\frac{1}{2}(E_r^V+E_r^A)(D_r^V-D_r^A)\right\}n_k\, dS\right].$$

In this last step we have used Gauss' theorem and the field equations (4.1.10) and (4.1.26). With a transformation very similar to that used in obtaining eqn. (4.1.23), we get

$$e_{ijk}\oint_V x_j t_k^0\, dS = e_{ijk}\int_V\left[\varrho\left(\frac{\partial\psi}{\partial x_{k;K}}\right)x_{j;K}+E_r D_j)\right)\right]dv$$

$$+e_{ijk}\oint_V x_j\left[\left\{E_r^V D_r^V-\frac{D_r^A D_r^A}{2\varepsilon_0} - \frac{1}{2}(E_r^V+E_r^A)(D_r^V-D_r^A)\right\}n_k - E_k^V D_r^V n_r\right]dS.$$

Also

$$e_{ijk}\oint_C x_j t_k^c\, dS = -e_{ijk}\oint_C x_j\left[\frac{D_k D_r}{\varepsilon_0}-\delta_{kr}\frac{D_s D_s}{2\varepsilon_0}\right]n_r\, dS$$

$$= e_{ijk}\oint_V x_j\left[\frac{D_k^A D_r^A}{\varepsilon_0}-\delta_{kr}\frac{D_s^A D_s^A}{2\varepsilon_0}\right]n_r\, dS.$$

Combining these equations we obtain

$$e_{ijk}\left[\oint_V x_j t_k^0\, dS+\oint_C x_j t_k^c\, dS\right] = e_{ijk}\int_V \varrho\left(\frac{\partial\psi}{\partial x_{k;K}}x_{j;K}+\frac{\partial\psi}{\partial D_k}D_j\right)dv,$$

where we have used the fact that the square bracket on the right-hand side of eqn. (4.1.25) is identically zero (see above) and the relation $E_k = \varrho(\partial\psi/\partial D_k)$.

Thus the net mechanical torque of the applied forces vanishes (i.e. angular momentum is conserved) if and only if

$$e_{ijk}\int_V \varrho\left(\frac{\partial\psi}{\partial x_{k;K}}x_{j;K}+\frac{\partial\psi}{\partial D_k}D_j\right)dv \equiv 0. \qquad (4.1.27)$$

Again, if ψ is an intrinsic property of the material making up the dielectric, then this identity is satisfied by ψ for any arbitrary shape and size of the dielectric. Under these circumstances the identity holds if and only if

$$e_{ijk}\left(\frac{\partial\psi}{\partial x_{k;K}}x_{j;K}+\frac{\partial\psi}{\partial D_k}D_j\right) = 0 \quad (i = 1, 2, 3) \qquad (4.1.28)$$

or

$$\left(\frac{\partial\psi}{\partial x_{k;K}}\right)x_{j;K}+\left(\frac{\partial\psi}{\partial D_k}\right)D_j = \left(\frac{\partial\psi}{\partial x_{j;K}}\right)x_{k;K}+\left(\frac{\partial\psi}{\partial D_j}\right)D_k$$

$$(j, k = 1, 2, 3). \quad (4.1.29$$

Thus the demands of conservation of angular momentum are such that ψ regarded as a function of $x_{i;K}$, D_i must satisfy the three partial differential equations described in eqn. (4.1.29). We reserve further discussion of these equations until we have described the connection between the requirements of conservation of angular momentum and the requirements of coordinate invariance.

4.1.6 Invariance I

In this and the next subsection we shall discuss the invariance of the energy density function ψ with respect to changes of the coordinate axes.

We introduce another set of rectangular axes x_1', x_2', x_3' to describe the final states of the system. Let the two sets of axes x_1, x_2, x_3 and x_1', x_2', x_3' possess a common origin and let the coordinate transformation between these axes be of the form $x_i' = R_{ij}x_j$, where $R_{ij}R_{kj} = \delta_{ik}$ and det $R_{ij} = +1$. Thus R_{ij} is the matrix representation of the rotation

operator R which when applied to the axes x_1, x_2, x_3 brings them into coincidence with the axes x_1', x_2', x_3' ($x_1 \to x_1'$, etc.). The set of all rotations R forms the Rotation group. Let us assume that the energy density is both *value* and *form* invariant with respect to the operations of the Rotation group, i.e.

$$\psi(x_{i;\,K} D_i) \equiv \psi(x'_{i;\,K}, D'_i). \tag{4.1.30}$$ †

where

$$x'_{i;\,K} = R_{ij} x_{j;\,K} \ddagger \quad \text{and} \quad D'_i = R_{ij} D_j .$$

In other words we are assuming that the value and form of the energy density ψ are the same for the set of all coordinate axes related by rotations. On physical grounds the first condition is reasonable, for the value of the energy density at a point in the dielectric cannot alter because the state of the system is described by a new set of coordinate axes, but the second condition, that the form remains unchanged, appears, at first sight, to be too stringent; however, as we now show, the demands of conservation of angular momentum are fully equivalent to the assumption that ψ is invariant in the sense defined above.

To explore the analytical implication of the identity in eqn. (4.1.30) it is sufficient to consider only the case where the rotation is infinitesimal, i.e. R_{ij} is of the form

$$R_{ij} = \delta_{ij} + \Omega_{ij} \tag{4.1.31}$$

with $\Omega_{ij} = -\Omega_{ji}$ and $|\Omega_{ij}| \ll 1$. Combining eqns. (4.1.30) and (4.1.31) we have

$$\psi[x_{i;\,K}, D_i] \equiv \psi[(\delta_{ij} + \Omega_{ij}) x_{j;\,K}, (\delta_{ij} + \Omega_{ij}) D_j].$$

Expand to terms linear in Ω_{ij}; thus

$$\left[\frac{\partial \psi}{\partial x_{i;\,K}} x_{j;\,K} + \frac{\partial \psi}{\partial D_i} D_j\right] \Omega_{ij} = 0.$$

Since Ω_{ij} is arbitrary and anti-symmetric the last equation holds if and only if

$$e_{ijk}\left[\frac{\partial \psi}{\partial x_{j;\,K}} x_{k;\,K} + \frac{\partial \psi}{\partial D_j} D_k\right] = 0 \quad (i = 1, 2, 3). \tag{4.1.32}$$

† We recall that the demands of the conservation of linear momentum require that ψ is not an explicit function of the x_i, § 4.1.5.

‡ Because the coordinates describing the initial and final states are referred to different axes we can transform one set of coordinates without affecting the other.

These equations are precisely those obtained from the condition that angular momentum is conserved (4.1.28). Thus the conservation of angular momentum and the value and form invariance of ψ with respect to rotational changes of the x_1, x_2, x_3 axes are fully equivalent. To discuss the constraints imposed on the functional form of ψ by the three partial differential equations (4.1.32) we shall assume that ψ may be expressed as a polynomial in $x_{i;K}$, D_i.† Then eqn. (4.1.32) expresses the condition that ψ is a polynomial scalar invariant, under the Rotation group, of the vectors whose components in the coordinate system x_1, x_2, x_3 are $x_{i;1}$, $x_{i;2}$, $x_{i;3}$ and D_i. From a theorem due to Cauchy we immediately conclude that ψ can be written as a polynomial of the invariants, under the Rotation group, of these twelve vectors (see, for example, Weyl, 1939). The invariants in this case are $x_{i;K}x_{i;L}$, $x_{i;K}D_i$, D_iD_i, $e_{ijk}x_{i;K}x_{j;L}x_{k;M}$, $e_{ijk}x_{i;K}x_{j;L}D_k$, $e_{ijk}x_{i;K}D_jD_k$ and $e_{ijk}D_iD_jD_k$. Since the invariance condition (4.1.32) is fully equivalent to the three partial differential equations of (4.1.29), we conclude that any arbitrary polynomial function of these invariants is automatically a solution to eqn. (4.1.29). We shall discuss what constitutes a minimal set of invariants (a set of functionally independent invariants) below.

In the proof connecting the conservation of angular momentum and the invariance of ψ we showed that, provided the two sets of axes x_i, x_i' could be brought into coincidence by a rotation, ψ must have the same value and form measured in these two frames if angular momentum is to be conserved. This proof does not apply to two frames related by an inversion ($x_i \to -x_i$) plus a rotation (i.e. a change in handedness of the axes). Thus the question of the invariance of ψ with respect to inversion is independent of the conservation of angular momentum. We shall make the additional assumption‡ that ψ is value and form invariant with respect to inversion. With this assumption and the assumption of conservation of angular momentum we immediately

† The analysis of the behaviour of real dielectrics is usually based on a representation of ψ, consisting of a truncated Taylor series in the field variables.

‡ The Hamiltonian describing the behaviour of the electrons and nuclei making up the solid is invariant with respect to inversion and as a result it seems to us to be extremely implausible to expect any scalar property of a system such as the energy density field not to display this invariance property.

conclude that the value and form of the energy density ψ is independent of the choice of axes used to describe the final state; in other words ψ is value and form invariant under the operations of the Full Orthogonal group ($\equiv$ Rotation group $\otimes$ Inversion).

If ψ is a polynomial scalar invariant under the full orthogonal group, then ψ is an explicit function just of the invariants of the vectors $x_{i;1}$, $x_{i;2}$, $x_{i;3}$, D_i under the full orthogonal group. To determine these invariants we note that the rotation group is a sub-group of the full orthogonal group and hence the invariants we require are the invariants of the rotation group which are also invariants of the inversion group. From an inspection of the list of invariants quoted above, we deduce that the invariants of the full orthogonal group are $x_{i;K}x_{i;L}$, $x_{i;K}D_i$ and D_iD_i. For convenience in the discussion in § 4.2 we choose to consider the equivalent set

$$\eta_{KL} = \tfrac{1}{2}(x_{i;K}x_{i;L} - \delta_{KL}), \tag{4.1.33}$$

$$\Pi_K = x_{i;K}D_i, \tag{4.1.34}$$

$$a = D_iD_i. \tag{4.1.35}$$

This is a minimal set of invariants since they are functionally independent. η_{KL} is the Lagrangian strain tensor introduced in § 4.1.1.

We may summarize the discussion given above thus; if a polynomial energy density function $\psi = \psi(x_{i;K}, D_i)$ is to be independent of the choice of axes used to discuss the final state (and hence satisfy the law of conservation of angular momentum) then any polynomial representation for ψ must be a function only of the arguments η_{KL}, Π_K and a.

We write $\psi = \chi(\eta_{KL}, \Pi_K, a)$ to distinguish between the polynomial $\psi(x_{i;K}, D_i)$ with arguments $x_{i;K}$, D_i and the polynomial $\chi(\eta_{KL}, \Pi_K, a)$ with arguments η_{KL}, Π_K, a.

We now deduce expressions for the quantities $(\partial\psi/\partial x_{i;K})x_{j;K}$ and $\partial\psi/\partial D_i$ valid when ψ is a function only of the arguments η_{KL}, Π_K, a. First of all the following identities are easily proved:

$$\frac{\partial \eta_{KL}}{\partial x_{i;\,M}} x_{j;\,M} = \tfrac{1}{2}(x_{i;\,K} x_{j;\,L} + x_{i;\,L} x_{j;\,K}),$$

$$\frac{\partial \Pi_K}{\partial x_{i;\,L}} x_{j;\,L} = x_{j;\,K} D_i,$$

$$\frac{\partial \Pi_K}{\partial D_i} = x_{i;\,K}.$$

Hence

$$\left(\frac{\partial \psi}{\partial x_{i;\,K}}\right) x_{j;\,K} = \frac{1}{2}\left(\frac{\partial \chi}{\partial \eta_{KL}}\right)_{\Pi,\,a} (x_{i;\,K} x_{j;\,L} + x_{i;\,L} x_{j;\,K}) + \left(\frac{\partial \chi}{\partial \Pi_K}\right)_{\eta,\,a} x_{j;\,K} D_i \tag{4.1.36†}$$

and

$$\left(\frac{\partial \psi}{\partial D_i}\right) = \left(\frac{\partial \chi}{\partial \Pi_K}\right)_{\eta,\,a} x_{i;\,K} + 2\left(\frac{\partial \chi}{\partial a}\right)_{\eta,\,\Pi} D_i. \tag{4.1.37}$$

4.1.7 Invariance II

In this subsection we shall discuss the invariance of ψ under changes of the axes X_1, X_2, X_3 which describe the initial state of the system.

We introduce another set of rectangular axes X_1', X_2', X_3' related to X_1, X_2, X_3 by an operation R of the full orthogonal group. We denote the matrix representation of R by R_{KL}, where $R_{KL}R_{JL} = \delta_{KJ}$ and det $R_{KL} = \pm 1$. Thus the coordinate transformation takes the form $X_K' = R_{KL}X_L$. The transformation properties of η_{KL}, Π_K and a under R are‡

$$\eta'_{KL} = R_{KM}R_{LN}\eta_{MN}, \quad \Pi'_K = R_{KL}\Pi_L, \quad a' = a. \tag{4.1.38}$$

† In differentiating ψ with respect to the strain components, we formally treat η_{LK}, η_{KL} as independent variables.

‡ *Proof.* The components D_i are unchanged hence $a' = a$. Consider Π'_K.

$$\Pi'_K = \frac{\partial x_i}{\partial X'_K} D_i = \frac{\partial x_i}{\partial X_L}\frac{\partial X_L}{\partial X'_K} D_i = R_{KL}\Pi_L.$$

Similarly,

$$\begin{aligned}\eta'_{KL} &= \frac{1}{2}\left(\frac{\partial x_i}{\partial X'_K}\frac{\partial x_i}{\partial X'_L} - \delta_{KL}\right)\\ &= \frac{1}{2}\left(\frac{\partial x_i}{\partial X_M}\frac{\partial x_i}{\partial X_N}\frac{\partial X_M}{\partial X'_K}\frac{\partial X_N}{\partial X'_L} - \delta_{KL}\right)\\ &= R_{KM}R_{LN}\eta_{MN} + \frac{1}{2}(\delta_{MN}R_{KM}R_{LN} - \delta_{KL}).\end{aligned}$$

The round bracket vanishes since $\delta_{MN}R_{KM}R_{LN} = R_{KM}R_{LM} = \delta_{KL}$.

Thus η_{KL}, Π_K, a are tensors of rank two, one and zero respectively.

The polynomial $\chi(\eta_{KL}, \Pi_K, a)$ representing the energy density is not form invariant under this transformation; that is to say the coefficients in the polynomial χ are functions of the choice of coordinate axes used to describe the initial state. We denote these coefficients by $H^{\alpha}_{KLM}\ldots$; thus, for example, we might write the term in χ which is a sum of products of η_{KL} and Π_K as $H^{6}_{KLM}\eta_{KL}\Pi_M$ (as usual repeated indices imply a sum) where the superscript 6 is taken to denote this type of term as opposed to another, for example of the form $H^{2}_{KL}\eta_{KL}$. The energy density can be regarded as a function of these coefficients;† to emphasize the functional dependence we write

$$\psi = g(H^{\alpha}_{KLM\ldots}, \eta_{KL}, \Pi_K, a),$$

where g is a homogeneous polynomial of degree one in the $H^{\alpha}_{KLM}\ldots$.

Now ψ is a scalar whose value cannot depend on the choice of coordinate axes. This applies to each type of sum appearing in the polynomial representation g, as, for example, $H^{3}_{KLM}\,\eta_{KL}\Pi_K$. Therefore‡ the $H^{\alpha}_{KLM}\ldots$ transform as tensors of rank n, where n is the number of subscripts $KLM\ldots$,

$$H^{\alpha'}_{KLM\ldots} = R_{KN}R_{LO}R_{MP}\ldots H^{\alpha}_{NOP\ldots} \qquad (4.1.39)$$

We see immediately that, as far as transformations of the axes X_1, X_2, X_3 are concerned, because of the special nature of g the demands of value invariance for ψ is sufficient to guarantee form invariance, namely,

$$g(H^{\alpha'}_{NOP\ldots}, \eta'_{NO}, \Pi'_N, a') \equiv g(H^{\alpha}_{KLM\ldots}, \eta_{KL}, \Pi_K, a). \qquad (4.1.40)$$

† Toupin (1956) refers to these coefficients as material descriptors.

‡ *Proof.* Consider the term $H^{3}_{KLM}\,\eta_{KL}\,\Pi_M$. Since ψ is a scalar

$$H^{3}_{KLM}\eta_{KL}\Pi_M \equiv H^{3'}_{NOP}\eta'_{NO}\Pi'_P$$
$$\equiv H^{3'}_{NOP}R_{NK}R_{OL}R_{PM}\eta_{KL}\Pi_M \qquad \text{from eqn. (4.1.38),}$$

i.e.
$$(H^{3}_{KLM}-H^{3'}_{NOP}R_{NK}R_{OL}R_{PM})\eta_{KL}\Pi_M \equiv 0$$

or
$$H^{3}_{KLM} = H^{3'}_{NOP}R_{NK}R_{OL}R_{PM}.$$

The inverse transformation is

$$H^{3'}_{NOP} = R_{NK}R_{OL}R_{PM}H^{3}_{KLM}.$$

In summary, then, the invariance condition satisfied by ψ under an arbitrary change of axes $X_K \to X'_K$ may be written in the form

$$\left.\begin{aligned}\psi &= g(H^{\alpha}_{KLM}\ldots, \eta_{KL}, \Pi_K, a)\\ &\equiv g(R_{KN}R_{LO}R_{MP}\ldots H^{\alpha}_{NOP}\ldots, R_{KN}R_{LO}\eta_{NO}, R_{KN}\Pi_N, a),\end{aligned}\right\} \quad (4.1.41)$$

where R_{KL} is the matrix representation of an operator belonging to the full orthogonal group. The polynomial g automatically satisfies the condition expressed by eqn. (4.1.41).

We now discuss *material invariance.* We shall assume that the coefficients $H^{\alpha}_{KLM}\ldots$ describing the elastoelectric properties of any given material are invariants of a sub-group of the full orthogonal group; that is to say

$$H^{\alpha}_{KLM}\ldots \equiv R_{KN}R_{LO}R_{MP}\ldots H^{\alpha}_{NOP}\ldots,$$

where R is an operation of the sub-group. This sub-group is referred to as the *symmetry group* of the material.† It follows immediately that ψ is value and form invariant with respect to transformations of the $H^{\alpha}_{KLM}\ldots$ under the operations of the symmetry group. Thus the so-called material invariance of ψ may be expressed

$$\psi = g(H^{\alpha}_{KLM}\ldots \eta_{KL}, \Pi_K, a) \equiv g(R_{KN}R_{LO}R_{MP}\ldots H^{\alpha}_{NOP}\ldots, \eta_{KL}, \Pi_K, a)$$

where R_{KL} is the matrix representation of any operator R of the symmetry group.‡

† If the energy density is not an intrinsic property of the material but depends, for example, on the shape of the dielectric, then the symmetry group will be characteristic of the particular specimen and not merely the material making up the specimen.

‡ We note that ψ is then also a polynomial invariant of the combinations of quantities η_{KL}, Π_K appearing in ψ under the operations of the material symmetry group.

Proof: Since each operator of the symmetry group is also an operator of the full orthogonal group we may write for any operator R of the symmetry group [eqn. (4.1.41)],

$$g(H^{\alpha}_{KLM}\ldots, \eta_{KL}, \Pi_K, a) \equiv g(R_{KN}R_{LO}R_{MP}\ldots H^{\alpha}_{NOP}\ldots, R_{KN}R_{LO}\eta_{NO}, R_{KL}\Pi_L, a)$$

or

$$g(H^{\alpha}_{KLM}\ldots, \eta_{KL}, \Pi_K, a) \equiv g(H^{\alpha}_{KLM}\ldots, R_{KN}R_{LO}\,\eta_{NO}, R_{KL}\Pi_L, a).$$

In the last step we have used the fact that the $H^{\alpha}_{KLM}\ldots$ are invariants of the symmetry group.

The symmetry group of isotropic materials is the full orthogonal group. The symmetry groups of the crystalline dielectrics are the 32 crystallographic point groups.

We conclude this sub-section by summarizing the various transformations which are assumed to leave the energy density both value and form invariant.

For any operator R of the Full Orthogonal group:

(i) Change of axes x_i, x_2, x_3:

$$x_i \to R_{ij} x_j.$$
$$X_K \to X_K.$$
$$D_i \to R_{ij} D_j.$$
$$x_{i;\,K} \to R_{ij} x_{j;\,K}.$$
$$\eta_{KL} \to \eta_{KL}.$$
$$\Pi_K \to \Pi_K.$$
$$H^{\alpha}_{KLM\ldots} \to H^{\alpha}_{KLM\ldots}$$

(ii) Change of axes X_1, X_2, X_3:

$$x_i \to x_i.$$
$$X_K \to R_{KL} X_L.$$
$$D_i \to D_i.$$
$$x_{i;\,K} \to R_{KJ} x_{i;\,J}.$$
$$\eta_{KL} \to R_{KM} R_{LN} \eta_{MN}.$$
$$\Pi_K \to R_{KL} \Pi_L.$$
$$H^{\alpha}_{KLM\ldots} \to R_{KN} R_{LO} R_{MP} \ldots H^{\alpha}_{NOP}.$$

For any operation R of the material symmetry group:

$$x_i \to x_i.$$
$$X_K \to X_K.$$
$$D_i \to D_i.$$
$$x_{i;\,K} \to x_{i;\,K}.$$
$$\eta_{KL} \to \eta_{KL}.$$
$$H^{\alpha}_{KLM\ldots} \equiv R_{KN} R_{LO} R_{MP} \ldots H^{\alpha}_{NOP\ldots}.$$

The assumption of the invariance of ψ under these transformations is consistent with the present body of experimental data.

4.1.8 Summary

To recapitulate, we have derived the following field equations and boundary conditions to describe the non-linear behaviour of an initially homogeneous elastic dielectric in static equilibrium in the presence of a charged conductor. (The extension to the case of several conductors is trivial.)

$$\left(\varrho\frac{\partial\psi}{\partial x_{i;K}}x_{j;K}\right)_{;j} - E_j D_{j;i} = 0 \quad \text{in } V, \tag{4.1.42}$$

$$D_{i;i} = 0 \quad \text{in } V \text{ and } A, \tag{4.1.43}$$

$$e_{ijk}E_{j;k} = 0 \quad \text{in } V \text{ and } A,^{\dagger} \tag{4.1.44}$$

$$t_i^0 = \left(\varrho\frac{\partial\psi}{\partial x_{i;K}}x_{j;K}\right)n_j - \left[\frac{D_j^A D_j^A}{2\varepsilon_0} + \frac{1}{2}(E_j^V + E_j^A)(D_j^V - D_j^A)\right]n_i \quad \text{on } V,^{\ddagger} \tag{4.1.45}$$

$$D_i^V n_i = D_i^A n_i \quad \text{on } V, \tag{4.1.46}$$

$$e_{ijk}n_j E_k^V = e_{ijk}n_j E_k^A \quad \text{on } V, \tag{4.1.47}$$

$$\oint_C D_i n_i \, dS = Q. \tag{4.1.48}$$

The energy density Ψ for the system is

$$\begin{aligned} \Psi &= \psi(x_{i;K}, D_i)\varrho && \text{in } V,\\ &= D_j D_j/2\varepsilon_0 && \text{in } A,\\ &\equiv 0 && \text{in } C, \end{aligned}$$

and the electric field E_i is generated from ψ, thus $E_i = \partial\Psi/\partial D_i$ with

$$E_i = \varrho\frac{\partial\psi}{\partial D_i} \quad \text{in } V, \tag{4.1.49}$$

$$E_i = D_i/\varepsilon_0 \quad \text{in } A, \tag{4.1.50}$$

$$E_i \equiv 0 \quad \text{in } C. \tag{4.1.51}$$

† This follows from the identity $e_{ijk}\phi_{;jk} \equiv 0$ with $E_i = -\phi_{;i}$.

‡ We emphasize that $\boldsymbol{t}^0$ is the force per unit deformed area and $\boldsymbol{n}$ the normal to the deformed surface. This convention is followed throughout this chapter.

The net mechanical and electrical work is

$$\delta \mathcal{W} = \oint_V t_i^0 \delta x_i \, dS + \phi_0 \oint_C \delta D_i n_i \, dS,$$

where $\boldsymbol{t}^0$ is the applied stress and ϕ_0 the potential of the conductor. The energy density ψ cannot be an explicit function of either the initial or final position coordinates of the material points; the first condition holds because the dielectric is assumed to be homogeneous in the initial state and the second condition follows from the demand that linear momentum be conserved. ψ is assumed to have arguments $x_{i;K}$ and D_i. If ψ is represented by a polynomial in these twelve arguments then the demands that ψ be both value and form invariant with respect to arbitrary changes of the axes describing the deformed state impose the condition that this representation is reducible to a polynomial in the ten arguments

$$\eta_{KL} = \tfrac{1}{2}(x_{i;K}x_{i;L} - \delta_{KL}), \quad \Pi_K = x_{i;K}D_i \quad \text{and} \quad a = D_i D_i.$$

The assumption of the invariance of ψ with respect to arbitrary changes of the axes describing the initial state trivially yields the transformation properties of the coefficients appearing in the polynomial representation. For any given material these coefficients are also invariants under the operations of the material symmetry group. It should be noted that the boundary conditions listed above are not valid if the dielectric is in contact with the conductor (no provision was made for this situation in carrying out the variational principle). We discuss a special case of conductor-dielectric contact in § 4.3.

4.1.9 Linear Theory

We now show that the non-linear theory described above yields the static case of the linear theory presented in § 2.5.

Let us suppose that the two sets of axes X_1, X_2, X_3 and x_1, x_2, x_3 coincide, i.e.

$$x_i = X_I, \quad (i = I = 1, 2, 3)^\dagger$$

† We recall that we have used the same symbols to denote the axes and the position coordinates; thus, while $x_i = X_I$ for the axes, in general $x_i \neq X_I$ for the position coordinates of material points.

Then $x_{i;K}-\delta_{iK}$ are the displacement gradients. To linearize the theory we treat the fields $x_{i;K}-\delta_{iK}$, D_i and E_i as small quantities and retain in the various expressions of the non-linear theory only the leading non-vanishing terms in these fields. Consider first the arguments appearing in the energy density $\psi(\eta_{KL}, \Pi_K, a)$; by definition we have

$$\begin{aligned}\eta_{KL} &= \tfrac{1}{2}[(x_{L;K}-\delta_{LK})+(x_{K;L}-\delta_{LK})+(x_{i;K}-\delta_{iK})(x_{i;L}-\delta_{iL})]\\ \Pi_K &= (x_{i;K}-\delta_{iK})D_i+\delta_{iK}D_i\\ a &= D_iD_i.\end{aligned}$$

Retaining only leading non-vanishing terms we obtain

$$\eta_{KL} = S_{KL}, \quad \Pi_K = D_K, \quad a = D_iD_i, \qquad (4.1.52)^\dagger$$

where S_{KL} is the infinitesimal strain tensor (§ 2.2.1).

The approximation procedure cannot affect the transformation properties (with respect to changes in the single set of coordinate axes). It is easy to verify that the a, Π_K and η_{KL} given in eqn. (4.1.52) are still tensors of rank zero, one and two respectively.

Clearly a is now redundant in a list of arguments for ψ. With our choice of initial state the leading non-vanishing terms in a polynomial representation for ψ are quadratic in S_{KL} and D_i; we write‡

$$\varrho_0\psi = \tfrac{1}{2}H^1_{KLMN}\eta_{KL}\eta_{MN}+H^2_{KLM}\eta_{KL}\Pi_M+\tfrac{1}{2}H^3_{KL}\Pi_K\Pi_L+H^4a,$$

where the coefficients H^1_{KLMN}, etc., are invariants under the operations of the material symmetry group. Substitute for the linearized forms for η_{KL} and Π_K [eqn. (4.1.52)],

$$\varrho_0\psi = \tfrac{1}{2}H^1_{KLMN}S_{KL}S_{MN}+H^2_{KLM}S_{KL}D_M+\tfrac{1}{2}(H^3_{KL}+2H^4\delta_{KL})D_KD_L. \qquad (4.1.53)$$

† Another linearization procedure of interest consists of treating $(x_{i;K}x_{i;L}-\delta_{KL})$ and D_i E_i as small quantities; then eqn. (4.1.52) is replaced by $\eta_{KL} = S_{KL}$, $\Pi_K = x_{i;K}D_i$, $a = D_iD_i$. Thus while the strain tensor S_{KL} is assumed to be infinitesimal, the associated displacement gradients $\partial u_i/\partial X_K$ can be finite. This condition will occur, for example, in the case of a crystal subjected to a simultaneous small strain and finite rotation.

‡ By introducing the factor ϱ_0 at this point we avoid an unnecessary proliferation of similar factors in later expressions for the equation of state. No terms linear in η_{KL} and Π_K appear in this representation because of the properties assumed for the initial state, viz. $\boldsymbol{t}^0 = 0$, $\boldsymbol{D} = 0$ and $\boldsymbol{E} = 0$.

Thus the coefficients of the quadratic function ψ (S_{KL}, D_i), which represents the energy density in the linear theory, are invariants under the operations of the material symmetry group (§ 3.3.7).

Turning now to the field equations (4.1.42) to (4.1.44), we note that

$$(\partial\psi/\partial x_{i;\,K})x_{j;\,K} = (\partial\psi/\partial S_{LM})(\partial S_{LM}/\partial x_{i;\,K})x_{j;\,K}{}^{\dagger} \cong \partial\psi/\partial S_{ij}$$

is a small quantity of order one in the variables S_{ij}, D_i, whereas E_jD is a small quantity of order two. The latter may be neglected in eqn. (4.1.42), and hence

$$\frac{\partial}{\partial x_j}\left(\varrho_0\frac{\partial\psi}{\partial S_{ji}}\right) = 0 \quad \text{in } V. \tag{4.1.54}$$

The form of the field equations (4.1.43) and (4.1.44) are not changed by the linearization; thus

$$\nabla_x\times\boldsymbol{E} = 0, \quad \nabla_x\cdot\boldsymbol{D} = 0 \quad \text{in } V \text{ and } A. \tag{4.1.55}$$

Of the three boundary conditions (4.1.45), (4.1.46) and (4.1.47), only the first is affected; by inspection we get

$$t_i^0 = \varrho_0\left(\frac{\partial\psi}{\partial S_{ij}}\right)n_j \quad \text{on } V, \tag{4.1.56}$$

$$\boldsymbol{n}\times(\boldsymbol{E}^V-\boldsymbol{E}^A) = 0 \quad \text{on } V, \tag{4.1.57}$$

$$\boldsymbol{n}\cdot(\boldsymbol{D}^V-\boldsymbol{D}^A) = 0 \quad \text{on } V. \tag{4.1.58}$$

If we set $\varrho_0(\partial\psi/\partial S_{ji}) = T_{ji}$, then eqns. (4.1.54) to (4.1.58) are precisely the equations for the time-independent linear theory [see eqns. (2.5.2) and (2.5.4) to (2.5.6)]. Thus T_{ji} is the stress tensor. The linearized form of the expression generating the electric field components, eqn. (4.1.49), is

$$E_i = \varrho_0\left(\frac{\partial\psi}{\partial D_i}\right). \tag{4.1.59}$$

Consider the linearized equations of state. From eqns. (4.1.53) and (4.1.59) and the definition of T_{ji} we obtain

$$T_{KL} = H^1_{KLMN}S_{MN}+H^2_{KLM}D_M, \tag{4.1.60}$$

$$E_K = H^2_{MNK}S_{MN}+(H^3_{KL}+2H^4\,\delta_{KL})D_L. \tag{4.1.61}$$

† See first footnote on p. 149.

Introducing the Voigt notation 11 → 1, 22 → 2, etc., and the infinitesimal strain S_α ($S_\alpha = S_{ij}\ i = j$, $S_\alpha = 2S_{ij}\ i \neq j$), we write these equations in the form

$$T_\alpha = c^D_{\alpha\beta} S_\beta - h_{j\alpha} D_j, \tag{4.1.62}$$

$$E_i = -h_{i\beta} S_\beta + \beta^S_{ij} D_j, \tag{4.1.63}$$

where $c^D_{\alpha\beta} = H^1_{KLMN}$, $h_{j\alpha} = -H^2_{KLj}$ and $\beta^S_{ij} = (H^3_{ij} + 2H^4\,\delta_{ij})$ represent the linear elastic stiffness, the linear piezoelectric coefficient and the linear dielectric permeability respectively (§ 2.5). Thus $c^D_{\alpha\beta}$, $h_{j\alpha}$, β^S_{ij} are (a) tensors (of rank four, three and two respectively) with respect to transformations of the coordinate axes, and (b) invariants under the operations of the material symmetry group (provided, of course, that the Voigt notation is replaced by the coordinate subscript notation). In terms of $c^D_{\alpha\beta}$, $h_{i\alpha}$ and β^S_{ij}, the linearized energy density satisfies the equation

$$\varrho_0\psi = \tfrac{1}{2} c^D_{\alpha\beta} S_\alpha S_\beta - h_{i\alpha} S_\alpha D_i + \tfrac{1}{2} \beta^S_{ij} D_i D_j. \tag{4.1.64}$$

The results are in agreement with the linear theory of §§ 2.5 and 3.3.

4.2 Approximation†

The mathematical apparatus of the static non-linear theory presented in the previous section is considerably more general than that required to analyse the non-linear dielectric properties which have been measured so far. It is the purpose of this section to present an approximation to this apparatus which is more in keeping with experimental needs. The approach we shall use is based on the following observation. It appears from measurements on real crystals that a good first-order non-linear approximation to the equations of state is one in which (a) the applied stress components are represented by the sum of two polynomials, one of order unity in the displacement gradients and the other of order two in the displacement $\boldsymbol{D}$ components, and (b) the electric field components are represented by the sum of three polynomials, one of order unity in the displacement gradients, one bilinear in the

† The Einstein summation convention is used throughout this section.

displacement gradients and displacement $\boldsymbol{D}$ components and one of arbitrary order in the displacement $\boldsymbol{D}$ components (cf. discussion of the quasi-linear theory in § 2.5.5). Thus the approximation amounts to a linearization in the displacement gradients alone. In practice, static non-linear measurements are almost invariably carried out using homogeneous states. For such states the field equations are satisfied identically in the dielectric; hence we shall discuss the effect of the linearization only for the case of the equations of state and boundary conditions.

4.2.1 Equations of State

Some of the conventions and notations of the previous section are no longer necessary. We now let the coordinate axes describing the final state coincide with those describing the initial state, $x_i = X_i$ ($i = 1, 2, 3$). We also drop (a) the convention involving small and capital Roman letters for subscripts and (b) the semicolon notation. For the displacement gradients we write $u_{KL} = \partial u_K/\partial X_L = \partial x_K/\partial X_L - \delta_{KL}$; then the strain tensor takes the form

$$\eta_{KL} = \tfrac{1}{2}(u_{KL} + u_{LK} + u_{MK}u_{ML}). \tag{4.2.1}$$

Consider the equation of state (4.1.49)

$$E_i = \varrho\left(\frac{\partial\psi}{\partial D_i}\right) \qquad \text{in } V \tag{4.2.2}$$

and the boundary condition (4.1.45). In terms of the displacement gradient this last equation now reads

$$t_i^0 = \varrho\left(\frac{\partial\psi}{\partial u_{iK}}\right)_D (u_{jK} + \delta_{jK})\, n_j - \left[\frac{D_j^A D_j^A}{2\varepsilon_0} + \frac{1}{2}\,(E_j^V + E_j^A)\,(D_j^V - D_j^A)\right] n_i. \tag{4.2.3}$$

Since a stress tensor was not defined for the non-linear theory, there is no equation of state in this theory which is analogous to that for T_{jl} in the linear theory.† However, eqn. (4.2.3) can be regarded as an equation of state in the sense that it relates the *applied* stress to the

† This is not to say that one could not introduce some tensor and call it a stress tensor. However, such a quantity has not proved necessary in the present analysis.

displacement gradients and displacement components. With this in mind we set

$$\varrho\left(\frac{\partial\psi}{\partial u_{iK}}\right)(u_{kK}+\delta_{kK})-\left[\frac{D_j^A D_j^A}{2\varepsilon_0}+\frac{1}{2}(E_j^V+E_j^A)(D_j^V-D_j^A)\right]\delta_{ik}=\tau_{ki}, \tag{4.2.4}$$

and hence from eqn. (4.2.3)

$$t_i^0=\tau_{ji}n_j. \tag{4.2.5}$$

We shall regard eqn. (4.2.4) as an equation of state, satisfied by τ_{ji}. The field τ_{ji} is defined only on the surface of the dielectric. From eqn. (4.2.5) we note that at any point on the surface at which the normal is parallel to one of the coordinate axes, X_K say, we have $t_i^0=\tau_{Ki}$, i.e. at such points τ_{KK} is a normal stress and τ_{Ki} ($i\neq K$) a shear stress. It is clear that τ_{ji} possesses some of the properties of the stress tensor field T_{jl} introduced in § 2.3.3. However, in general $\tau_{ji}\neq\tau_{ij}$.

The remaining boundary conditions (4.1.46) and (4.1.47) are not affected by our approximation and hence will not enter directly in the discussion which follows.

4.2.2 Approximation

Let us assume that the energy density $\psi(\partial x_i/\partial X_K, D_i)$ can be represented by a polynomial in the arguments $\partial x_i/\partial X_K$, D_i. Then the demands of invariance with respect to changes of axes are satisfied if and only if ψ is a function of these arguments through the variables η_{KL}, Π_K and a (see § 4.1.6).

From an inspection of eqns. (4.2.2) and (4.2.4) and the definitions of η_{KL} and Π_K we may verify that we get the polynomial representations for τ_{ji} and E_i described at the end of the introduction to this section from the following polynomial representation for ψ:†

$$\left.\begin{aligned}\varrho_0\psi=&\tfrac{1}{2}H^1_{KLMN}\eta_{KL}\eta_{MN}+H^2_{KLM}\eta_{KL}\Pi_M+\tfrac{1}{2}H^3_{KL}\Pi_K\Pi_L\\&+H^4a+H^5_{KLMN}\eta_{KL}\Pi_M\Pi_N+H^6_{KL}\eta_{KL}a+g(\Pi_K,a),\end{aligned}\right\} \tag{4.2.6}$$

where g is a polynomial containing terms of at least order two in a and of at least order three in Π_K. The $H^\alpha_{KLM}\ldots$'s are tensors charac-

† See second footnote on p. 155.

teristic of the dielectric of interest (§ 4.1.7), and since they are coefficients in a polynomial, each one may be invariant under certain interchanges of subscripts, for example $H^3_{KL} = H^3_{LK}$. In addition these coefficients are invariant under the operations of the material symmetry group (§ 4.1.7). While the tensor η_{KL} is symmetric, it is convenient in algebraic manipulations to use the convention that η_{KL} and η_{LK} be treated as independent variables.

In the first order non-linear approximation described above we have, from eqns. (4.2.2) and (4.2.6),

$$\left.\begin{aligned} E_M = {} & H^2_{KLM}u_{KL}+(H^3_{MN}+2H^4\,\delta_{MN})\,D_N+2\Big[H^5_{KLMN} \\ & +H^6_{KL}\,\delta_{MN}+\frac{1}{2}(H^3_{ML}\,\delta_{NK}+H^3_{LN}\,\delta_{MK}-H^3_{MN}\,\delta_{KL}) \\ & -H^4\,\delta_{MN}\,\delta_{KL}\Big]\,u_{KL}D_N+\frac{dg^0}{dD_M}, \end{aligned}\right\} \quad (4.2.7)$$

where $g^0 = g(\Pi_K, a)\,\big|_{u_{KL}=0}$.

The term $[-(H^3_{MN}\,\delta_{KL}+2H^4\,\delta_{MN}\,\delta_{KL})\,u_{KL}D_N]$ occurs because of the presence of the factor ϱ in eqn. (4.2.2).

Also

$$\left.\begin{aligned} \varrho\left(\frac{\partial\psi}{\partial u_{Kj}}\right)(u_{Lj}+\delta_{Lj}) = {} & H^1_{KLMN}\,u_{MN}+H^2_{KLM}D_M \\ +\Big[H^5_{KLMN}+H^6_{KL}\,\delta_{MN}+\frac{1}{2}(H^3_{ML}\,\delta_{NK}+H^3_{LN}\,\delta_{MK})\Big]\,D_MD_N. & \end{aligned}\right\} \quad (4.2.8)$$

Since g forms part of the total energy density ψ, all the coefficients appearing in a polynomial representation for g must be invariant under the operations of the material symmetry group. Thus we conclude that all the coefficients appearing on the right-hand sides of eqns. (4.2.7) and (4.2.8) are invariants of the material symmetry group. Within the approximation embodied in eqns. (4.2.7) and (4.2.8) the right-hand side of eqn. (4.2.6) has the form

$$\left.\begin{aligned} \varrho_0\,\psi = {} & \tfrac{1}{2}H^1_{KLMN}u_{KL}u_{MN}+H^2_{KLM}\,u_{KL}D_M+\tfrac{1}{2}(H^3_{MN} \\ & +2H^4\,\delta_{MN})\,D_MD_N+[H^5_{KLMN}+H^6_{KL}\,\delta_{MN} \\ & +\tfrac{1}{2}(H^3_{ML}\,\delta_{NK}+H^3_{LN}\,\delta_{MK})]\,u_{KL}D_MD_N+g^0. \end{aligned}\right\} \quad (4.2.9)$$

In other words the difference between the right-hand side of eqn. (4.2.6) and the right-hand side of eqn. (4.2.9) consists of terms which are neglected to obtain eqns. (4.2.7) and (4.2.8).

We now consider the Maxwell stress, i.e. the term in square brackets on the left of eqn. (4.2.4). Using the boundary conditions (4.1.46) and (4.1.47), the free space relationship $D_i^A = \varepsilon_0 E_i^A$ and eqn. (4.2.7), we can express the nine components of the three vectors $\boldsymbol{D}^A$, $\boldsymbol{E}^A$, $\boldsymbol{E}^V$ in terms of the components D_i^V and u_{KL}. Thus the leading term in the Maxwell stress is a quadratic function of D_i and u_{KL}. However, within our approximation scheme we neglect all but the quadratic terms in D_i. Hence we set

$$\left[\frac{D_j^A D_j^A}{2\varepsilon_0} + \frac{1}{2}(E_j^V + E_j^A)(D_j^V - D_j^A)\right] = J_{MN} D_M^V D_N^V, \qquad (4.2.10)$$

where the J_{MN} are functions of ε_0 and the $H^{\alpha}_{KLM}\ldots$. Explicit expressions for J_{MN} are derived in the following subsection.

Combining eqns. (4.2.4), (4.2.8) and (4.2.10) we get

$$\left.\begin{aligned}\tau_{KL} = {} & H^1_{KLMN} u_{MN} + H^2_{KLM} D_M^V + [H^5_{KLMN} + H^6_{KL}\,\delta_{MN} \\ & + \tfrac{1}{2}(H^3_{ML}\,\delta_{NK} + H^3_{NL}\,\delta_{MK}) - J_{MN}\,\delta_{KL}]\, D_M^V D_N^V.\end{aligned}\right\} \qquad (4.2.11)$$

Equations (4.2.7) and (4.2.11) are the most general forms the non-linear equations of state may take within the framework of our approximation. They are to be compared with the equations of state of the quasi-linear theory, (2.5.44) and (2.5.45). H^1_{KLMN}, H^2_{KLM} and $(H^3_{KL} + 2H^4\,\delta_{MN})$ are the linear elastic stiffness, the linear piezoelectric coefficient and the linear dielectric permeability respectively of the linear theory (see also § 4.1.9). The quantities in square brackets in eqns. (4.2.7) and (4.2.11) play the role of electrostrictive coefficients. However, they differ in two respects from their counterparts in the quasi-linear theory. First of all these quantities are not in general symmetric under the interchange of the subscripts K and L; as a result (i) $\tau_{KL} \not\equiv \tau_{LK}$, and (ii) the electric field components E_M are functions of the displacement gradient tensor u_{KL} and not merely the infinitesimal strain tensor $\frac{1}{2}(u_{KL} + u_{LK})$. Secondly, the electrostrictive coefficients appearing in the expression for τ_{KL} differ from the analogous

coefficients in the relation for E_M. The difference is

$$(J_{MN}-\tfrac{1}{2}H^3_{MN}-H^4\,\delta_{MN})\,\delta_{KL}.$$

$J_{MN}\,\delta_{KL}$ is associated with the Maxwell stress and the remaining term $\frac{1}{2}(H^3_{MN}+2H^4\delta_{MN})\,\delta_{KL}$ occurs because of the factor ϱ in eqn. (4.2.2). We reserve the discussion of the numerical significance of these quantities until after the effect of dielectric-conductor contact has been described (see § 4.3.3).

4.2.3 Maxwell Stress

We now determine expressions for the J_{MN} in terms of H^3_{KL} and H^4. Consider a point on the surface of the dielectric. For simplicity let the normal to this point be parallel to the X_1-axis. From the boundary conditions for $\boldsymbol{D}$ and $\boldsymbol{E}$, we have $D^A_1 = D^V_1$, $E^A_2 = E^V_2$ and $E^A_3 = E^V_3$. Since $\boldsymbol{D}^A = \varepsilon_0\boldsymbol{E}^A$, then $E^A_1 = D^A_1/\varepsilon_0$, $D^A_2 = \varepsilon_0 E^V_2$ and $D^A_3 = \varepsilon_0 E^V_3$. Within our approximation, to calculate J_{MN} we need only consider E^V_M as a linear function of D^V_M, that is

$$E^V_M = (H^3_{MN}+2H^4\,\delta_{MN})\,D^V_N$$

[see eqn. (4.2.7)]. Setting $(H^3_{MN}+2H^4\,\delta_{MN}) = \beta_{MN}$ and combining the results listed above, we get

$$J_{MN}D^V_M D^V_N = \frac{(D^V_1)^2}{2\varepsilon_0}-\frac{\varepsilon_0}{2}(\beta_{2M}\beta_{2N}+\beta_{3M}\beta_{3N})\,D^V_M D^V_N$$
$$+(\beta_{2M}D^V_2 D^V_M+\beta_{3M}D^V_3 D^V_M).$$

Hence

$$\left.\begin{aligned}
J_{11} &= \frac{1}{2}\left[\frac{1}{\varepsilon_0}-\varepsilon_0(\beta^2_{12}+\beta^2_{13})\right],\\
J_{12} &= \tfrac{1}{2}[\beta_{12}-\varepsilon_0(\beta_{12}\,\beta_{22}+\beta_{13}\,\beta_{23})],\\
J_{22} &= [\beta_{22}-\tfrac{1}{2}\varepsilon_0(\beta^2_{22}+\beta^2_{23})],\\
J_{23} &= [\beta_{23}-\tfrac{1}{2}\varepsilon_0(\beta_{22}+\beta_{33})\,\beta_{23}].
\end{aligned}\right\} \qquad (4.2.12)$$

J_{33}, J_{13} can be generated from J_{22}, J_{12} by replacing 2 with 3.

4.2.4 *m*3*m* Material Symmetry

In this subsection we shall write out the equations of state (4.2.7) and (4.2.11) in a form which displays the full symmetry of the crystallo-

graphic point group $m3m$. We may then compare them directly with the corresponding equations of state based on the quasi-linear theory, (3.3.90) to (3.3.92).

For material symmetry belonging to the group $m3m$ the $H^{\alpha}_{KLM}\cdots$ appearing in eqn. (4.2.6) have the properties listed below.† The axes used here are the standard cubic crystallographic axes.

$$\text{(i)}\ H^{\alpha}_{1111} = H^{\alpha}_{2222} = H^{\alpha}_{3333}$$
$$H^{\alpha}_{1122} = H^{\alpha}_{2211} = H^{\alpha}_{2233} = H^{\alpha}_{3322} = H^{\alpha}_{3311} = H^{\alpha}_{1133}$$
$$H^{\alpha}_{1212} = H^{\alpha}_{2112} = H^{\alpha}_{2121} = H^{\alpha}_{1221} = H^{\alpha}_{2323} = H^{\alpha}_{3223} = H^{\alpha}_{3232} = H^{\alpha}_{2332}$$
$$= H^{\alpha}_{1313} = H^{\alpha}_{3113} = H^{\alpha}_{3131} = H^{\alpha}_{1331}.$$

The remaining H^{α}_{KLMN} are identically zero ($\alpha = 1, 5$).

$$\text{(ii)}\ H^{2}_{KLM} = 0 \qquad \text{all } K, L, M.$$
$$\text{(iii)}\ H^{\alpha}_{11} = H^{\alpha}_{22} = H^{\alpha}_{33}$$
$$H^{\alpha}_{12} = H^{\alpha}_{21} = H^{\alpha}_{23} = H^{\alpha}_{32} = H^{\alpha}_{31} = H^{\alpha}_{13} \equiv 0 \quad (\alpha = 3,6).$$

Using these results we find that the energy density [eqn. (4.2.9)] and the equations of state (4.2.7) and (4.2.11) take the following forms:

$$\left.\begin{aligned}\varrho_0\psi = {} & \tfrac{1}{2}c^D_{11}(u^2_{11}+u^2_{22}+u^2_{33})+c^D_{12}(u_{11}u_{22}+u_{22}u_{33}+u_{33}u_{11})\\ & +\tfrac{1}{2}c^D_{44}[(u_{12}+u_{21})^2+(u_{23}+u_{32})^2+(u_{31}+u_{13})^2]\\ & +\tfrac{1}{2}\beta(D^2_1+D^2_2+D^2_3)+q_{11}(u_{11}D^2_1+u_{22}D^2_2+u_{33}D^2_3)\\ & +q_{12}[u_{11}(D^2_2+D^2_3)+u_{22}(D^2_3+D^2_1)+u_{33}(D^2_1+D^2_2)]\\ & +q_{44}[(u_{21}+u_{12})D_1D_2+(u_{32}+u_{23})D_2D_3\\ & +(u_{13}+u_{31})D_3D_1]+g^0,\end{aligned}\right\} \quad (4.2.13)$$

$$\tau_{11} = c^D_{11}u_{11}+c^D_{12}(u_{22}+u_{33})+q_{11}(D^V_1)^2+q_{12}[(D^V_2)^2+(D^V_3)^2]-J_{MN}D^V_MD^V_N, \quad (4.2.14)$$

† Most of these results were obtained from tables (see, for example, Huntington, 1958). When a tensor invariant of a crystallographic point group is not available in the literature, its properties are easily deduced using the "direct inspection" method of Fumi (1952) and the generators listed by Birss (1963). H^{5}_{KLMN} is tabulated by Forsbergh (1956).

$$\tau_{12} = c_{44}^D(u_{21}+u_{12})+q_{44}D_1^V D_2^V, \tag{4.2.15}$$

etc.

$$\left.\begin{aligned} E_1 = \beta D_1+(2q_{11}-\beta)u_{11}D_1+(2q_{12}-\beta)(u_{22}+u_{33})D_1 \\ +q_{44}[D_2(u_{12}+u_{21})+D_3(u_{31}+u_{13})]+\frac{dg^0}{dE_1}^{\dagger}, \end{aligned}\right\} \tag{4.2.16}$$

etc., where

$$c_{11}^D = H_{1111}^1, \quad c_{12}^D = H_{1122}^1, \quad c_{44}^D = H_{1212}^1, \quad \beta = (H_{11}^3+2H^4),$$
$$q_{11} = (H_{1111}^5+H_{11}^6+H_{11}^3), \quad q_{12} = (H_{1122}^5+H_{11}^6),$$
$$q_{44} = (2H_{1212}^5+H_{11}^3).$$

Thus for $m3m$ symmetry $\tau_{KL} \equiv \tau_{LK}$ and the components E_M are functions only of the symmetric part of the tensor u_{KL}.

For material symmetry $m3m$, $\beta_{KL} = \beta\,\delta_{KL}$ and therefore [eqn. (4.2.12)]

$$J_{11} = \frac{1}{2\varepsilon_0}, \quad J_{22} = J_{33} = \beta(1-\varepsilon_0\beta/2), \quad J_{12} = J_{23} = J_{31} = 0. \tag{4.2.17}$$

Now the expressions for J_{MN} in eqn. (4.2.12) hold for a surface point at which the normal lies parallel to the X_1-axis, hence at such a point on a dielectric with material symmetry $m3m$ the normal stress satisfies the relation

$$\begin{aligned} \tau_{11} = c_{11}^D u_{11}+c_{12}^D(u_{22}+u_{33})+\left(q_{11}-\frac{1}{2\varepsilon_0}\right)(D_1^V)^2 \\ +[q_{12}-\beta(1-\varepsilon_0\beta/2)][(D_2^V)^2+(D_3^V)^2]. \end{aligned} \tag{4.2.18}$$

The shear stress components τ_{12}, τ_{13} do not involve the Maxwell stress. We reserve discussion of this equation until we have described the effect of dielectric-conductor contact (see § 4.3.3).

† We recall that g^0 is a polynomial in D_i whose terms are at least of order three in the D_i. A g^0 which displays the material symmetry $m3m$ is given in eqn. (3.3.89).

4.3 Dielectric Slab

4.3.1 Introduction

In this section we derive again from first principles the boundary conditions satisfied by the applied stress in the case of small deformations. The derivation is based on a system consisting of a dielectric slab fitted tightly between the plates of a condenser. Our reasons for describing this calculation are threefold. First the system is sufficiently simple that much of the cumbersome mathematics characteristic of the previous two sections is avoided and we hope that as a result this derivation provides some physical insight into the origin and nature of the terms appearing in the applied stress boundary condition. Secondly, part of the surface of the dielectric is assumed to be in contact with the conductor and hence we obtain a result not contained in the previous sections, namely an expression for the applied stress boundary condition at a dielectric-conductor interface. Thirdly, this system is commonly used in the experimental investigation of the elasto-electric properties of crystals; because of this we are able to apply the results of the analysis to the case of barium titanate and hence determine directly the values of some of the coefficients appearing in the polynomial representation for the energy density of this crystal. From these values we are able to obtain some indication of the role of the Maxwell stress in the behaviour of barium titanate. Since the following analysis is considerably simpler than that of the previous sections we have attempted to make this section as self-contained as possible.

4.3.2 Boundary Conditions

Consider the case of a thin rectangular slab of homogeneous crystal with square faces placed symmetrically between two tightly fitting square metal plates of a condenser (Fig. 4.3.1). Let the slab thickness be z, the side of the square face be l and the side of the square plate be L. To ensure that there are no fringe effects (i.e. no non-uniform fields) in or near the slab, we choose the dimensions of the slab to be such that $L-l \gg z$ and $l \gg z$. The state in which the condenser is uncharged and no stresses are applied to the crystal is taken to be the

initial state. We suppose that in the initial state the displacement field D_i and the macroscopic electric field are everywhere zero.

Let X_1, X_2, X_3 be a set of rectangular axes fixed in the laboratory and oriented so that each axis is parallel to a slab edge. In particular let X_3 lie parallel to the plate normal.

For the sake of mathemetical simplicity we suppose that the properties of the dielectric are such that when a charge is placed on the condenser and uniform *normal* surface tractions are applied to the exposed slab

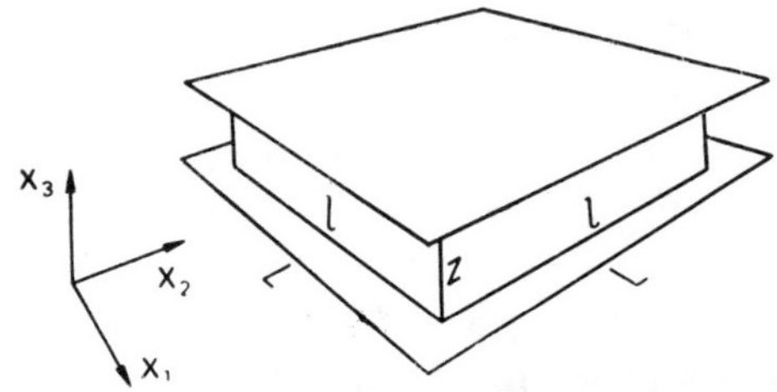

FIG. 4.3.1. A condenser consisting of a rectangular parallelepiped of dielectric tightly fitted between two rigid parallel conductor plates. Each coordinate axis is chosen to lie parallel to one of the slab edges.

faces and the outsides of the condenser plates (assumed rigid) then the displacement and the electric fields induced are normal to the plates and no shear strains occur,† i.e. the deformed dielectric is in the shape of a rectangular parallelepiped. Since the dimensions of the system have been chosen to prevent fringing effects in the slab and the applied stresses are uniform, the fields within and at the surface of the slab are uniform. Thus the equilibrium state of the slab is described by the following field *values*, the applied stress τ_{ii}, where τ_{ii} is the force (per unit deformed area) acting normally on the planes perpendicular to the X_i-axis, the three normal displacement gradients $S_i = \partial u_i/\partial X_i$ (no sum) and the components E and D respectively of the electric field and displacement field in the X_3-direction.

From the definition of displacement gradient, the dimensions of the uniformly deformed slab are $l(1+S_1)$, $l(1+S_2)$, $z(1+S_3)$. The electric

† For the case of material symmetry $m3m$ discussed in a later subsection this supposition is valid.

field in the vacuum is parallel to X_3 and of magnitude E (this last result comes from the demands of the boundary condition satisfied by the electric field at a dielectric-vacuum interface). Hence the charge density on the upper plate (Fig. 4.3.2) is D at the dielectric-conductor surface and $\varepsilon_0 E$ at the vacuum-conductor surface; since fringing effects are negligible the net charge Q on the upper plate is

$$Q = l^2(1+S_1)(1+S_2)D+[L^2-l^2(1+S_1)(1+S_2)]\varepsilon_0 E. \qquad (4.3.1)$$

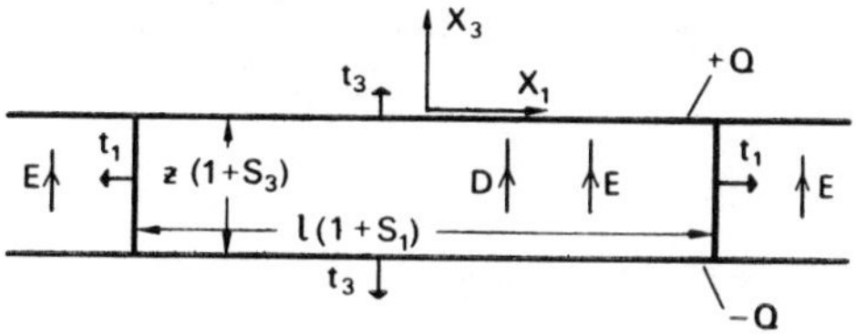

FIG. 4.3.2. A cross-section of the condenser in Fig. 4.3.1 showing the magnitudes and directions of most of the fields describing the mechanical and dielectric properties of the system. $t_i = \tau_{ii}$

We denote the energy density (energy/unit mass) in the dielectric by $\psi = \psi(S_1, S_2, S_3, D)$ and we set the energy density (energy/unit volume) in the vacuum $\varepsilon_0 E^2/2 = \chi$. The net energy of the system is thus

$$\mathcal{U} = V_0[(1+S_1)(1+S_2)(1+S_3)\varrho\psi+(1+S_3)\{(L/l)^2-(1+S_1)(1+S_2)\}\chi], \qquad (4.3.2)$$

where ϱ is the density of the slab in the deformed state and $V_0 = l^2 z$ is the volume of the slab in the initial state.

Suppose an additional infinitesimal charge δQ is placed on the condenser and the stresses are altered by infinitesimal amounts so that changes δS_i, δD, δE, $\delta\psi$ and $\delta\chi$ occur in S_i, D, E, ψ and χ respectively. If $\delta\mathcal{W}$ is the work done and $\delta\mathcal{U}$ the energy stored during this transformation, then by the conservation of energy

$$\delta\mathcal{U} = \delta\mathcal{W}. \qquad (4.3.3)$$

From the definitions of mechanical and electrical work

$$\delta\mathcal{W} = V_0[\tau_{11}(1+S_2)(1+S_3)\,\delta S_1+\tau_{22}(1+S_1)(1+S_3)\,\delta S_2 + \tau_{33}(1+S_1)(1+S_2)\,\delta S_3]+Ez(1+S_3)\,\delta Q. \qquad (4.3.4)$$

From eqn. (4.3.2)

$$\left.\begin{aligned}\delta\mathcal{U} = V_0[&(1+S_1)(1+S_2)(1+S_3)\varrho\delta\psi+(1+S_3)\{(L/l)^2\\&-(1+S_1)(1+S_2)\}\delta\chi-(1+S_3)\{\delta S_1(1+S_2)\\&+\delta S_2(1+S_1)\}\chi+\delta S_3\{(L/l)^2-(1+S_1)(1+S_2)\}\chi]\end{aligned}\right\} \quad (4.3.5)$$

since $\delta\,[(1+S_1)(1+S_2)(1+S_3)\varrho] = \delta$ (mass of slab) $= 0$. The changes $\delta\psi$, $\delta\chi$ are

$$\delta\psi = \left(\frac{\partial\psi}{\partial D}\right)\delta D+\sum_{i=1}^{3}\left(\frac{\partial\psi}{\partial S_i}\right)\delta S_i, \quad (4.3.6)$$

$$\delta\chi = \varepsilon_0 E\,\delta E. \quad (4.3.7)$$

Equation (4.3.1) imposes a constraint on the infinitesimals; thus

$$\left.\begin{aligned}\delta Q = l^2[D-\varepsilon_0 E]&[(1+S_2)\,\delta S_1+(1+S_1)\,\delta S_2]\\&+[L^2-l^2(1+S_1)(1+S_2)]\varepsilon_0\,\delta E\\&+l^2(1+S_1)(1+S_2)\,\delta D.\end{aligned}\right\} \quad (4.3.8)$$

Substituting $\delta\mathcal{U}$, $\delta\mathcal{W}$ and δQ from eqns. (4.3.5), (4.3.4) and (4.3.8) respectively into eqn. (4.3.3) and treating δS_1, δS_2, δS_3, δD and δE as independent, arbitrary variations, we get the following conditions;

from δD
$$\varrho\left(\frac{\partial\psi}{\partial D}\right) = E, \quad (4.3.9)$$

from δE
$$E = E, \quad (4.3.10)$$

from δS_1
$$\tau_{11} = \varrho\left(\frac{\partial\psi}{\partial S_1}\right)(1+S_1)-\left[\frac{\varepsilon_0 E^2}{2}+E(D^V-\varepsilon_0 E)\right]. \quad (4.3.11)$$

from δS_2
$$\tau_{22} = \varrho\left(\frac{\partial\psi}{\partial S_2}\right)(1+S_2)-\left[\frac{\varepsilon_0 E^2}{2}+E(D^V-\varepsilon_0 E)\right], \quad (4.3.12)$$

from δS_3
$$\left.\begin{aligned}&\tau_{33}[l^2(1+S_1)(1+S_2)]\\&\quad= \varrho\left(\frac{\partial\psi}{\partial S_3}\right)(1+S_3)\,[l^2(1+S_1)(1+S_2)]\\&\qquad+\frac{\varepsilon_0 E^2}{2}[L^2-l^2(1+S_1)(1+S_2)].\end{aligned}\right\} \quad (4.3.13)$$

The quantity on the left-hand side of (4.3.13) is the net force applied normally to the outside of each condensor plate and the terms on the right-hand side are the balancing forces, the first from the dielectric and the second from the electrostatic field in the free space between the plates. (We recall that $\varepsilon_0 E^2/2$ is the stress acting on a charged conduction.) Hence we deduce that the normal stress acting on the slab face in contact with the conductor is

$$\tau_{33} = \varrho \left(\frac{\partial \psi}{\partial S_3} \right) (1+S_3), \qquad (4.3.14)$$

where τ_{33} is now the force (per unit deformed area) acting normally on the large faces of the slab. Equation (4.3.9) shows that the electric field can be generated from the energy density in a form which agrees with the results in § 4.1.8. Equations (4.3.11) and (4.3.12) refer to dielectric-vacuum boundaries, and it is a simple exercise to show that they are special cases of eqn. (4.1.45). The terms in square brackets in eqns. (4.3.11) and (4.3.12) describe the Maxwell stress acting on the conductor-free faces. Equation (4.3.14) is the boundary condition for the stress at a dielectric-conductor interface. We see that in this particular situation the Maxwell stress is suppressed. This is a new result, not contained in the previous sections.

In summary: we have treated the case of a dielectric slab in a parallel plate condensor and we have obtained an equation of state for the electric field and boundary conditions for the applied stress at a dielectric-free space interface and these are in agreement with the more general results of § 4.1. The additional boundary condition for the stress at a dielectric-conductor interface is new.

4.3.3 Comparison with the Quasi-linear Theory

We now compare the boundary conditions for the stress and the equations of state for the electric field in the approximation described in § 4.2 with the corresponding results of the quasi-linear theory (§ 3.3.8). The discussion is restricted to the case of material symmetry *m*3*m*; the results obtained are then directly applicable to barium titanate, the only crystal whose electrostrictive properties have so far been experimentally investigated in any detail.

In the approximation introduced in § 4.2 the normal stress τ_{11} at a dielectric-conductor boundary with normal parallel to the X_1-axis satisfies the boundary condition

$$\tau_{11} = c_{11}^{D} u_{11} + c_{12}^{D}(u_{22}+u_{33}) + q_{11}(D_1^V)^2 \quad (D_2^V = D_3^V = 0). \quad (4.3.15)$$

This result is obtained by substituting the polynomial representation for ψ given by eqn. (4.2.13) into eqn. (4.3.14) and relabelling the subscripts $3 \to 1$, $1 \to 3$. We recall that $S_1 = u_{11}$, $S_2 = u_{22}$, $S_3 = u_{33}$. The components D_2^V, D_3^V vanish at this interface since (i) $\boldsymbol{E}$ is normal to a conductor surface and (ii) for cubic material symmetry $\boldsymbol{E}$ and $\boldsymbol{D}$ are parallel. From eqns. (4.2.14) and (4.2.16) the analogous boundary condition at a dielectric-free space interface is

$$\tau_{11} = c_{11}^{D} u_{11} + c_{12}^{D}(u_{22}+u_{33}) + \left(q_{11} - \frac{1}{2\varepsilon_0}\right)(D_1^V)^2$$
$$+ [q_{12} - \beta(1-\varepsilon_0\beta/2)][(D_2^V)^2 + (D_3^V)^2], \quad (4.3.16)$$

and the electric field E_1 satisfies the equation of state

$$E_1 = \beta D + (2q_{11} - \beta)u_{11}D_1 + (2q_{12} - \beta)(u_{22}+u_{33})D_1$$
$$+ q_{44}[D_2(u_{12}+u_{21}) + D_3(u_{13}+u_{31})] + \frac{dg^0}{dD_1}. \quad (4.3.17)$$

In the quasi-linear theory with the *energy density representation used above*, namely eqn. (4.2.13), the corresponding equations read [cf. eqns. (3.3.90) and (3.3.92) with $\gamma_0 = 0$]

$$\tau_{11} = c_{11}^{D} u_{11} + c_{12}^{D}(u_{22}+u_{33}) + q_{11}(D_1^V)^2 + q_{12}[(D_2^V)^2 + (D_3^V)^2] \quad (4.3.18)$$

for both dielectric-conductor and dielectric-free space boundaries, and

$$E_1 = \beta D_1 + 2q_{11}u_{11}D_1 + 2q_{12}(u_{22}+u_{33})D_1$$
$$+ q_{44}[D_2(u_{12}+u_{21}) + D_3(u_{13}+u_{31})] + \frac{dg^0}{dD_1}. \quad (4.3.19)$$

The boundary conditions for the shears τ_{12}, τ_{13} are exactly the same in both theories.

By inspection the boundary conditions for the stress at a dielectric-

free space interface and the equation of state for the electric field in the quasi-linear theory are obtained as a special case from the approximate theory of § 4.2 if and only if

$$2\varepsilon_0 |q_{11}| \gg 1, \quad |q_{11}| \gg \frac{|\beta|}{2}, \quad |q_{12}| \gg \frac{|\beta|}{2}, \quad |q_{12}| \gg |\beta(1-\varepsilon_0\beta/2)|. \tag{4.3.20}$$

The boundary condition for the stress at a dielectric-conductor interface agrees with the quasi-linear expression [compare eqns. (4.3.15) and (4.3.18)].

If we suppose that the approximate theory of § 4.2 is valid, then the inequalities quoted here are necessary criteria for the validity of the quasi-linear theory. These criteria are also sufficient in the case where the strain, electric and displacement (***D***) fields in the dielectric are uniform, for then the field equations are satisfied identically and play no role in determining the values of the fields. As remarked before, this is a common situation experimentally when static measurements are being made.

4.3.4 Barium Titanate

Since there is more experimental data available for the non-linear properties of barium titanate than for any other crystal, we shall determine the relative sizes of q_{11}, q_{12}, β and $1/\varepsilon_0$ in the case of this crystal.

Let us suppose that the crystal in the system described in § 4.3.2 (see Fig. 4.3.1) is barium titanate. We further suppose that the cubic axes are parallel to the coordinate axes described in that subsection. At room temperatures the shorted single domain crystal will display a spontaneous polarization parallel to the X_3-axis. This state is described by an electric field, displacement field and deformation gradient tensor with the forms

$$\boldsymbol{E} = (0, 0, 0), \quad \boldsymbol{D} = (0, 0, P^0),$$

$$u_{11} = u_{22} = u_{11}^0, \quad u_{33} = u_{33}^0, \quad u_{12} = u_{21} = u_{23} = u_{32} = u_{31} = u_{13} = 0,$$

where P^0 is the spontaneous polarization and u_{11}^0, u_{33}^0 are respectively the spontaneous strain perpendicular and parallel to the spontaneous polarization vector. For the crystal free of applied mechanical stress,

the boundary condition at the dielectric-conductor interface reads [eqn. (4.3.15)]

$$0 = c_{11}^D u_{33}^0 + 2c_{12}^D u_{11}^0 + q_{11}(P^0)^2, \tag{4.3.21}$$

and the stress boundary condition at the dielectric-vacuum interface is [eqn. (4.3.16)]

$$0 = (c_{11}^D + c_{12}^D)u_{11}^0 + c_{12}^D u_{33}^0 + [q_{12} - \beta(1 - \varepsilon_0\beta/2)](P^0)^2. \tag{4.3.22}$$

Hence by measuring c_{11}^D, c_{12}^D, β, u_{11}^0, u_{33}^0 and P^0 one can determine q_{11} and q_{12}.

We note that if the spontaneous polarization is measured in a slab, free of conductor contacts, eqn. (4.3.21) must be replaced by the boundary condition [eqn. (4.3.16)]

$$0 = c_{11}^D u_{33}^0 + 2c_{12}^D u_{11}^0 + \left(q_{11} - \frac{1}{2\varepsilon_0}\right)(P^0)^2. \tag{4.3.23}$$

Solving eqn. (4.3.21) and eqn. (4.3.22) for u_{11}^0 and u_{33}^0 we find

$$u_{11}^0 = -\{(s_{11}^D + s_{12}^D)[q_{12} - \beta(1 - \varepsilon_0\beta/2)] + [s_{12}^D q_{11}]\}(P^0)^2, \tag{4.3.24}$$

$$u_{33}^0 = -\{2s_{12}^D[q_{12} - \beta(1 - \varepsilon_0\beta/2)] + s_{11}^D q_{11}\}(P^0)^2. \tag{4.3.25}$$

with

$$s_{11}^D = (c_{11}^D + c_{12}^D)/(c_{11}^D - c_{12}^D)(c_{11}^D + 2c_{12}^D), \quad s_{12}^D = -c_{12}^D/(c_{11}^D - c_{12}^D)(c_{11}^D + 2c_{12}^D).$$

The following room temperature ($\sim$ 25°C) values are reported in the literature:

$s_{11}^D = +7{\cdot}25 \times 10^{-12}$ MKS. $s_{12}^D = -3{\cdot}2 \times 10^{-12}$ MKS (Berlincourt and Jaffe, 1958).

$\beta = 6{\cdot}6 \times 10^5(\theta - \theta_0)$ MKS with $\theta_0 = 110$°C (Drougard and Young, 1954).

$u_{11}^0 = -3{\cdot}4 \times 10^{-3}$, $u_{33}^0 = 7{\cdot}5 \times 10^{-3}$ (Kay and Vousden, 1949),†

$P^0 = 2{\cdot}6 \times 10^{-1}$ MKS (Merz, 1953).

† These strains are measured relative to an initial state in the paraelectric phase of $BaTiO_3$ and hence strictly should not be used in eqns. (4.3.24) and (4.3.25). However, these experimental values differ from the correct values by the normal strain components [cf. eqns. (3.1.45) and (3.3.72)] and, since these latter provide only small corrections, we are justified in neglecting their effect in the order of magnitude calculation which follows.

The permittivity of free space $\varepsilon_0 = 8{\cdot}85 \times 10^{-12}$ MKS; hence, at 25°C,

$$\beta = -5{\cdot}6 \times 10^7 \text{ MKS} \quad \text{and} \quad \varepsilon_0\beta/2 = 2{\cdot}5 \times 10^{-4} \ll 1. \quad (4.3.26)$$

Kay and Vousden do not describe the nature of the boundaries of the crystal they used to determine the spontaneous strains. Let us assume that the plated slab is appropriate for these values; hence, using eqns. (4.3.24) and (4.3.25) and the data quoted above, we get

$$q_{11} = -1{\cdot}5 \times 10^{10} \text{ MKS}, \quad (4.3.27)$$

$$q_{12} - \beta(1 - \varepsilon_0\beta/2) = +1{\cdot}5 \times 10^9 \text{ MKS}. \quad (4.3.28)$$

From eqn. (4.3.26) the second term on the left-hand side of eqn. (4.3.28) is negligible; hence

$$q_{12} = 1{\cdot}5 \times 10^9 \text{ MKS}. \quad (4.3.29)$$

It is evident from these values that in the expressions for the electrostrictive coefficients, eqns. (4.3.16) and (4.3.17), the terms involving β are negligibly small; thus the equation of state for E [eqn. (4.3.17)] reduces to the quasi-linear expression. However, the quantity $(1/2\varepsilon_0)$ is not negligible and the dielectric-free space boundary condition [eqn. (4.3.16)] does not reduce to the quasi-linear expression [eqn. (4.3.18)]. (Referring to the inequalities in eqn. (4.3.20) we see that only the latter three are satisfied.)

To summarize: we have found that for the case of barium titanate at room temperatures the approximate theory developed in § 4.2 and the quasi-linear theory of § 2.5 agree as far as the equation of state for the electric field components and the stress boundary condition at a dielectric-conductor interface are concerned but disagree on the form of the stress boundary condition at a dielectric-free space interface. If the approximate theory is correct, then the forms of the stress boundary condition at the two types of faces, dielectric-free space and dielectric-conductor, are quite distinct—the term† $1/2\varepsilon_0$ is almost four

† This term can be tracted to the quantity $D_j^A D_j^A/2\varepsilon_0$ appearing in the Maxwell stress [eqn. (4.2.10)]. Hence, when $\boldsymbol{D}^A = 0$, as in the case of a conductor, $1/2\varepsilon_0$ does not appear in the expressions for the electrostrictive coefficients in the boundary condition for the stress.

times $|q_{11}|$. In particular this means that the *results of measurements of spontaneous strain as a function of spontaneous polarization are influenced by the nature of the boundaries of the specimen.* In the case of the spontaneous strain values used above we do not know which boundary condition is appropriate and hence we must regard the values calculated for q_{11} and q_{12} as merely order of magnitude estimates;† even so this does not affect the conclusions drawn at the beginning of this paragraph.

At the time of writing there have been no experiments carried out to check the validity of the approximate theory developed in § 4.2. The size of the quantity $(1/2\varepsilon_0)$ relative to q_{11} and q_{12} suggests that it should not prove difficult to distinguish experimentally between this theory and the quasi-linear theory.

Bibliography

The encyclopedic articles of Truesdell and Toupin (1960) and Truesdell and Noll (1965) provide background information on the non-linear field theories. Most of the material in this chapter was originally published in a series of articles Grindlay (1966a, b, 1967). In these articles we used an energy density (energy per unit volume) in contrast to the energy density (energy per unit mass) introduced above. The choice is a matter of taste and affects only the form of the field equations and boundary conditions and not the physical description of the theory. §§ 4.1.6 and 4.1.7 on invariance are based largely on Toupin's work (Toupin, 1956). There are other static non-linear theories of the elastic dielectric, we mention in particular Toupin (1956) and Eringen (1963). Hajicek (1967) has recently shown that Toupin's theory is equivalent to the one described in the present chapter.

Toupin (1963) and Grot and Eringen (1966) have developed time-dependent non-linear theories of the elastic dielectric. The latter authors include thermal effects in their discussion.

† These values, eqns. (4.3.27) and (4.3.29), agree reasonably well with the values obtained by other methods (see Jona and Shirane, 1962). However, a critical review of the experimental data on ferroelectrics lies outside the scope of this book and we shall not pursue this topic.

CHAPTER 5

Phenomenological Models for Ferroelectricity

In § 3.3.8 we showed that by introducing non-linear terms into the dielectric equation of state one may account for the occurrence of spontaneous polarization even in the case in which the material symmetry of the free energy is non-polar. In the present chapter we elaborate on this point. To continue the discussion using the general form of the equations of state introduced in § 3.3.8 would be cumbersome and, as far as the ferroelectrics are concerned, not necessarily fruitful; hence we follow the treatment of Ginzburg (1946) and Devonshire (1954) and specialize these equations to simple forms which, in addition to being amenable to straightforward algebraic analysis, are also appropriate in the case of a number of real ferroelectric crystals. These special forms provide us with phenomenological models which exhibit ferroelectricity.

The main characteristic of these models is the vanishing of one of the linear permeability coefficients at some non-zero temperature. This vanishing gives rise to a number of other anomalies in the dielectric, elastic and thermal properties of the models. The results obtained are based on two important assumptions: (i) the equations of state can be represented by polynomials, and (ii) these polynomials describe the dielectric, elastic and thermal properties of the system both above and below the transition temperature. The type of analysis introduced in this chapter is commonly applied in the literature to describe the properties of real ferroelectrics. While many numerical discrepancies

occur in these applications there is as yet no strong experimental evidence that their source lies in assumptions (i) or (ii).

The material in this chapter has been subdivided into five sections. Section 5.1 contains an analysis of the dielectric properties of the models. The effect of the dielectric anomalies on the thermal and elastic properties is described in §§ 5.2 and 5.3. In § 5.4 we gather the various results from the previous three sections and show that the phase transitions exhibited by these models fit into the Ehrenfest classification scheme. The discussion to this point is concerned solely with "one-dimensional" or uniaxial ferroelectric models. In § 5.5 we describe a "three-dimensional" model for the ferroelectric crystal $BaTiO_3$. Since the analysis of this model is merely a complicated extension of the analysis presented in previous sections, we have merely indicated how the calculation would "go". At various points in the text we have introduced numbers appropriate to some ferroelectrics to give the reader a feeling for the order of magnitudes involved. There is no attempt to consider the question of how well these models can account for the quantitative behaviour of real ferroelectrics.

5.1 Dielectric Properties

5.1.1 Stability and Constraints

Most of the material in this chapter deals with ferroelectric models which are subject to the constraint of constant electric field. Before starting this programme it is instructive to discuss other constraints. The discussion is not intended to be complete or exhaustive; we wish merely to give an indication of some of the problems involved.

Consider the circuit in Fig. 5.1.1a. A and B are parallel plate condensers in series connected to a cell of e.m.f. V, as shown. Suppose that A is empty and B is filled by a dielectric with the property that $\boldsymbol{E}$ and $\boldsymbol{D}$ are both uniform and normal to the plates when B is charged.† If we introduce a set of axes X_1, X_2, X_3 with X_3 perpendicular to the plates, then the electric and displacement fields have components

† Fringing effects are assumed to be negligible in both A and B.

(0, 0, E) and (0, 0, D) measured in this frame. We further suppose that the components E and D satisfy an equation of state of the form sketched in Fig. 5.1.1b. It is assumed that E is an odd function of D although this is not an essential point in the following discussion. For a given V the values of E and D are easily determined. First of all we note that since the condensers A and B carry the same charge the potential drop across A is proportional to D whereas the potential

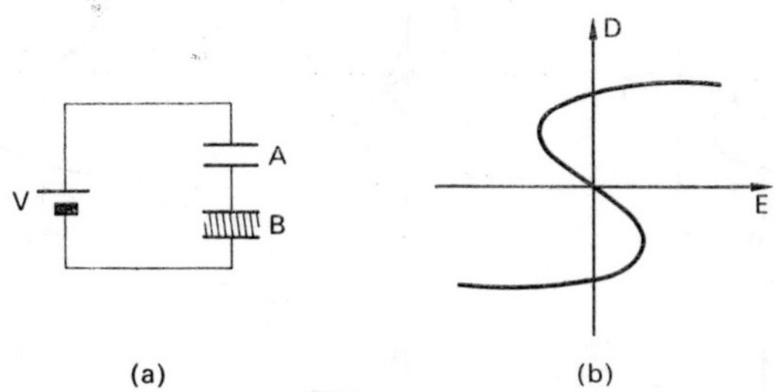

FIG. 5.1.1. (a) Circuit consisting of a cell of e.m.f. V connected across two parallel plate condensers, A and B, in series. A is empty and B is filled with dielectric. (b) Form assumed for the dielectric equation of state relating those components of $\mathbf{E}$ and $\mathbf{D}$ which are perpendicular to the plates of the condenser B.

drop across B is proportional to E. It is a simple exercise to determine the proportionality constants; hence

$$V = \left(\frac{a}{a'}\right) z' \frac{D}{\varepsilon_0} + zE, \tag{5.1.1}$$

where a', a, z', z are the plate areas and plate separations of A and B respectively. This is one relation determining D and E, the other is the equation of state. Thus for a given V the corresponding values of D and E are determined by the intersection of the straight line [eqn. (5.1.1)] with the equation of state in the D, E plane. We may think of the former as a load line. By varying V we change the intersections of the load line with the D and E axes and by varying the size of A we vary the slope of the load line. Equation (5.1.1) represents the constraint appropriate to this system. Consider the case in which $z' = 0$ (Fig. 5.1.2a). The load line is parallel to the D-axis; for large values of E

there is one intersection; the slope $\partial E/\partial D > 0$ in this region so that this solution yields a uniform state which is locally stable. By reducing the value of E, a regime is reached in which there are three intersections α, β, γ (Fig. 5.1.2a). From the sign of $\partial E/\partial D$ we deduce that the state corresponding to the intermediate intersection β is locally

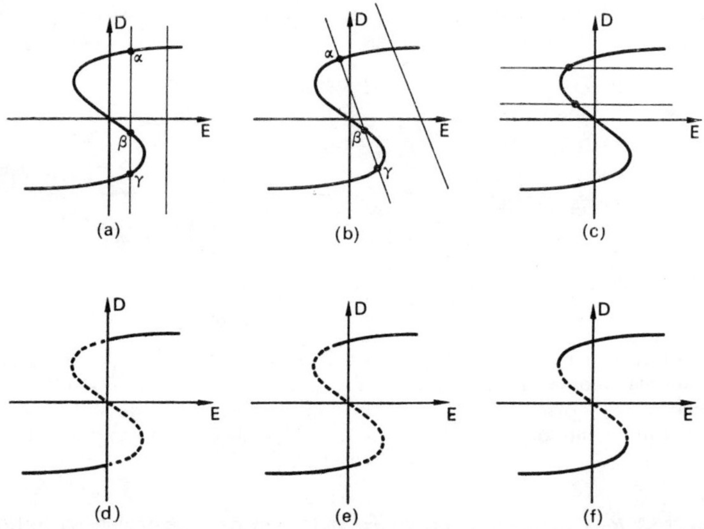

FIG. 5.1.2. A simultaneous plot of the equation of state (Fig. 5.1.1b) and the load line [eqn. (5.1.1)] for the general case (b) and the special cases (a) $z' = 0$ and (c) $(a/d') \ll (z/z')$. The lines of absolutely stable states (full lines) for these three cases (a), (b) and (c) are sketched in (d), (e) and (f) respectively.

unstable and the remaining two states, α, γ, stable. The question of the metastability and absolute stability of α and γ is solved by calculating the space integral of the relevant thermodynamic potential (G in this case). We show in the following subsections that the state α is absolutely stable and γ is metastable. Similarly, the upper part of the branch in the second quadrant refers to states which are metastable relative to those on the branch in the third quadrant. Hence for the constraint in which the load line is parallel to the D-axis the absolutely stable states lie on the curve sketched in Fig. 5.1.2d.

In the general case the load line has some arbitrary slope in the D, E plane. In the example sketched in Fig. 5.1.2b there are two regions—one in which there is a single intersection and one in which there are three. The discussion is qualitatively similar to that given above for the special case $z' = 0$. The absolutely stable states lie on the curve sketched in Fig. 5.1.2e. The point we wish to illustrate here is that in this general case, for small values of V the question of absolute stability and metastability arises for two states which, in the special case $z' = 0$, are both metastable (relative to other states). Thus the question of absolute stability and metastability is a function of the constraints applied to the system.

As the plate separation of A is increased from zero a regime is reached in which for all values of V the load line has merely one intersection with the equation of state. The extreme case occurs when the load line is parallel to the E-axis (Fig. 5.1.2c). Referring to this figure, we see that for a range of V-values in the vicinity of $V = 0$ the load line cuts the E, D curve at a single point and this point represents an unstable state, $\partial E/\partial D < 0$. Thus for these values of V and this constraint there are no uniform states which are locally stable. Under these circumstances the system *must* pass through non-uniform states. This result underlines the fact that any discussion of the dielectric response is incomplete if non-uniform states are not considered. It is also clear that the phrase "absolutely stable" used above refers to the relative global stability of two uniform states. It is conceivable that some of these "absolutely stable" states are in fact metastable relative to non-uniform states. In the case of the ferroelectric, non-uniform states occur; these states are characterized by domains† (§ 1.1.2); in the light of the remark made in the previous sentence it is of interest to note that domains *are* seen with constraints for which locally stable uniform states exist. The theory of ferroelectric domains in particular and non-uniform states in general lies outside the scope of this book.

The discussion given above dealt with the role of the constraints in determining the stability of the system. We conclude this subsection

† There is some experimental evidence that space charges occur in ferroelectrics, especially close to the surface of the sample. This charge would also give rise to non-uniform states (see, for example, Jona and Shirane, 1962).

by discussing the effect of the shape of the dielectric on the stability of the states. Consider the case of the homogeneous dielectric ellipsoid in a uniform applied electric field $\mathcal{E}$. Let us suppose that (a) one of the principal axes of the ellipsoid lies parallel to the applied field, (b) uniform fields $\boldsymbol{E}$ and $\boldsymbol{D}$ are induced in the ellipsoid and (c) these uniform fields $\boldsymbol{E}$ and $\boldsymbol{D}$ are parallel to $\mathcal{E}$. If we denote the magnitude of $\boldsymbol{E}$ by E, etc., then

$$E = \mathcal{E} - \frac{L}{\varepsilon_0}(D - \varepsilon_0 E), \tag{5.1.2}$$

where L is the depolarization factor appropriate to the principal axis parallel to $\mathcal{E}$ (§ 2.4.4). For given $\mathcal{E}$, eqn. (5.1.2) defines a load line in the D, E plane. By introducing the equation of state described by the sketch in Fig. 5.1.1b we get a situation very similar to the one described above where $\mathcal{E}$ now plays a role analogous to the e.m.f. V and changes of constraint are achieved by altering the depolarization factor L. The case of the isolated crystal is of special interest; in this case $\mathcal{E} = 0$ and the load line is determined by the equation

$$0 = E(1-L) + \frac{LD}{\varepsilon_0}.$$

In particular, for a slab $L = 1$ and for a needle $L = 0$; hence the constraints for these shapes are $D = 0$ and $E = 0$ respectively. Thus there is a stable uniform solution for the isolated needle but not for the isolated slab.

5.1.2 Dielectric Equation of State

The discussion in the remainder of this chapter will be restricted to the case of a homogeneous plated dielectric slab with a potential V applied across the plates. We shall suppose that in the charged condenser the electric and displacement fields are uniform and normal to the plates. Introducing rectangular axes X_1, X_2, X_3 with X_3 perpendicular to the plates, we write $\boldsymbol{E}$ and $\boldsymbol{D}$ in component form as $(0, 0, E)$ and $(0, 0, D)$ respectively. Now suppose that E and D satisfy an equation of state of the form

$$E = \beta D + \xi D^3 + \zeta D^5 \tag{5.1.3}$$

with $\beta = A(\theta - \theta_0)$. θ is the temperature of the dielectric and A, θ_0, ζ positive constants and ξ a positive or negative constant.† We shall further suppose that the dielectric properties in the plane parallel to the plates are such that the components of $\boldsymbol{E}$ and $\boldsymbol{D}$ in this plane vanish simultaneously under all circumstances.‡

In this and the following subsections we explore the isothermal response of this capacitor to an applied e.m.f. V. In other words we shall determine the values of E and D for different given values of V and θ. Since E is trivially determined, $V = Ez$, where z is the plate separation in the condenser, the only problem we have is the determination of D. As in the previous subsection, fringing effects are assumed to be negligible.

It is convenient to replace D, E and θ by the dimensionless variables

$$d = (2\zeta/|\xi|)^{\frac{1}{2}} D, \quad e = (2/|\xi|)^{\frac{5}{2}} \zeta^{\frac{3}{2}} E, \quad t = 4A(\zeta/\xi^2)(\theta - \theta_0). \quad (5.1.4)$$

The equation of state then takes the form

$$e = td + 2d^3 + d^5 \quad (\xi > 0)^{\S} \quad (5.1.5)$$

and

$$e = td - 2d^3 + d^5 \quad (\xi < 0). \quad (5.1.6)$$

5.1.3 Phase Transition, $\xi > 0$

Consider first the case $\xi > 0$. To sketch the family of curves e versus d for different t we note that (i) e is an odd function of d, (ii) $e \to \pm\infty$ when $d \to \pm\infty$ and (iii) $e(d, t) = 0$ when $d = 0$ and $d^2 = -1 \pm (1-t)^{\frac{1}{2}}$. By inspection $e = 0$ has one real root when $t > 0$ and three real roots when $t \leqslant 0$. The required family is sketched in Fig. 5.1.3. Since t is related to the temperature, this is a family of isotherms. For a given V across the condenser the load line is parallel to the d-axis. When $t \geqslant 0$ the load line cuts the equation of state in one point for which $\partial e/\partial d > 0$; hence this point corresponds to a stable state. When $t < 0$ the load line

† The *form* of the results obtained in this section does not depend on the mechanical constraints applied to the system and therefore we shall not burden these symbols with superscripts denoting the choice of mechanical constraint. However it must be emphasized that in a given system the numerical values of A, θ_0, ξ and ζ may depend critically on the choice of mechanical constraints (see § 5.3).

‡ We discuss briefly a case in which this condition does not hold in § 5.5.

§ The case $\xi = 0$ is described at the end of § 5.1.7.

cuts the equation of state at one or three points; in the latter case, two points correspond to stable states and one an unstable state. To determine which of the two stable states is absolutely stable we require the space integral of the appropriate thermodynamic potential. Since the potentials of the condenser plates and the temperature of the dielectric are held fixed, the requisite potential is the Gibbs function G.† Because the electric field vanishes outside the condenser the

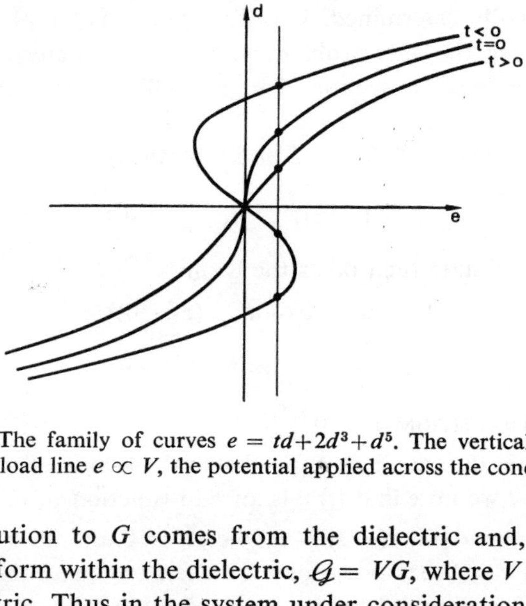

FIG. 5.1.3. The family of curves $e = td+2d^3+d^5$. The vertical line represents a load line $e \propto V$, the potential applied across the condenser.

only contribution to G comes from the dielectric and, because the fields are uniform within the dielectric, $\mathcal{G} = VG$, where V is the volume of the dielectric. Thus in the system under consideration the question of absolute stability of the uniform states is settled by the thermodynamic potential G. The state with the least value of G is absolutely stable. It is convenient to introduce the dimensionless Gibbs function $g = (8\zeta^2/|\xi|^3)G$. Since $D = -(\partial G/\partial E)_\theta$, it follows that $d = -(\partial g/\partial e)_t$,

† Strictly, G is only appropriate if the mechanical constraint is constant applied stress (§ 3.2.1). However, G differs from the potential for whatever choice of mechanical constraint is made by a Legendre transformation in the stress and strain variables and, as far as the dielectric properties are concerned, this difference is a constant which does not affect the general discussion given above. For a particular system the mechanical constraints do dictate what are the appropriate coefficients appearing in the dielectric equation of state, i.e. free, clamped, etc.

and hence we may determine g by integrating the equation of state $d = d(e)$.

The qualitative form of g is easily obtained without using numerical or analytical integration techniques. For example, consider the branch OC in the equation of state sketched in Fig. 5.1.4b; at each point on this branch d is positive; hence $\partial g/\partial e < 0$ which means that g increases from O to C. At C, g has a cusp and decreases with increasing e along the branch CBA of the d versus e curve. The branch CO of g lies above the branch CBA because the d values corresponding to these branches

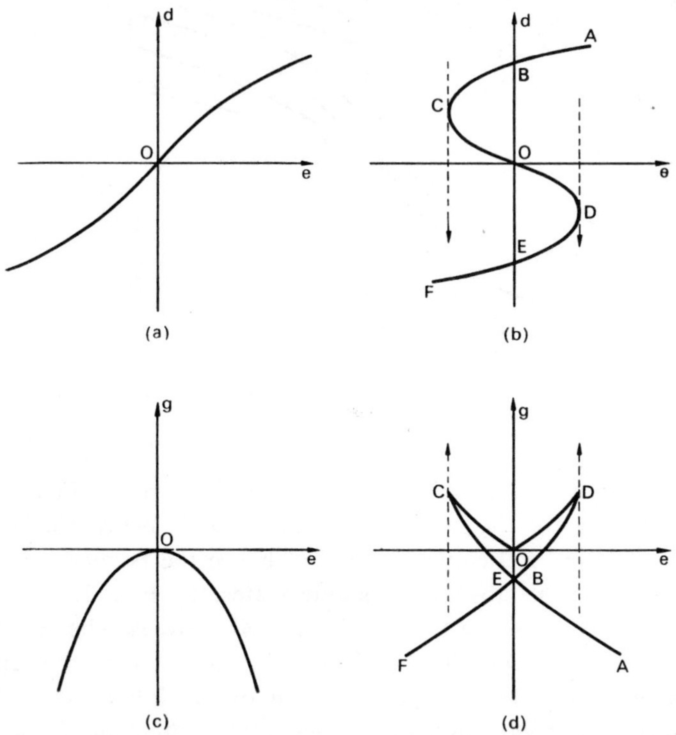

FIG. 5.1.4. Sketch of the two types of curves characteristic of the equation of state $e = td+2d^3+d^5$ and the corresponding integral $g = -\int_0^e d\,de$; (a) and (c) refer to the regime $t \geq 0$ and (b) and (d) to the regime $t < 0$. The same curves hold for two regimes of the equation of state $e = td-2d^3+d^5$; when $t > t_{cr}$, (a) and (c) apply, and when $t < 0$, (b) and (d) apply.

are smaller in the former branch; thus the curve *OCBA* integrates to the curve *OCBA* shown in Fig. 5.1.4d. Since e is an odd function of d the g curve is symmetric about $e = 0$ so that *ODEF* integrates to *ODEF* as shown. In the g–e plane the points B and E coincide. The value of $g(e = 0)$ is chosen quite arbitrarily to be zero. For comparison we have included a sketch of g versus e when $t \geqslant 0$ (Fig. 5.1.4c). The

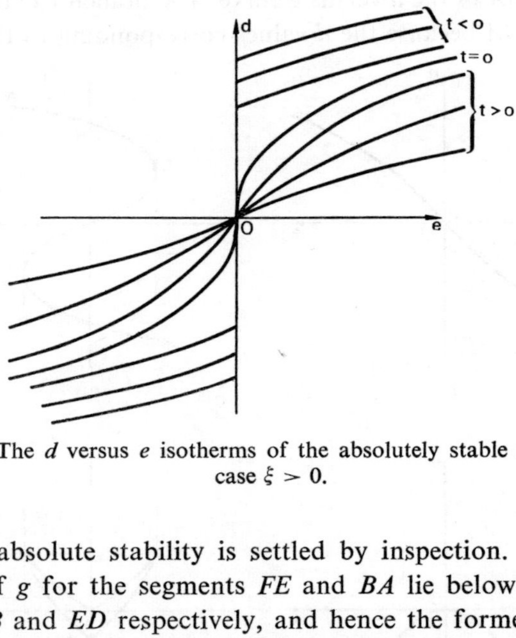

FIG. 5.1.5. The d versus e isotherms of the absolutely stable states in the case $\xi > 0$.

question of absolute stability is settled by inspection. For example, the values of g for the segments *FE* and *BA* lie below those for the segments *CB* and *ED* respectively, and hence the former pair of segments describe the absolutely stable states. In Fig. 5.1.5 we have sketched the isotherms of the absolutely stable states. The information given by Fig. 5.1.5 can also be displayed in a sketch of d versus t for different e values. The qualitative form of the latter family can be deduced from the following observations: (a) the intercept of an isotherm with a vertical line e = cont. (Fig. 5.1.5) is a continuous monotonic function of t except in the case $e = 0$, and (b) e is an odd function of d; hence Fig. 5.1.6. These two families describe the behaviour of the dielectric-filled condenser provided the system passes through only absolutely stable uniform states. It is clear from Fig. 5.1.5 that for

$t < 0$ the dielectric possesses a reversible spontaneous dipole moment†; hence, when $\xi > 0$, eqn. (5.1.3) describes a ferroelectric material. Since there is no spontaneous dipole moment for $t \geqslant 0$, this system undergoes a phase change at $t = 0$. From Fig. 5.1.6 we see that $d(e = 0, t)$ is a continuous function of t, whereas $(\partial d/\partial e)|_{e=0}$ is discontinuous at

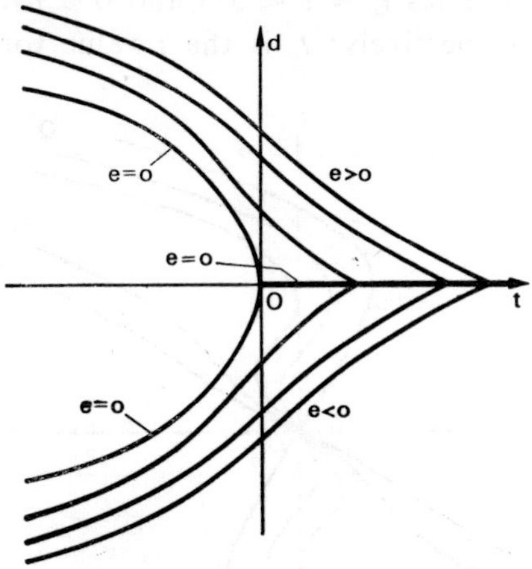

FIG. 5.1.6. The family of d versus t curves with parameter e for the absolutely stable states in the case $\xi > 0$.

$t = 0$. Since $d = -(\partial g/\partial e)_t$, this last result suggests that this is a second-order phase transition (see § 5.4.3 for further details).

5.1.4 Phase Transition, $\xi < 0$

A similar discussion applies to the case $\xi < 0$. However, the results are somewhat more complicated. From eqn. (5.1.6), $e(d)$ has turning values when

$$\left(\frac{\partial e}{\partial d}\right)_t = t - 6d^2 + 5d^4 = 0.$$

The roots of this equation are

$$d = \pm\sqrt{\tfrac{1}{5}\left[3 \pm \sqrt{9-5t}\right]}.$$

† $e(d, t)$ is an odd function of d and hence the non-zero roots d of $e = 0$ occur in pairs of opposite sign and equal magnitudes.

By inspection there are (i) no real roots when $t > 9/5$, (ii) four real roots when $0 \leqslant t \leqslant 9/5$ and (iii) two real roots when $t < 0$. The family d versus e is sketched in Fig. 5.1.7. Figure 5.1.4a and b apply to the regimes (i) and (iii) respectively. In (ii) the four turning values in the equation of state give rise to four cusps in the function g. Here there are two sub-regimes $t_c < t < 9/5$ and $0 \leqslant t \leqslant t_c$ illustrated by Fig. 5.1.8a and b respectively; t_c is the t value for which the lower

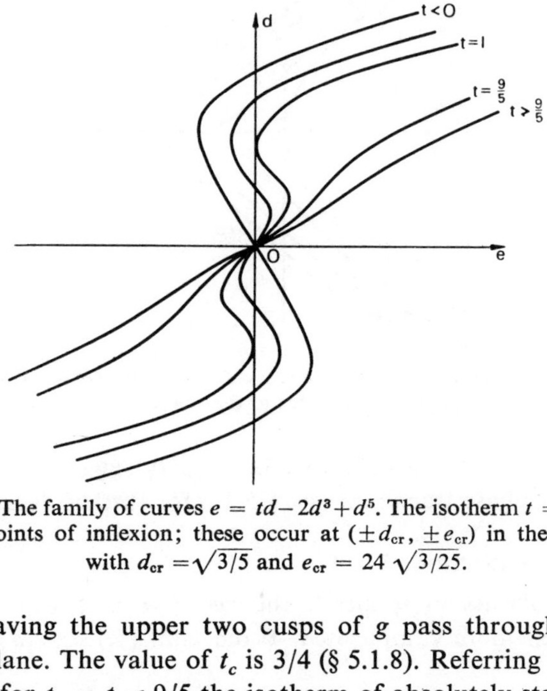

FIG. 5.1.7. The family of curves $e = td - 2d^3 + d^5$. The isotherm $t = t_{cr} = 9/5$ has two points of inflexion; these occur at $(\pm d_{cr}, \pm e_{cr})$ in the d–e plane with $d_{cr} = \sqrt{3/5}$ and $e_{cr} = 24\sqrt{3/25}$.

branches leaving the upper two cusps of g pass through the origin in the g–e plane. The value of t_c is 3/4 (§ 5.1.8). Referring to Fig. 5.1.8 we see that for $t_c < t \leqslant 9/5$ the isotherm of absolutely stable uniform states suffers a discontinuity for non-zero values of e, whereas for $0 \leqslant t \leqslant t_c$ the discontinuity occurs for $e = 0$ just as in regime (iii). The complete set of isotherms of absolutely stable uniform states is sketched in Fig. 5.1.9. By inspection the dielectric is ferroelectric for $t \leqslant t_c$. When $t = 9/5$, these isotherms have points of inflexion; we denote the value of t, d and e at these points by t_{cr}, $\pm d_{cr}$ and $\pm e_{cr}$ respectively. The subscripts "cr" are derived from "critical" (see below);

from above, $t_{cr} = 9/5$; it is a simple exercise to show that $d_{cr} = \sqrt{3/5}$ and $e_{cr} = 24/25\sqrt{3/5}$ (§ 5.1.10). The family d versus t with parameter e is easily derived from Fig. 5.1.9; the results are sketched in Fig. 5.1.10. Referring to this figure we see that both $d(t)$ and the slope $(\partial d/\partial t)_e$ have discontinuities when $0 \leqslant |e| < e_{cr}$; this suggests that the system undergoes a first-order phase transition in this range of e values (§ 5.4.2). For each value of e in this range the transition occurs

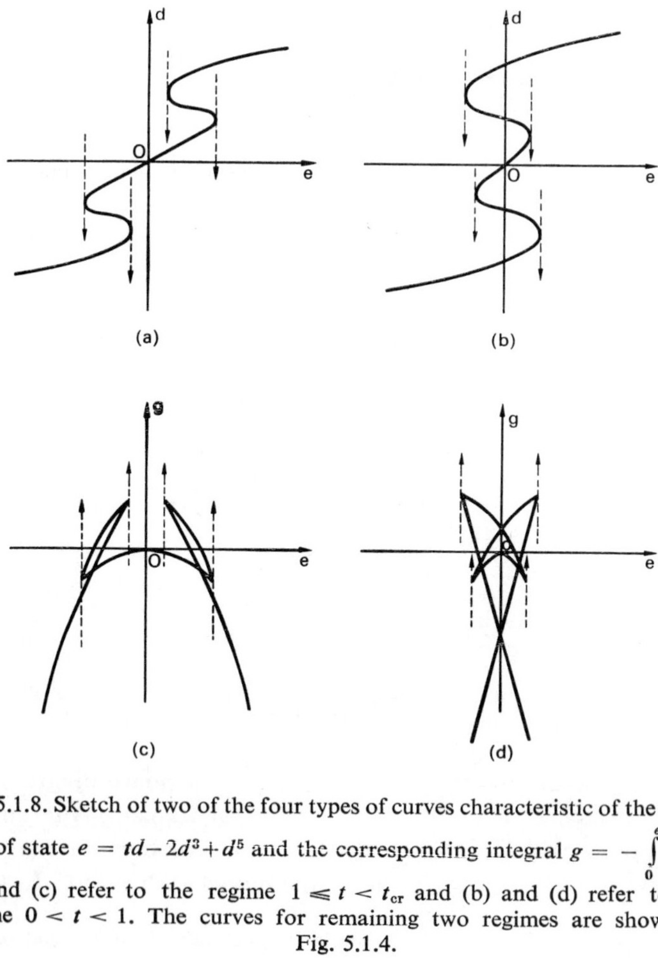

FIG. 5.1.8. Sketch of two of the four types of curves characteristic of the equation of state $e = td-2d^3+d^5$ and the corresponding integral $g = -\int_0^e d\, de$; (a) and (c) refer to the regime $1 \leqslant t < t_{cr}$ and (b) and (d) refer to the regime $0 < t < 1$. The curves for remaining two regimes are shown in Fig. 5.1.4.

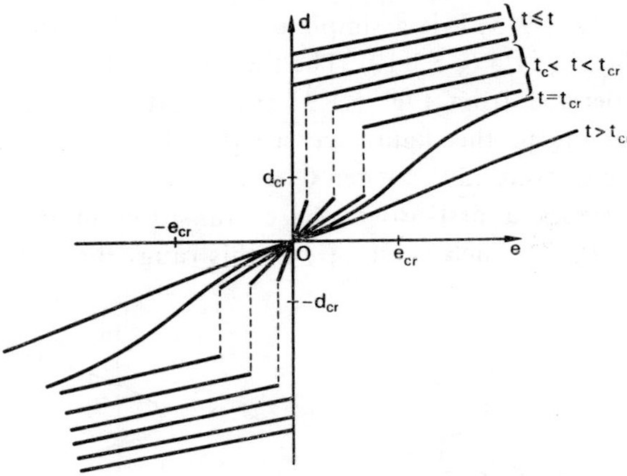

FIG. 5.1.9. The d versus e isotherms of the absolutely stable states in the case $\xi < 0$.

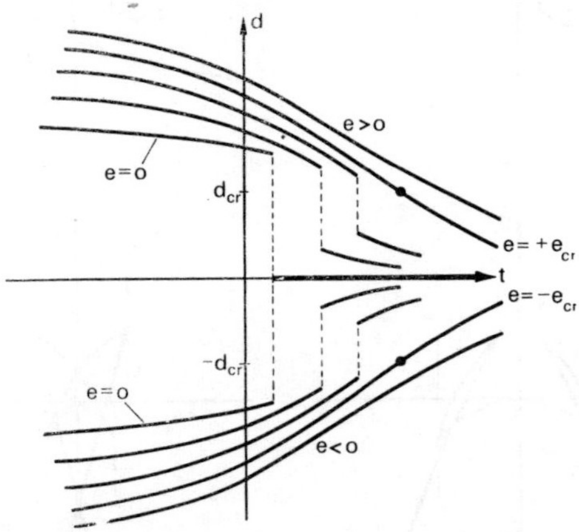

FIG. 5.1.10. The family of d versus t curves with parameter e for the absolutely stable states in the case $\xi < 0$.

at a value t in the range $t_c \leqslant t < t_{cr}$. The e, t phase diagram for the system is sketched in Fig. 5.1.11. The line of transitions terminates at critical points with coordinates $(\pm e_{cr}, t_{cr})$. For comparison we have also included the phase diagram for the model with $\xi \geqslant 0$.† In this case

† The case $\xi = 0$ is qualitatively similar to the case $\xi > 0$ (see end of § 5.1.7).

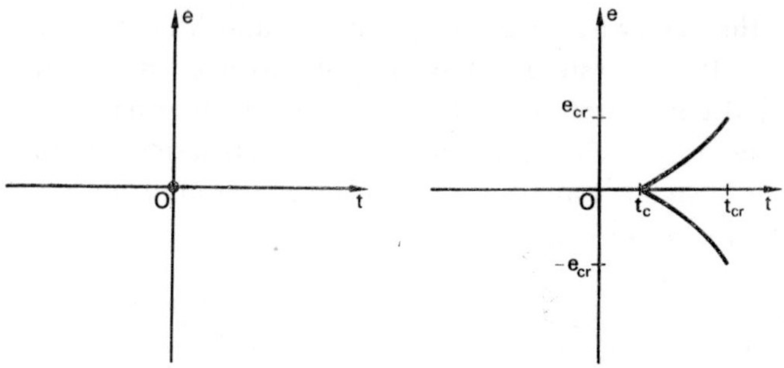

FIG. 5.1.11. e versus t phase diagrams (a) $\xi \geqslant 0$, (b) $\xi < 0$. In the latter case, $t_{cr} = 9/5\ e_{cr} = 24\sqrt{3/25}$ and $t_c = 3/4$ (see text).

there is only a phase transition for $e = 0$, and this appears as an isolated point in the e, t plane.

To summarize: we find that eqn. (5.1.3) describes a system which exhibits a phase transition between a ferroelectric and non-ferroelectric phase. The nature of the transition depends critically on the sign of ξ.

5.1.5 Dielectric Hysteresis

In the previous subsections we have discussed the response of a dielectric filled condenser to a voltage applied across the plates on the assumption that the crystal passes through only absolutely stable (uniform) states. If we admit metastable states in our determination of the response we obtain hysteresis effects. In Fig. 5.1.12 we have sketched the isothermal paths on the assumption that the switchover from one branch of locally stable states to another occurs at the points at which local instability sets in. (i.e. the turning values of the e versus d curve). Referring to Fig. 5.1.12 we see that in the case $\xi < 0$ single loops, double loops and broken loops occur depending on the value of t. Examples of such loops have been observed experimentally; however, the criterion for the switchover from one branch to another is not always the onset of local instability. This is demonstrated by the fact that switchovers can occur at electric field values several orders of magnitudes smaller than those at the turning values of the e versus d curve. The theory of the switching process is complicated.

We refer the reader to the book by Fatuzzo and Merz (1967) for further discussion. If the hysteresis loop is not traversed quasistatically then the thermal constraint is adiabatic and not isothermal. In the case of barium titanate the change in temperature associated with the adiabatic constraint is small. Equation (5.2.33) gives the appropriate expression for the temperature change.

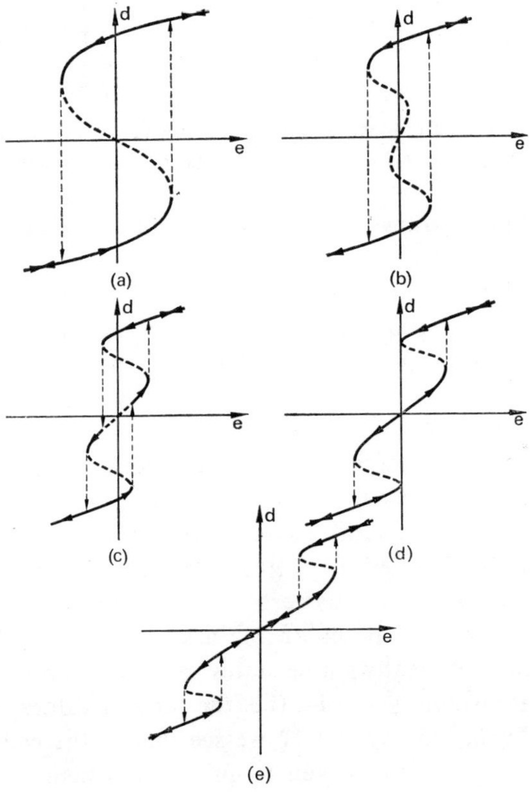

FIG. 5.1.12. The d versus e isotherms (full lines) for the plated slab when the criterion for the switchover from one branch to another is the onset of local instability, i.e. $(\partial e/\partial d)_t < 0$. The dotted lines indicate those states on curve $e(d, t)$ which are not accessible to this system. The vertical dashed lines connect the states at which local instability occurs and the states which the system switches to. (a) refers to the case $\xi > 0$ with $t < 0$, and (a) to (e) refer to the case $\xi < 0$ with $t < t_{cr}$. These isotherms should be compared with the isotherms for absolutely stable states in Figs. 5.1.5. and 5.1.9.

5.1.6 Numerics

Equation (5.1.3) provides a good representation of the dielectric behaviour of barium titanate, ($\xi < 0$), in the vicinity of the Curie–Weiss temperature. In this subsection, as an example, we shall use the experimentally determined values of the coefficients of the dielectric equations of state of this crystal to calculate the scaling factors which appear in eqn. (5.1.4).

In the literature the experimental results of dielectric measurements on ferroelectrics are usually expressed in terms of the electric field and polarization. Let us write the $E(P, \theta)$ representation of the dielectric equation of state given in eqn. (5.1.3) in the form

$$E = \eta P + BP^3 + CP^5. \qquad (5.1.7)$$

Drougard and Huibgreste (1957) report for $BaTiO_3$.

$\eta = 6{\cdot}7 \times 10^5\,(\theta - \theta_0)$ MKS, $B = -9{\cdot}7 \times 10^8$ MKS,† $C = 3{\cdot}9 \times 10^{10}$ MKS.

It is a simple exercise to show that the coefficients in eqns. (5.1.7) and (5.1.3) satisfy the following identities:

$$\beta = \frac{\eta}{1+\varepsilon_0\eta}, \quad \xi = \frac{B}{(1+\varepsilon_0\eta)^4}, \quad \zeta = \frac{C}{(1+\varepsilon_0\eta)^6} - \frac{3\varepsilon_0 B^2}{(1+\varepsilon_0\eta)^7}. \qquad (5.1.8)$$

Now $\varepsilon_0 = 8{\cdot}9 \times 10^{-12}$ MKS, and hence for $BaTiO_3$, $\varepsilon_0\eta = 6 \times 10^{-6}(\theta - \theta_0)$ Thus $|\varepsilon_0\eta| \ll 1$ over a wide range of temperatures about θ_0. We deduce immediately that to an excellent approximation in this crystal, $\beta = \eta$ and $\xi = B$ in the vicinity of θ_0. Further, $3\varepsilon_0 B^2/C = 6{\cdot}4 \times 10^{-4}$; hence ζ and C differ by a negligibly small quantity.

With the values quoted above, eqn. (5.1.4) reads

$$D = 0{\cdot}11\, d\ (\mathrm{C/m^2}), \quad E = 6{\cdot}9 \times 10^5\, e\ (\mathrm{V/m}), \quad \theta - \theta_0 = 11\, t(^\circ\mathrm{C})$$

for barium titanate. At the Curie point in barium titanate the spontaneous polarization $P^0 = 0{\cdot}18$ C/m² (Merz, 1953). Thus the following observations hold in the case of barium titanate. (i) For typical experimental electric field values, $|e| < 1$. (ii) Within about 10 degrees

† In barium titanate B is a linear function of θ. For simplicity we have taken the value $B(\theta_0)$.

on either side of θ_0, $|t| < 1$. (iii) The reduced spontaneous polarization values are of order of magnitude unity or larger.

5.1.7 $E = 0, \xi \gtrless 0$

In this and the following subsection we obtain explicit expressions for some of the dielectric properties of the system introduced in § 5.1.2. We restrict the discussion to the case of the shorted condenser, i.e. $E = 0$.

First of all we calculate the reduced free energy $f = (8\zeta^2/|\xi|^3)F$. Since $E = (\partial F/\partial D)_\theta$, it follows that $e = (\partial f/\partial d)_t$; hence from eqns. (5.1.5) and (5.1.6)

$$f = f_0 + \frac{td^2}{2} + \frac{1}{2}d^4 + \frac{1}{6}d^6 \qquad (\xi > 0) \tag{5.1.9}$$

and

$$f = f_0 + \frac{td^2}{2} - \frac{1}{2}d^4 + \frac{1}{6}d^6 \qquad (\xi < 0), \tag{5.1.10}$$

where f_0 is some function of t (temperature). These functions f are even functions of d and have turning values at the roots $e(d, t) = 0$. Thus from Figs. 5.1.3 and 5.1.7 we may immediately sketch the families of curves f–f_0 versus d for different values of t in the two cases $\xi > 0$ and $\xi < 0$ (Fig. 5.1.13).

The reduced values of the potentials F and G are related by the Legendre transformation $g = f - ed$. Now when $e = 0$, (i) $e = (\partial f/\partial d) = 0$ and (ii) $g = f$, that is the roots of $e(d) = 0$ occur at turning values of f and these turning values are equal to the value of $g(e = 0)$. Moreover, at a maximum turning value $\partial^2 f/\partial d^2 < 0$, i.e. $\partial e/\partial d < 0$, and at a minimum turning value $\partial^2 f/\partial d^2 > 0$, i.e. $\partial e/\partial d > 0$. Therefore a root of $e(d) = 0$ is locally unstable if it corresponds to a maximum in f and locally stable for a minimum. Referring to Fig. 5.1.13 we see that a phase transition occurs in the shorted condenser when $(f - f_0)$ touches the d-axis at three points.† In the case $\xi > 0$ this condition occurs when $t = 0$, in agreement with the discussion given in § 5.1.3, whereas when

† As in § 5.1.4, we are assuming here that the system passes through only absolutely stable states (see discussion of thermal hysteresis at end of § 5.1.8).

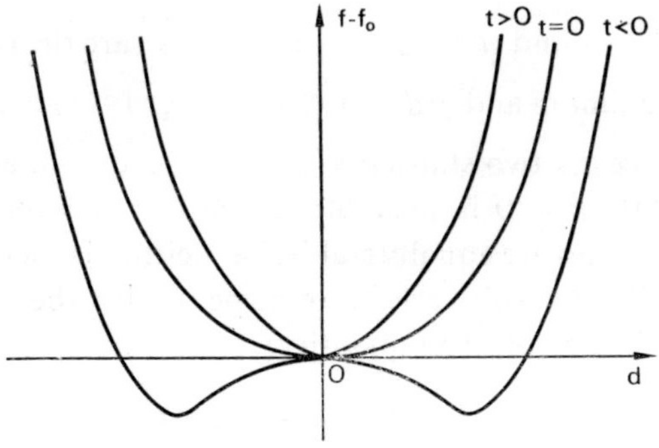

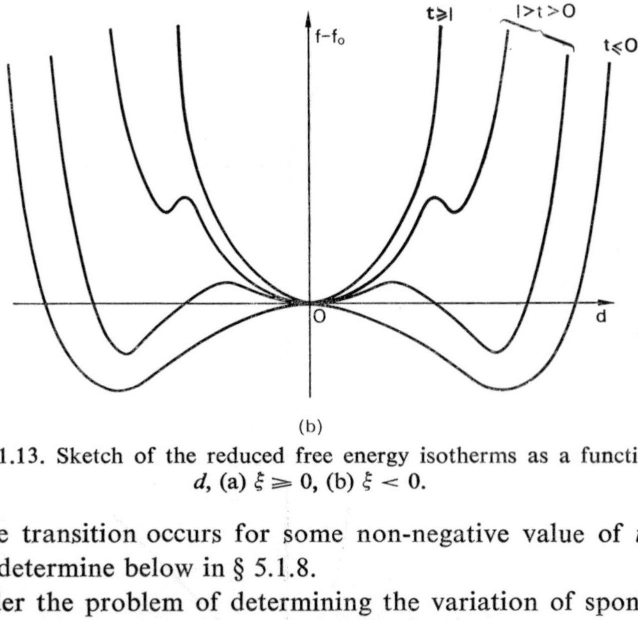

FIG. 5.1.13. Sketch of the reduced free energy isotherms as a function of d, (a) $\xi \geqslant 0$, (b) $\xi < 0$.

$\xi < 0$ the transition occurs for some non-negative value of t, which we shall determine below in § 5.1.8.

Consider the problem of determining the variation of spontaneous polarization and permeability with temperature, for the case $\xi > 0$. When $e = 0$, d is a root of the equation $0 = d[t+2d^2+d^4]$. The roots are

$$d = 0, \quad d = \pm\sqrt{-1\pm\sqrt{1-t}}.$$

When (a) $t \geqslant 0$, $d = 0$ is the only real root; $f(d = 0)$ is a minimum and hence the state described by $d = 0$, $e = 0$ is stable.

(b) $t < 0$, $d = 0$ and $d = \pm\sqrt{-1+\sqrt{1-t}}$ are the only real roots; $f(d = 0)$ is a maximum and $f(d = \pm\sqrt{-1+\sqrt{1-t}})$ are minima (Fig. 5.1.13), and hence the two states $d = \pm\sqrt{-1+\sqrt{1-t}}$ are locally stable whereas the state $d = 0$ is unstable. Since f is an even function of d the former two states are equally stable, i.e. neither is metastable relative to the other. These results are in agreement with the qualitative conclusion drawn in § 5.1.4. The quantity

$$p^0 = \sqrt{-1+\sqrt{1-t}} \tag{5.1.11}$$

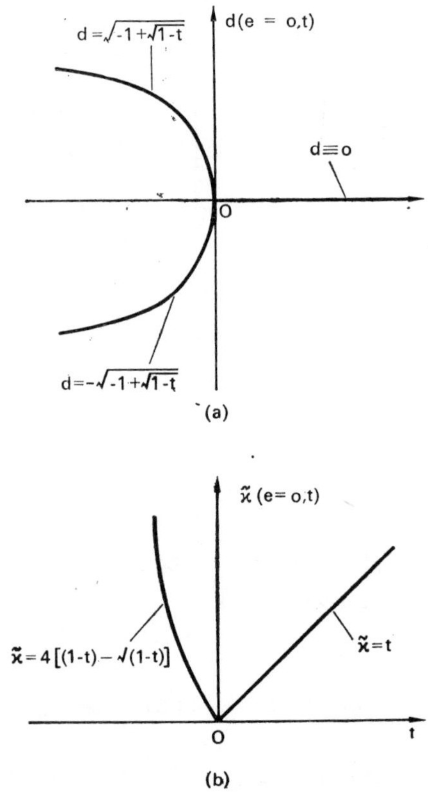

FIG. 5.1.14. Sketch of the t (temperature) dependence of the zero electric field values of the reduced displacement d and the reduced permeability $\tilde{\varkappa} = (\partial e/\partial d)_t$ in the case $\xi > 0$.

is the reduced value of the spontaneous polarization.

The function $p^0(t) = \sqrt{-1+\sqrt{1-t}}$ has the following properties:

(i) $p^0(t \to 0^-) \to 0$.

(ii) $\frac{dp^0}{dt}(t \to 0^-) \to -\infty$.

(iii) $(p^0)^2 = -\frac{t}{2}\left[1+\frac{t}{4}+0(t^2)\right]^\dagger \quad (-1 < t \leqslant 0)$.

(iv) $p^0\left(\frac{dp^0}{dt}\right) = -\frac{1}{4}-\frac{t}{8}+0(t^2) \quad (-1<t\leqslant 0)$.

From (iv) $p^0(dp^0/dt) \to -\frac{1}{4}$ as $t \to 0^-$. $p^0(t)$ is sketched in Fig. 5.1.14. We note that as $t \to -\infty$, $p^0 \to \infty$, i.e. the function $p^0(t)$ does not account for the saturation observed in the spontaneous polarization of real ferroelectric crystals. In terms of the variables P^0 and θ, where P^0 is the spontaneous polarization, the results in (iii) and (iv) read

$$(P^0)^2 = [-\xi+\sqrt{\xi^2-4\zeta A(\theta-\theta_0)}]/2\zeta$$
$$= \frac{A(\theta_0-\theta)}{|\xi|}\left[1+\frac{A\zeta}{\xi^2}(\theta-\theta_0)+0[(\theta-\theta_0)^2]\right]^\ddagger \qquad (5.1.12)$$

† The radius of convergence of this series expansion is unity. This means that the closer the value of $|t|$ to unity the greater the number of terms required in the expansion to obtain an accurate polynomial representation for the function $-1+\sqrt{1-t}$. For example, when $t = -0{\cdot}5$

$$-1+\sqrt{1-t} = 0{\cdot}226; \quad -t/2 = 0{\cdot}250; \quad -t/2-t^2/8 = 0{\cdot}219.$$

Thus when $t = -0{\cdot}5$ the leading term provides a representation which is accurate to within 11 per cent.

‡ This equation may be used to determine ξ and ζ experimentally. To second order in $(\theta-\theta_0)$ we have

$$(P^0)^2 = \frac{A(\theta_0-\theta)}{|\xi|} - \frac{A^2\zeta}{|\xi|^3}(\theta-\theta_0)^2.$$

Thus a least square fitting of the experimental curve $(P^0)^2$ versus $(\theta_0-\theta)$ to this form should yield the values of ξ and ζ provided A is known. However, the following *caveat* is in order; this expression for $(P^0)^2$ is based on two assumptions: (a) ξ and ζ are independent of temperature, and (b) $|\theta-\theta_0| < \xi^2/4A\zeta$. The last condition is a consequence of the fact that the radius of convergence of the power series expansion in (iii) is unity for t or equivalently $\xi^2/4A\zeta$ for the variable $(\theta_0-\theta)$.

and

$$P^0 \frac{dP^0}{d\theta} = -\frac{A}{2|\xi|}\left[1+\frac{2A\zeta}{\xi^2}(\theta-\theta_0)+0\{(\theta-\theta_0)^2\}\right]. \quad (5.1.13)$$

The isothermal dielectric permeability $\varkappa = (\partial E/\partial D)_\theta|_{E=0}$. In reduced form $\tilde{\varkappa} = (\partial e/\partial d)_t$. Since $e = td+2d^3+d^5$,

$$\tilde{\varkappa}(d, t) = t+6d^2+5d^4. \quad (5.1.14)$$

From the discussion above, when $e = 0$ and $t \geqq 0$, the only stable root is $d = 0$. Therefore the zero field reduced permeability is

$$\left(\frac{\partial e}{\partial d}\right)_t\Bigg|_{e=0} = \tilde{\varkappa}(0, t) = t. \quad (5.1.15)$$

When $e = 0$ and $t < 0$ the stable roots are $d = \pm p^0$, where p^0 is given by eqn. (5.1.11); hence

$$\left(\frac{\partial e}{\partial d}\right)_t\Bigg|_{e=0} = \tilde{\varkappa}(\pm p^0, t) = 4[(1-t)-\sqrt{1-t}]. \quad (5.1.16)$$

The functions $d(e = 0, t)$ and $\tilde{\varkappa}(e = 0, t)$ describing the absolutely stable states are sketched in Fig. 5.1.14.

For $|t| < 1$,

$$4[(1-t)-\sqrt{(1-t)}] = -2t+\tfrac{1}{2}t^2+0(t^3) \quad (t < 0)$$

When $|t|$ is sufficiently small that we may truncate the series after the leading term, we have

$$\tilde{\varkappa}(\pm p^0, t) = -2t \quad (t \lesssim 0). \quad (5.1.17)$$

Referring to eqns. (5.1.15) and (5.1.17) we see that the reduced zero field permeability is continuous through the transition and its slope $d\tilde{\varkappa}/dt$ changes magnitude by a factor of two (see Fig. 5.1.14). In terms of the unreduced variables,

$$\left(\frac{\partial E}{\partial D}\right)\Bigg|_{E=0} = A(\theta-\theta_0) \quad (\theta \geqslant \theta_0); \qquad \left(\frac{\partial E}{\partial D}\right)\Bigg|_{E=0} = 2A(\theta_0-\theta) \quad (\theta \lesssim \theta_0). \quad (5.1.18)$$

For completeness we quote the first few terms in the expansion of $(p^0)^2$ and $\tilde{\varkappa}|_{e=0}$ when $t < -1$:

$$(p^0)^2 = (-t)^{\frac{1}{2}} - 1 - \tfrac{1}{2}(-t)^{-\frac{1}{2}} - \tfrac{1}{8}(-t)^{-\frac{3}{2}} + \ldots, \qquad (5.1.19)$$

$$\tilde{\varkappa}\Big|_{e=0} = 4\left[-t - (-t)^{\frac{1}{2}} + 1 + \tfrac{1}{2}(-t)^{-\frac{1}{2}} + \ldots\right]. \qquad (5.1.20)$$

To summarize: in the case $\xi > 0$ the shorted, plated dielectric undergoes a transition when $\theta = \theta_0$. Above the Curie temperature $\theta_c (= \theta_0)$ the spontaneous polarization vanishes and the linear dielectric permittivity $\varepsilon = (\partial D/\partial E)|_{E=0}$ obeys a Curie–Weiss law, $1/[A(\theta - \theta_0)]$. Below the Curie temperature there is a non-zero spontaneous polarization and the linear dielectric permittivity has the form $\varepsilon = 1/[2A(\theta_0 - \theta)]$, provided $\theta \gtrsim \theta_0$. Both the spontaneous polarization and the zero field permeability vanish at the transition whereas their temperature derivatives are discontinuous. The continuity of P^0 and discontinuity in $dP^0/d\theta$ characterize a second-order phase transition in the Ehrenfest classification (see § 5.4.3).

We close this subsection with a brief discussion of the model with $\xi = 0$. In this case the equation of state reads

$$E = \beta D + \zeta D^5.$$

The roots of $E = 0$ are $D = 0$ and $D = \pm\sqrt[4]{-\beta/\zeta}$. Thus the curve $E(D)$ cuts the D-axis once when $\beta \geq 0$ and thrice when $\beta < 0$. Therefore the family of isotherms E versus D, and hence the associated phase transition, are qualitatively similar to those for $\xi > 0$. There is a second-order ferroelectric phase transition at the temperature θ_0. There are, however, quantitative differences; consider the spontaneous polarization and zero field permeability. With $E = 0$,

$$P^0 = \pm\sqrt[4]{\frac{A}{\zeta}(\theta_0 - \theta)}$$

and

$$\varkappa(0, \theta) = A(\theta - \theta_0) \quad (\theta \geq \theta_0),$$
$$\varkappa(P^0, \theta) = 4A(\theta_0 - \theta) \quad (\theta < \theta_0).$$

These are to be compared with the results in eqns. (5.1.12) and (5.1.18).

5.1.8 $E = 0, \xi < 0$

We first determine the value of t at the transition. When $e = 0$, d is a root of $0 = d(t-2d^2+d^4)$. The roots are

$$d = 0, \quad d = \pm\sqrt{1+\sqrt{1-t}}, \quad d = \pm\sqrt{1-\sqrt{1-t}}. \quad (5.1.21)$$

The last pair of solutions are locally unstable (see Fig. 5.1.13b). The phase transition occurs between a state with $d = 0$ and a state with $d = \pm\sqrt{1+\sqrt{1-t}}$. Thus the reduced value of the spontaneous polarization has the form

$$p^0(t) = \sqrt{1+\sqrt{1-t}}. \quad (5.1.22)$$

At the transition the g values of the two phases are equal. Since $e = 0$, $g = f$; hence at the transition,

$$f(p^0, t) = f(0, t)^\dagger$$

or $$f_0+\tfrac{1}{2}t(p^0)^2-\tfrac{1}{2}(p^0)^4+\tfrac{1}{6}(p^0)^6 = f_0 \quad \text{[see eqn. (5.1.10)]}. \quad (5.1.23)$$

Equations (5.1.22) and (5.1.23) serve to determine the values of t and p^0 at the transition. The roots are $t_c = \frac{3}{4}$ and $p^0(t_c) = \sqrt{3/2}$. Expanding the right-hand side of eqn. (5.1.22) in terms of $(t-t_c)$ we get

$$(p^0)^2 = \tfrac{3}{2}+(t_c-t)-(t_c-t)^2+0[(t_c-t)^3].^\ddagger$$

Hence

$$p^0\frac{dp^0}{dt} = -\frac{1}{2}+(t_c-t)+0[(t_c-t)^2]$$

and

$$p^0\frac{dp^0}{dt}\bigg|_{t\to t_c} = -\tfrac{1}{2}.$$

The function $p^0 = \sqrt{1+\sqrt{1-t}}$ is sketched in Fig. 5.1.15. In terms of the unreduced variables,

$$\begin{aligned}(P^0)^2 &= [|\xi|+\sqrt{\xi^2-4\zeta A(\theta-\theta_0)}]/2\zeta \\ &= \frac{3}{4}\frac{|\xi|}{\zeta}\left[1+\frac{8A\zeta}{3\xi^2}(\theta_c-\theta)-\frac{2}{3}\left(\frac{4A\zeta}{\xi^2}\right)^2(\theta_c-\theta)^2+\ldots\right] \quad (5.1.24)\end{aligned}$$

† Mathematically, this is the condition for the three minima in f to touch the d-axis.

‡ The radius of convergence is 0·25.

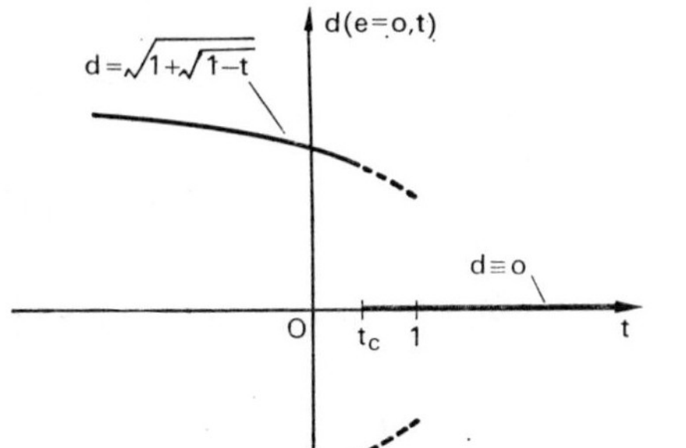

FIG.5.1.15. Sketch of the t (temperature) dependence of the zero electric field values of the reduced displacement d and the reduced permeability $\tilde{\varkappa} = (\partial e/\partial d)_t$ for absolutely stable states in the case $\xi < 0$.

and
$$P^0 \frac{dP^0}{d\theta} = -\frac{A}{|\xi|} + \frac{8A^2\zeta}{|\xi|^3}(\theta_c - \theta) +, \tag{5.1.25}$$

where
$$\theta_c = \theta_0 + \tfrac{3}{16}\xi^2/A\zeta. \tag{5.1.26}$$ †

Since $e = td - 2d^3 + d^5$, the reduced permeability is

$$\tilde{\varkappa} = t - 6d^2 + 5d^4.$$

† From this expression we deduce that a ferroelectric phase transition will occur when θ_0 is negative, provided $\theta_0 + \frac{3}{16}\xi^2/A\zeta > 0$ and $\xi < 0$.

Thus when $e = 0$ and $t > t_c$, the absolutely stable states have $d = 0$ and

$$\tilde{\varkappa}\Big|_{e=0} = \tilde{\varkappa}(0, t) = t\,; \qquad (5.1.27)$$

whereas when $e = 0$ and $t < t_c$, the absolutely stable states have $d = \pm p^0$ and

$$\tilde{\varkappa}\Big|_{e=0} = \tilde{\varkappa}(p^0, t) = 4[(1-t)+\sqrt{1-t}]. \qquad (5.1.28)$$

The functions $d(e = 0, t)$ and $\tilde{\varkappa}(e = 0, t)$ describing the absolutely stable states are sketched in Fig. 5.1.15.

The series expansion in terms of $(t_c - t)$ for $\varkappa(p^0, t)$ is

$$\tilde{\varkappa}(p^0, t) = 3+8(t_c-t)-4(t_c-t)^2+\dots. \qquad (5.1.29)$$

When t is close to the value t_c, the reduced value of the zero field permeability is given by the approximate form

$$\tilde{\varkappa}(p^0, t) = 3+8(t_c-t) \quad (t \sim t_c). \qquad (5.1.30)$$

Referring to eqns. (5.1.27) and (5.1.30) we deduce that the reduced permeability and its slope $d\tilde{\varkappa}/dt$ are discontinuous through the transition. In terms of the unreduced variables (5.1.27) and (5.1.30), read

$$\left(\frac{\partial E}{\partial D}\right)\Big|_{E=0} = A(\theta-\theta_0) \quad (\theta \geqslant \theta_c),$$

$$\left(\frac{\partial E}{\partial D}\right)\Big|_{E=0} = \frac{\xi^2}{4\zeta}\left[3+\frac{32A\zeta}{\xi^2}(\theta_c-\theta)\right] \quad (\theta \gtrsim \theta_c). \qquad (5.1.31)^\dagger$$

For completeness we quote the first few terms in the expansion of $(p^0)^2$ and $\tilde{\varkappa}(p^0, t)$ for $t < -1$:

$$(p^0)^2 = \left[1+(-t)^{\frac{1}{2}}-\tfrac{1}{2}(-t)^{-\frac{1}{2}}-\tfrac{1}{8}(-t)^{-\frac{3}{2}}+\dots\right], \qquad (5.1.32)$$

$$\tilde{\varkappa}(p^0, t) = 4\left[-t+(-t)^{\frac{1}{2}}+1-\tfrac{1}{2}(-t)^{-\frac{1}{2}}+\dots\right]. \qquad (5.1.33)$$

† When $\theta \to \theta_c^-$, $P^0 \to \sqrt{(\frac{3}{4})(|\xi|/\zeta)}$ (see eqn. 5.1.24) and $(\partial E/\partial D)\big|_{E=0} \to 3\xi^2/4\zeta$. Thus by measuring P^0 and $(\partial E/\partial D)\big|_{E=0}$ at the transition one could in principle determine the values of ξ and ζ. However, in practice, because of thermal hysteresis (see below) this method cannot yield accurate values of these quantities.

To summarize: in the case $\xi<0$, the shorted, plated dielectric slab undergoes a transition between absolutely stable states when $\theta = \theta_c = \theta_0 + \frac{3}{16}(\xi^2/A\zeta)$. Above the Curie temperature θ_c the spontaneous polarization vanishes and the linear dielectric permittivity $\varepsilon = 1/A(\theta-\theta_0)$. Below θ_c there is a non-zero spontaneous polarization and the linear dielectric permittivity has the form $\varepsilon \simeq 1/8A(\theta_b-\theta)$ for $\theta \sim \theta_b$, with $\theta_b = \theta_c+(3/32)(\xi^2/A\zeta)$. The spontaneous polarization and the zero field permeability are discontinuous at θ_c. The former fact suggests that this is a first-order phase transition in the Ehrenfest classification (see § 5.4.2). It should be noted that $\theta_0 \neq \theta_c$, i.e. the Curie–Weiss temperature does not equal the Curie temperature.

We conclude this subsection with a brief discussion of thermal hysteresis. Let us assume that the shorted crystal described above is cooled from some paraelectric state $D = 0$, $\theta>\theta_c$. If the system can only pass through absolutely stable states, the previous analysis predicts that a transition to the ferroelectric state occurs at θ_c. In practice this restriction is too strong and the system would continue to pass through the metastable paraelectric states below θ_c until it switched over at some temperature $\theta_c^l<\theta_c$ to the absolutely stable ferroelectric states $D = \pm P^0$. This effect is known as *supercooling*. In real systems θ_c^l is determined by such factors as rate of cooling, impurity content of the specimen, etc., which lie outside the scope of the present analysis. However, this analysis does provide a lower bound to θ_c^l, namely θ_0, the temperature at which the paraelectric states $D = 0$ become locally unstable;† thus $\theta_0 \leqslant \theta_c^l \leqslant \theta_c$. Similarly if the shorted crystal is heated from some ferroelectric state $D = \pm P^0$, $\theta<\theta_c$, the system may pass through metastable ferroelectric states before undergoing a transition at $\theta_c^u>\theta_c$ to the absolutely stable paraelectric states. This is referred to as *superheating*. θ_c^u is bounded above because the function $P^0(\theta)$ becomes imaginary when $\theta>\theta_c+\xi^2/4A\zeta$; thus $\theta_c \leqslant \theta_c^u \leqslant \theta_c+\xi^2/4A\zeta$. When superheating and/or supercooling occur, the past history of the system will determine whether it occupies a paraelectric or a ferroelectric state in the temperature range $\theta_c^l<\theta<\theta_c^u$. This effect is called *thermal hystere-*

†While the demands of absolute stability are not always met in practice, there is no experimental evidence that the demands of local stability are ever violated.

sis. We note that thermal hysteresis cannot occur for the case $\xi > 0$, because in this model there are no states which are both metastable and locally stable (see Fig. 5.1.13).

5.1.9 Field Dependent Permeability $\theta \gtrless \theta_c$.

In this subsection we derive expressions for the field dependent permeability of the model introduced in § 5.1.2. Unless otherwise stated, the expressions quoted below are valid when ξ and ζ are functions of temperature. We shall continue to suppose that $\beta = A(\theta - \theta_0)$.

In reduced form the equation of state reads

$$e = td + 2\alpha d^3 + d^5,$$

where $\alpha = \pm 1$ according as $\xi \gtrless 0$. The reduced permeability is

$$\tilde{\varkappa}(d, t) = \left(\frac{\partial e}{\partial d}\right)_t = t + 6\alpha d^2 + 5d^4. \qquad (5.1.34)$$

If $d = 0$ when $e = 0$, then the right-hand side of eqn. (5.1.34) can be regarded as the first part of a power series expansion for $\tilde{\varkappa}(d, t)$ in terms of the small quantity d. On the other hand, if $d = p^0 \neq 0$ when $e = 0$, we treat $(d - p^0)$ as a small quantity to obtain the following expansion:

$$\tilde{\varkappa}(d, t) = \tilde{\varkappa}(p^0, t) + 4[3\alpha p^0 + 5(p^0)^3](d - p^0) + 6[\alpha + 5(p^0)^2](d - p^0)^2 + \ldots \qquad (5.1.35)$$

In terms of the unreduced variables, eqns. (5.1.34) and (5.1.35) read

$$\varkappa(D, \theta) = \varkappa(0, \theta) + 3\xi D^2 + 5\xi D^4 \quad (\theta \geqslant \theta_c) \qquad (5.1.36)$$

and

$$\varkappa(D, \theta) = \varkappa(P^0, \theta) + [6\xi + 20\zeta(P^0)^2]P^0(D - P^0) + [3\xi + 30\zeta(P^0)^2](D - P^0)^2 + \quad (\theta \leqslant \theta_c), \qquad (5.1.37)$$

where $\varkappa(0, \theta) = \beta$

and $$\varkappa(P^0, \theta) = \beta + 3\xi(P^0)^2 + 5\zeta(P^0)^4 = (1/\zeta)[(\xi^2 - 4\beta\zeta) - \xi\sqrt{(\xi^2 - 4\beta\zeta)}]\,. \qquad (5.1.38)$$

$\varkappa(D, \theta)$ is the isothermal permeability. To obtain the adiabatic permeability we use the identity [eqn. (3.1.21)]

$$\left(\frac{\partial E}{\partial D}\right)_\sigma^{-1} = \left(\frac{\partial E}{\partial D}\right)_\theta^{-1} - \frac{\theta}{\varrho_0 C_E}\left(\frac{\partial D}{\partial \theta}\right)_E^2,$$

where C_E is the specific heat at constant E and ϱ_0 the mass density in the initial state. If ξ and ζ are independent of temperature, $(\partial D/\partial\theta)_E = -AD(\partial D/\partial E)_\theta$ and hence

$$\left(\frac{\partial E}{\partial D}\right)_\sigma = \varkappa(D, \theta) + \left(\frac{\theta A^2}{\varrho_0 C_D}\right) D^2 + O(D^4) \quad (\theta \lesseqgtr \theta_c), \qquad (5.1.39)$$

where C_D is the specific heat at constant displacement. Power series expansions for $(\partial E/\partial D)_\sigma$ in terms of D and in terms of $D-P^0$ are readily obtained by combining eqn. (5.1.39) with eqns. (5.1.36) and (5.1.37) respectively. For the special case $E = 0$, we find, using eqns. (5.1.12), (5.1.18), (5.1.24) and (5.1.31), that for $\xi > 0$

$$\left.\left(\frac{\partial E}{\partial D}\right)_\sigma\right|_{E=0} = A(\theta-\theta_c), \quad \theta > \theta_c;$$

$$\left.\left(\frac{\partial E}{\partial D}\right)_\sigma\right|_{E=0} = 2A(\theta_c-\theta)\left[1+\frac{\theta_c A^2}{2\varrho_0 C_D(\theta_c)}\frac{1}{\xi}\right] \quad \theta \lesssim \theta_c, \qquad (5.1.40)$$

and for $\xi < 0$

$$\left.\left(\frac{\partial E}{\partial D}\right)_\sigma\right|_{E=0} = A(\theta-\theta_c), \quad \theta > \theta_c;$$

$$\left.\left(\frac{\partial E}{\partial D}\right)_\sigma\right|_{E=0} = \frac{3}{4}\frac{\xi^2}{\zeta}\left[1+\frac{A^2\theta_c}{\varrho_0 C_D(\theta_c)}\frac{1}{|\xi|}\right] + 8A(\theta_c-\theta)\left[1+\frac{A}{8\varrho_0 C_D(\theta_c)}\left\{\frac{2A\theta_c}{|\xi|} - \frac{3|\xi|}{4\zeta} - \frac{3\theta_c}{4C_D}\frac{|\xi|}{\zeta}\left(\frac{dC_D}{d\theta}\right)\right\}\right], \quad \theta \lesssim \theta_c, \qquad (5.1.41)$$

Again we emphasize that these results are valid only when ξ and ζ are independent of temperature. Equation (5.1.40) should be compared

with eqn. (5.1.18) and eqn. (5.1.41) with eqn. (5.1.31). In particular

$$(\partial E/\partial D)_\sigma|_{E=0}/(\partial E/\partial D)_\theta|_{E=0} = 1 \quad \text{for} \quad \theta > \theta_c\,(\xi \lessgtr 0) \qquad (5.1.42)$$

and

$$(\partial E/\partial D)_\sigma|_{E=0}/(\partial E/\partial D)_\theta|_{E=0} = \left[1+\frac{\theta_c A^2}{2\varrho_0 C_D(\theta_c)}\frac{1}{|\xi|}\right]$$

$$\text{for} \quad \theta \gtrsim \theta_c \quad (\xi > 0). \qquad (5.1.43)$$

The permeability may also be expressed as a power series in the electric field E. Inverting the representation $e = td+2\alpha d^3+d^5$ we get

$$d = t^{-1}e-2\alpha t^{-4}e^3+(12t^{-7}-t^{-6})e^5+\ldots \qquad (5.1.44)$$

if $d = 0$ when $e = 0$, and

$$d = p^0+[e/\tilde{\varkappa}(p^0, t)]-2[3\alpha+5(p^0)^2]p^0e^2/\tilde{\varkappa}^3(p^0, t)+\ldots \qquad (5.1.45)$$

if $d = p^0 = 0$ when $e = 0$.†

In terms of the unreduced variables, these series read

$$D = \frac{E}{A(\theta-\theta_0)}-\frac{\xi E^3}{A^4(\theta-\theta_0)^4}+\frac{[3\xi^3-\zeta A(\theta-\theta_0)]E^5}{A^7(\theta-\theta_0)^7}+\ldots \quad (\theta > \theta_c) \qquad (5.1.46)$$

and

$$D = P^0+\frac{E}{\varkappa(P^0, \theta)}-[3\xi+10\zeta(P^0)^2]\frac{P^0E^2}{\varkappa^3(P^0, \theta)}+\ldots \quad (\theta < \theta_c) \qquad (5.1.47)$$

where P^0, $\varkappa(P^0, \theta)$ are the spontaneous polarization and zero field permeability respectively.

Substituting for d from eqn. (5.1.44) into eqn. (5.1.34) and for $(d-p^0)$ from eqn. (5.1.45) into eqn. (5.1.35) we get

$$\tilde{\varkappa} = t+6\alpha t^{-2}e^2-(24t^{-5}-5t^{-4})e^4+\ldots \quad (t > t_c) \qquad (5.1.48)$$

† The radii of convergence of these series are not known; however, we can obtain some idea of the mathematical range of validity of the inversion from the following observation. If $\alpha = +1$, e is a monotonic increasing function of d and hence eqn. (5.1.41) is not valid if $|t^{-1}e|-|2t^{-4}e^3| < 0$, i.e. $e^2 > (|t|^3/2)$. This condition of course not a satisfactory substitute for a precise knowledge of the radius of convergence.

and

$$\tilde{\varkappa} = \tilde{\varkappa}(p^0, t) + 4[3\alpha + 5(p^0)^2]\, p^0\, e/\tilde{\varkappa}(p^0, t) + \dots \quad (t < t_c) \quad (5.1.49)$$

In terms of the unreduced variables these equations read

$$\left(\frac{\partial E}{\partial D}\right)_\theta = \beta + (3\xi/\beta^2)\, E^2 + [(5\zeta\beta - 6\xi^2)/\beta^5]\, E^4 + \dots \quad (\theta > \theta_c) \quad (5.1.50)$$

and

$$\left(\frac{\partial E}{\partial D}\right)_\theta = \varkappa(P^0, \theta) + 2[3\xi + 10\zeta(P^0)^2]\, P^0 E/\varkappa(P^0, \theta) + \dots \quad (\theta < \theta_c). \quad (5.1.51)$$

Thus† by measuring the permeability of the crystal as a function of D or E for different temperatures the expressions obtained above enable one to obtain the temperature dependence of the non-linear coefficients ξ, ζ.

5.1.10 The Critical Point

In this subsection we derive some of the dielectric properties of the system in the region of the critical point described in § 5.1.4.

For $\xi < 0$, the reduced equation of state is

$$e = td - 2d^3 + d^5.$$

By definition, at the critical point $(\partial e/\partial d)_t = 0$ and $(\partial^2 e/\partial d^2)_t = 0$; these equations serve to determine the critical values $\pm d_{cr}$ and t_{cr}. Thus $t_{cr} - 6d_{cr}^2 + 5d_{cr}^4 = 0$ and $-12d_{cr} + 20d_{cr}^3 = 0$. The solution is $t_{cr} = 9/5$, $d_{cr} = \sqrt{3/5}$. Substituting these values into the equation of state we find $e_{cr} = (24/25)\sqrt{3/5}$. It is convenient to express e as a function of $(d - d_{cr})$; thus

$$e - e_{cr} = (t - t_{cr})\, d_{cr} + (t - t_{cr})(d - d_{cr}) + 4(d - d_{cr})^3 + 5\sqrt{3/5}\,(d - d_{cr})^4 + (d - d_{cr})^5.$$

† In the only case in which this approach has been attempted, Triebwasser (1960) found evidence of space charge effects in a barium titanate crystal at the field strengths required to observe the term involving ζ and hence was not able to determine its value by this method.

Let us suppose that $(d-d_{cr})$ is sufficiently small that we may neglect the fourth and fifth order terms in this representation; hence

$$e-e_{cr} = (t-t_{cr})\,d_{cr}+(t-t_{cr})\,(d-d_{cr})+4(d-d_{cr})^3. \tag{5.1.52}$$

The following expressions for the reduced dielectric permeability and the reduced pyroelectric coefficient $(\partial d/\partial t)_e$ are easily obtained from this equation;

$$\left(\frac{\partial e}{\partial d}\right)_t = (t-t_{cr})+12(d-d_{cr})^2, \tag{5.1.53}$$

$$\left(\frac{\partial d}{\partial t}\right)_e = -\frac{d}{(t-t_{cr})+12(d-d_{cr})^2}\,. \tag{5.1.54}$$

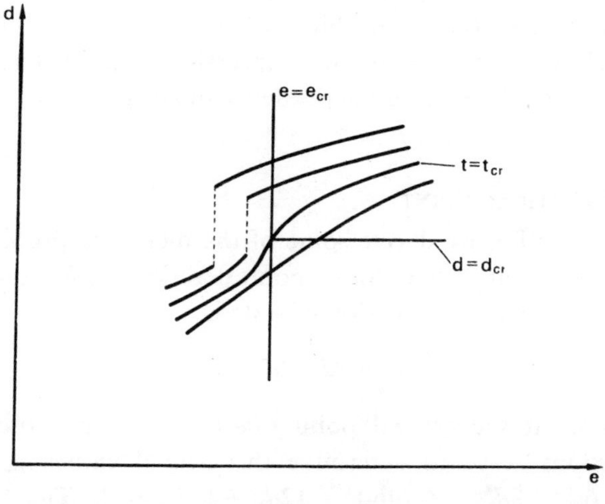

FIG. 5.1.16. Sketch of the isotherms $e = td-2d^3+d^5$ in the region of the critical point $(+d_{cr}, +e_{cr})$ (see Fig. 5.1.9). The paths $e = e_{cr}$ and $d = d_{cr}$ are straight lines in this diagram and the path $t = t_{cr}$ is the critical isotherm.

Consider the line of states $e = e_{cr}$ in the (e, d) plane (Fig. 5.1.16). In the neighbourhood of the critical point (d_{cr}, e_{cr}), d and t are related by the equation obtained by putting $e = e_{cr}$ in eqn. (5.1.52); thus

$$0 = (t-t_{cr})\,d_{cr}+(t-t_{cr})\,(d-d_{cr})+4(d-d_{cr})^3.$$

If $(t-t_{cr})>0$, this equation has one real root. In a first approximation

the root is

$$d-d_{cr} = -\left[\frac{d_{cr}}{4}(t-t_{cr})\right]^{\frac{1}{3}}. \qquad (5.1.55)$$

Substituting this last expression for $(d-d_{cr})$ into eqns. (5.1.53) and (5.1.54) we obtain as a first approximation

$$\left(\frac{\partial e}{\partial d}\right)_t = 12\left(\frac{t-t_{cr}}{4}\right)^{\frac{2}{3}} d_{cr}^{2/3}, \quad \left(\frac{\partial d}{\partial t}\right)_e = -\frac{d_{cr}^{1/3}}{12}\left(\frac{4}{t-t_{cr}}\right)^{\frac{2}{3}}. \qquad (5.1.56)$$

Thus, in particular, the dielectric permeability goes to zero as $(\theta-\theta_{cr})^{\frac{1}{3}}$ along the line $E = E_{cr}$, for $\theta > \theta_{cr}$. Similarly, for paths (a) $d = d_{cr}$ ($e \geqslant e_{cr}$) and (b) $t = t_{cr}$ we find the following expressions:

(a) if $d = d_{cr}$,

$$e-e_{cr} = (t-t_{cr})^1 d_{cr}, \qquad (5.1.57)$$

$$\left(\frac{\partial e}{\partial d}\right)_t = (t-t_{cr})^1, \quad \left(\frac{\partial d}{\partial t}\right)_e = -\frac{d_{cr}}{(t-t_{cr})^1}, \qquad (5.1.58)$$

and (b) if $t = t_{cr}$,

$$d-d_{cr} = \left(\frac{e-e_{cr}}{4}\right)^{\frac{1}{3}}, \qquad (5.1.59)$$

$$\left(\frac{\partial e}{\partial d}\right)_t = 12\left(\frac{e-e_{cr}}{4}\right)^{\frac{2}{3}}, \quad \left(\frac{\partial d}{\partial t}\right)_e = -\frac{d_{cr}}{12}\left(\frac{4}{e-e_{cr}}\right)^{\frac{2}{3}}. \qquad (5.1.60)$$

It is clear from these results that for some properties of the model the critical point is also a singular point in the terminology of mathematical analysis.

Let us formally set $d_{cr} = 0$, $e_{cr} = 0$ and $t_{cr} = 0$ in the equation of state given by (5.1.52); then

$$e = td + 4d^3. \qquad (5.1.61)$$

Aside from the factor 2 (which may be absorbed by a judicious choice of scaling factors), this expression for e coincides with the reduced

equation of state ($\xi > 0$) in the case where d is sufficiently small that the terms in d^5 may be neglected, eqn. (5.1.5). Thus we may formally regard the isolated transition point which occurs in the (E, θ) phase diagram for the case $\xi > 0$, as a critical point. The results given above for the behaviour of the functions $(\partial e/\partial d)_t$ and $(\partial d/\partial t)_e$ in the neighbourhood of the critical point are not valid when d_{cr} vanishes. In this special case, [eqn. (5.1.61)], when $t > 0$, d and e vanish simultaneously; hence the paths $d = 0$ and $e = 0$ coincide; for this path we find

$$\left(\frac{\partial e}{\partial d}\right)_t = t, \quad \left(\frac{\partial d}{\partial t}\right)_e = 0 \tag{5.1.62}$$

[cf. (eqn. 5.1.15)].

When $t = 0$,

$$d = (e/4)^{\frac{1}{3}} \tag{5.1.63}$$

and

$$\left(\frac{\partial e}{\partial d}\right)_t = 12\left(\frac{e}{4}\right)^{\frac{2}{3}}, \quad \left(\frac{\partial d}{\partial t}\right)_e = -\frac{1}{12}\left(\frac{4}{e}\right)^{\frac{1}{3}}. \tag{5.1.64}$$

The various powers appearing in eqns. (5.1.55) to (5.1.60) and eqns. (5.1.62) to (5.1.64) are referred to as *critical indices.* The values quoted above are the so-called classical values derived from a polynomial representation of the dielectric equation of state. Another set of classical values can be obtained by considering the behaviour of the system along the paths $d = d_{cr}$ and $e = e_{cr}$ for $t < t_c$. These critical indices are important because they provide a measure of the suitability of the polynomial representation of the dielectric equation of state. Thus, for example, if in some ferroelectric the permeability $(\partial E/\partial D)_\theta$ is measured along the path $D = D_{cr}$ for $\theta > \theta_{cr}$ and it is found that close to θ_{cr} the critical index is not unity, eqn. (5.1.58) [i.e. $(\partial E/\partial D)_\theta$ does not go to zero as $(\theta - \theta_{cr})^1$], then this result indicates that a polynomial in integral powers of D does not provide a suitable representation for E close to θ_{cr}. The only reported case† of a non-classical critical index in a ferroelectric is that of the permeability of Rochelle salt below the lower Curie temperature (Craig, 1966).

† See Fisher (1966) for a discussion of critical indices in other systems.

5.2 Electrocaloric Coupling

5.2.1 Entropy and Specific Heats

The entropy density σ and temperature θ of the model introduced in § 5.1.2 are related to D and E by the Maxwell relation (§ 3.1.2)

$$\left(\frac{\partial \sigma}{\partial D}\right)_{\theta,\,X} = -\left(\frac{\partial E}{\partial \theta}\right)_{D,\,X}, \tag{5.2.1}$$

where X represents the set of mechanical variables (stress or strain) which are held fixed during the differentiation process. If the isothermal equation of state $E = E(\theta, D, X)$ is known, then we may insert this function into the right-hand side of eqn. (5.2.1) and integrate with respect to D to obtain

$$\sigma(\theta, D, X) = -\int \frac{\partial E}{\partial \theta}(\theta, D, X)\, dD + \sigma_0(\theta, X). \tag{5.2.2}$$

$\sigma_0(\theta, X)$ represents the constant of the D integration in which θ and X are held fixed; thus as has been indicated σ_0 may be a function of θ and X. From this expression for the entropy density we see that σ has two components; one is linked with the isothermal dielectric equation of state and the other is independent of this equation.† Therefore σ_0 and its temperature derivatives can exhibit discontinuities and, in the case of the derivatives,‡ singularities which have no counterparts in $E(\theta, D, X)$ and vice versa. In other words discontinuities, etc., in σ and its temperature derivatives are not necessarily detectable by means of isothermal dielectric measurements.

Let us set $X = 0$ (i.e. the crystal is either clamped or free) and put

$$E = A(\theta - \theta_0)\, D + \xi D^3 + \zeta D^5 \tag{5.2.3}$$

as in § 5.1.2. Substituting this representation for E in eqn. (5.2.2) and

† Experimentally σ_0 may be obtained from a series of either calorimetric measurements or *adiabatic* dielectric measurements [see eqns. (5.2.5), (5.2.6) and (5.1.39) respectively].

‡ Physically the entropy function must remain finite.

assuming that ξ and ζ are independent of θ, we find

$$\sigma(\theta, D) = -\frac{AD^2}{2} + \sigma_0(\theta). \tag{5.2.4}$$

The specific heats C_E and C_D then have the form [eqn. (3.1.18)]

$$C_E = -\frac{\theta AD}{\varrho_0}\left(\frac{\partial D}{\partial \theta}\right)_E + \frac{\theta}{\varrho_0}\left(\frac{d\sigma_0}{d\theta}\right) \tag{5.2.5}$$

and

$$C_D = \frac{\theta}{\varrho_0}\left(\frac{d\sigma_0}{d\theta}\right). \tag{5.2.6}$$

Subtracting,

$$C_E - C_D = -\frac{\theta AD}{\varrho_0}\left(\frac{\partial D}{\partial \theta}\right)_E. \tag{5.2.7}$$

From eqn. (5.2.3) the pyroelectric coefficient takes the form

$$\left(\frac{\partial D}{\partial \theta}\right)_E = -\frac{AD}{[A(\theta-\theta_0)+3\xi D^2+5\zeta D^4]}. \tag{5.2.8}$$

Equations (5.2.4) to (5.2.6) are thermodynamic identities relating the calorimetric and dielectric properties of the system. These identities are not completely general because we have assumed a special representation for E.

5.2.2 $E = 0$

Above the Curie point θ_c, $D = 0$ when $E = 0$ and hence, by eqns. (5.2.4) and (5.2.7),

$$\sigma^0 - \sigma_0 = 0 \quad (\theta > \theta_c), \tag{5.2.9}$$

$$C_E^0 - C_D = 0 \quad (\theta > \theta_c), \tag{5.2.10}$$

where the superscript 0 denotes the value of the function when $E = 0$. Below the Curie point $D = \pm P^0$ when $E = 0$, where

$$(P^0)^2 = \left\{-\xi + \sqrt{[\xi^2 - 4A\zeta(\theta-\theta_0)]}\right\}/2\zeta$$

[cf. eqns. (5.1.12) and (5.1.24)].

Hence

$$\sigma^0 - \sigma_0 = \frac{A}{4\zeta}\{\xi - \sqrt{[\xi^2 - 4A\zeta(\theta-\theta_0)]}\} \quad (\theta < \theta_c) \qquad (5.2.11)$$

and

$$C_E^0 - C_D = \frac{1}{2}\frac{\theta A^2/\varrho_0}{\sqrt{\xi^2 - 4A\zeta(\theta-\theta_0)}} \quad (\theta < \theta_c) \qquad (5.2.12)$$

When $\xi < 0$, eqns. (5.2.11) and (5.2.12) hold only if there is no thermal hysteresis.

If $\theta \gtrsim \theta_c$ we may expand in terms of $(\theta_c - \theta)$ to get

$$\sigma^0 - \sigma_0 = -\frac{A^2(\theta_c-\theta)}{2\xi} + \frac{A^3\zeta}{2\xi^3}(\theta_c-\theta)^2 + \ldots \quad (\xi > 0) \quad (5.2.13)$$

[see eqn. (5.1.12)] and

$$\sigma^0 - \sigma_0 = \frac{3}{8}\frac{A\xi}{\zeta} + \frac{A^2}{\xi}(\theta_c-\theta) - \frac{4A^3\zeta}{\xi^3}(\theta_c-\theta)^2 + \ldots \quad (\xi < 0) \quad (5.2.14)$$

[see eqn. (5.1.24)].

Hence

$$C_E^0 - C_D = \frac{A^2\theta_c}{2\varrho_0\xi} - \frac{A^2}{2\varrho_0\xi}\left[\frac{2A\zeta\theta_c}{\xi^2} + 1\right](\theta_c-\theta) + \quad (\xi > 0) \quad (5.2.15)$$

$$C_E^0 - C_D = -\frac{A^2\theta_c}{\xi\varrho_0} + \frac{A^2}{\varrho_0\xi}\left[\frac{8A\zeta\theta_c}{\xi^2} + 1\right](\theta_c-\theta) + \quad (\xi < 0) \quad (5.2.16)$$

By inspection, as $\theta \to \theta_c^-$,

$$\sigma^0 - \sigma_0 \to 0, \quad C_E^0 - C_D \to \frac{A^2\theta_c}{2\varrho_0\xi} \quad (\xi > 0) \qquad (5.2.17)$$

and

$$\sigma^0 - \sigma_0 \to \frac{3A\xi}{8\zeta}, \quad C_E^0 - C_D \to -\frac{A^2\theta_c}{\varrho_0\xi} \quad (\xi < 0). \qquad (5.2.18)$$

The temperature variation of the quantities $\sigma^0 - \sigma_0$ and $C_E^0 - C_D$ in the neighbourhood of θ_c are sketched in Fig. 5.2.1. Based on the calorimetric properties listed in eqns. (5.2.17) and (5.2.18) we should classify the transition for the case $\xi > 0$ as second order and that for $\xi < 0$ as

first order in agreement with the classification based on the dielectric properties (see §§ 5.1.3 and 5.1.4 and §§ 5.4.2 and 5.4.3). In the first case, $\xi > 0$, the assignment is valid only if σ_0 is a smooth function of temperature near θ_c.

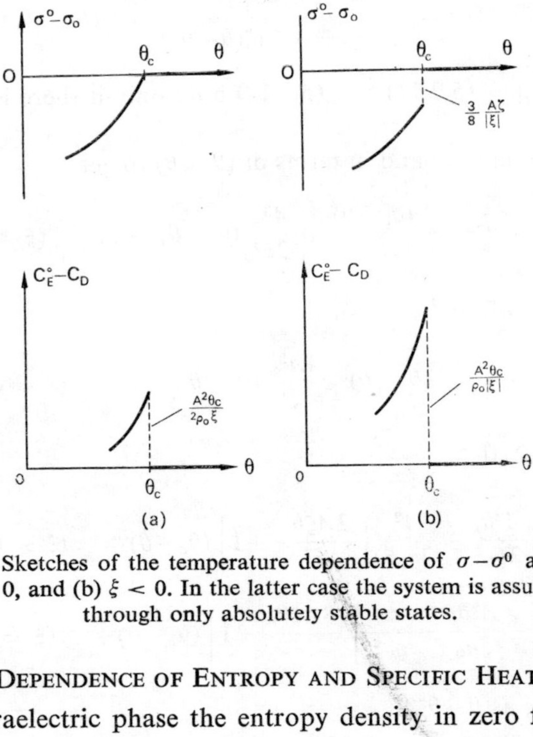

FIG. 5.2.1. Sketches of the temperature dependence of $\sigma - \sigma^0$ and $C_E^0 - C_D$ for (a) $\xi > 0$, and (b) $\xi < 0$. In the latter case the system is assumed to pass through only absolutely stable states.

5.2.3 Field Dependence of Entropy and Specific Heats

In the paraelectric phase the entropy density in zero field $\sigma^0 = \sigma_0$, and hence from eqn. (5.2.4)

$$\sigma = \sigma^0 - \frac{AD^2}{2} \qquad (\theta \geqslant \theta_c), \tag{5.2.19}$$

whereas in the ferroelectric phase $\sigma^0 = \sigma_0 - A(P^0)^2/2$ and therefore

$$\sigma = \sigma^0 - \frac{A}{2}[D^2 - (P^0)^2] \quad (\theta \leqslant \theta_c). \tag{5.2.20}$$

On differentiating both sides of eqns. (5.2.19) and (5.2.20) with respect

to θ, we obtain expressions for C_E; thus

$$C_E = C_E^0 - \frac{A\theta D}{\varrho_0}\left(\frac{\partial D}{\partial \theta}\right)_E \qquad (\theta \geqslant \theta_c) \qquad (5.2.21)$$

and

$$C_E = C_E^0 - \frac{\theta A}{\varrho_0}\left[D\left(\frac{\partial D}{\partial \theta}\right)_E - P^0\left(\frac{dP^0}{d\theta}\right)\right] \qquad (\theta \leqslant \theta_c). \qquad (5.2.22)$$

Substituting for $(\partial D/\partial\theta)_E$ from eqn. (5.2.8) and expanding in terms of D and $(D-P^0)$ for $\theta > \theta_c$ and $\theta < \theta_c$ respectively, we have

$$C_E = C_E^0 + \frac{\theta A D^2}{\varrho_0(\theta-\theta_0)} - \frac{3\theta\xi}{\varrho_0(\theta-\theta_0)^2} D^4 + \dots \qquad (\theta > \theta_c) \quad (5.2.23)$$

and

$$C_E = C_E^0 + \frac{\theta A^2}{\varrho_0}\left[\frac{1}{\varkappa(P^0, \theta_0)}\right]\left[2P^0 - \frac{6\xi(P^0)^3 + 20\xi(P^0)^5}{\varkappa(P^0, \theta)}\right](D-P^0) + \dots$$
$$(\theta < \theta_c), \qquad (5.2.24)$$

where $\varkappa(P^0, \theta) = A(\theta-\theta_0) + 3\xi(P^0)^2 + 5\zeta(P^0)^4$ is the permeability in zero electric field. These expressions are not valid if $\theta \sim \theta_0$. Equations (5.2.19) (5.2.20), (5.2.23) and (5.2.24) describe the variations of the entropy density σ and specific heat C_E about the corresponding zero electric field values. These expressions are written explicitly in terms of D or $(D-P^0)$, whichever is appropriate. It is a simple exercise to replace D and $(D-P^0)$ by E using eqns. (5.1.46) and (5.1.47) respectively.

The behaviour of $C_E - C_D$ along the isotherm $\theta = \theta_0$ is of some interest in the case $\xi > 0$. From eqns. (5.2.5), (5.2.6) and (5.2.8)

$$C_E - C_D = \frac{A^2\theta}{\varrho_0}\,\frac{D^2}{[A(\theta-\theta_0) + 3\xi D^2 + 5\zeta D^4)]}.$$

Provided $E \not\equiv 0$ we may put $\theta = \theta_0$ in this expression to obtain

$$C_E - C_D = \frac{A^2\theta_0}{\varrho_0}\,\frac{1}{(3\xi + 5\zeta D^2)}. \qquad (5.2.25)$$

Thus, if $5\zeta D^2 \ll 3\xi$, $C_E - C_D$ is independent of D and hence the family of curves $(C_E - C_D)$ versus θ with parameter D pass through a common point determined by the temperature θ_0.

5.2.4 Transition Heat and Transition Entropy

The identity eqn. (5.2.7) provides a test of the validity of the form of the equation of state given in eqn. (5.2.3). In this subsection we derive two integrated forms of this identity which are commonly used in the literature.

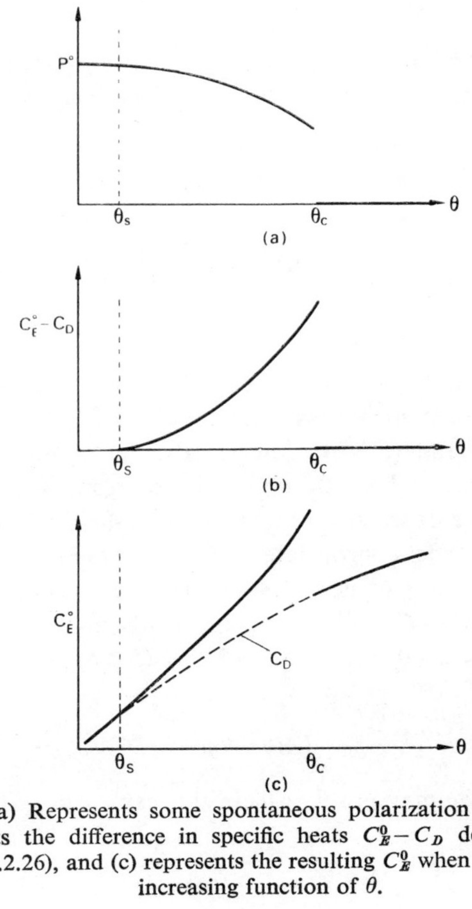

FIG. 5.2.2. (a) Represents some spontaneous polarization function $P^0(\theta)$, (b) represents the difference in specific heats $C_E^0 - C_D$ derived from P^0 using eqn. (5.2.26), and (c) represents the resulting C_E^0 when C_D is a smooth increasing function of θ.

Consider the case in which the spontaneous polarization P^0 has the form sketched in Fig. 5.2.2a. Below θ_s, P^0 is assumed to be independent of θ. This saturation effect is not described by the equation of state

(5.2.3); nevertheless, let us suppose that the identity (5.2.7) [based on eqn. (5.2.3)] holds over the range of temperature displayed in Fig. 5.2.2a. In particular we have

$$C_E^0 - C_D = -\frac{A\theta}{\varrho_0} P^0 \left(\frac{dP^0}{d\theta}\right). \tag{5.2.26}$$

From this identity and the form assumed for P^0 we deduce that $C_E^0 - C_D$ has the appearance sketched in Fig. 5.2.2b. Clearly $C_E^0 - C_D$ is non-zero only in the temperature range (θ_s, θ_c). Now let us suppose that $C_D(\theta)$ behaves normally, that is to say C_D is a monotonic increasing function of θ.† Referring to eqn. (5.2.26) and Fig. 5.2.2b we then see that C_E^0 consists of a peak superimposed on this monotonic function C_D (Fig. 5.2.2c).‡ If the spontaneous polarization in the vicinity of θ_c is correctly described by the polynomial representation of eqn. (5.2.3), the peak is finite.

The *transition heat* Q and *transition entropy* Σ are defined

$$Q = \int_{\theta_s}^{\theta_c} (C_E^0 - C_D)\, d\theta, \tag{5.2.27}$$

$$\Sigma = \int_{\theta_s}^{\theta_c} (C_E^0 - C_D) \frac{d\theta}{\theta}\,; \tag{5.2.28}$$

† To our knowledge C_D has never been measured directly in a ferroelectric crystal. Recent work on the ferroelectric crystals KH_2PO_4, KH_2AsO_4 and TGS throws doubt on the general validity of the assumption that C_D is a smooth function of θ. The specific heats of these crystals (C_E^0 in the case of TGS and tentatively C_E^0 in the case of the other two crystals) appear to have logarithmic singularities at θ_c (§ 1.2); provided the polynomial representation eqn. (5.2.3) is valid for these crystals, $C_E^0 - C_D$ is finite at θ_c (Fig. 5.2.1), and hence C_D must share the singular properties, if any, of C_E^0.

‡ The specific heat curve drawn in Fig. 5.2.2c is qualitatively similar to the experimental specific heat curves observed for some ferroelectrics (see, for example, Fig. 1.2.7). However, for the most part these experiments were carried out without prescribing the electrical boundary conditions and as a result it is not known whether these measurements refer to specific heat at constant E, specific heat at constant D or some other specific heat. In no case has C_E and C_D been measured, and hence he identities eqn. (5.2.26) has not been subjected to a rigorous test.

Q provides a measure of the area between the curves C_E^0 and C_D. Inserting the expression for $C_E^0 - C_D$ from eqn. (5.2.26) we get

$$Q = -\frac{A}{\varrho_0} \int_{\theta_s}^{\theta_c} \theta P^0 \frac{dP^0}{d\theta} d\theta \tag{5.2.29}$$

and

$$\Sigma = -\frac{A}{\varrho_0} \int_{\theta_s}^{\theta_c} P^0 \frac{dP^0}{d\theta} d\theta$$

or

$$\Sigma = \frac{A}{2\varrho_0} [P^0(\theta_s)^2 - P^0(\theta_c)^2]. \tag{5.2.30}$$

If (θ_s, θ_c) is a small temperature interval,

$$Q \cong \frac{A\theta_c}{2\varrho_0} [P^0(\theta_s)^2 - P^0(\theta_c)^2]. \tag{5.2.31}$$

Equations (5.2.29) and (5.2.30) are integrated forms of the identity equation (5.2.26). They are useful in practice only if the circumstances envisaged above occur.

5.2.5 The Electrocaloric Effect

Consider the following transformation: suppose an electric field is applied adiabatically to the plated slab causing a transformation from a state $E = 0$, $\theta = \theta_i$, $D = P^0(\theta_i)$, to a state $E \neq 0$, $\theta = \theta_f$, $D = D_f$ (Fig. 5.2.3). The change in θ associated with the change in D or E when σ is held fixed is referred to as the *electrocaloric effect*. From eqn.(5.2.4) the entropy density remains constant if

$$\sigma_0(\theta_i) - \frac{A}{2} [P^0(\theta_i)]^2 = \sigma_0(\theta_f) - \frac{A}{2} D_f^2$$

or

$$\sigma_0(\theta_f) - \sigma_0(\theta_i) = \frac{A}{2} [D_f^2 - P^0(\theta_i)^2]. \tag{5.2.32}$$

This equation determines the isentropic curve in the D, θ plane; the curves terminates at the point $[P^0(\theta_i), \theta_i]$. If σ_0 is a smooth function of

θ we may expand $\sigma_0(\theta_f)$ in terms of $(\theta_f-\theta_i)$; as a first approximation

$$\left[\left(\frac{\partial\sigma_0}{\partial\theta}\right)\bigg|_{\theta=\theta_i}\right](\theta_f-\theta_i) = \frac{A}{2}[D_f^2-[P^0(\theta_i)]^2]$$

or

$$(\theta_f-\theta_i) = \frac{A\theta_i}{2\varrho_0 C_D(\theta_i)}[D_f^2-\{P^0(\theta_i)\}^2]. \qquad (5.2.33)$$

Thus for fixed θ_i a plot of $(\theta_f-\theta_i)$ versus D_f^2 is a straight line with slope $A\theta_i/2\varrho_0 C_D(\theta_i)$ and intercept on the D_f^2-axis, $[P^0(\theta_i)]^2$ (Fig. 5.2.4). In

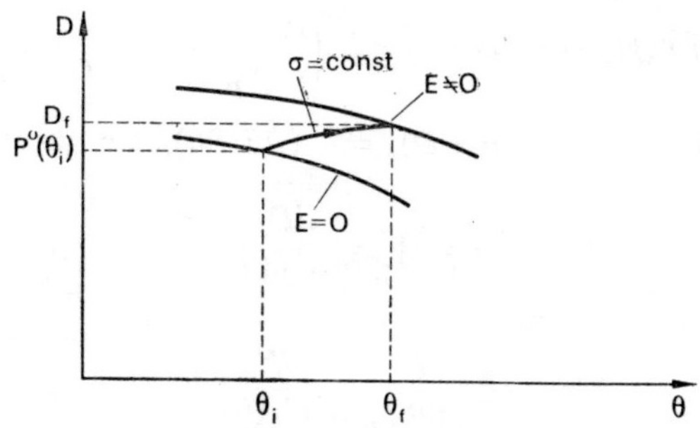

FIG. 5.2.3. The long full lines represent paths of constant E and the short full line represents an adiabatic path.

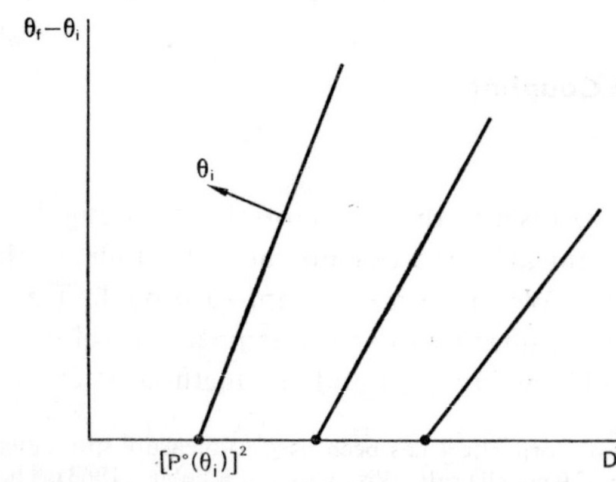

FIG. 5.2.4. The family of straight lines $(\theta_f-\theta_i)$ versus D^2 [(eqn. (5.2.33)]. The arrow indicates the direction of increasing θ_i, assuming that $(dP^0/d\theta) < 0$.

practice the linear portion of the $(\theta_f - \theta_i)$ versus D_f^2 curve appears to be greatest in extent when θ_i is close to the Curie point; thus the electrocaloric effect should provide a good method of measuring the spontaneous polarization in this temperature region,† *provided of course that* C_D *is a slowly varying function of* θ *near* θ_c.

The results obtained above refer to a particular phenomenological model, described by eqn. (5.2.3). More general expressions for the electrocaloric effect may be obtained as follows. Consider the function $\sigma(\theta, D)$. For an infinitesimal isentropic process in which $\theta \rightarrow \theta + d\theta$, $D \rightarrow D + dD$, then

$$\sigma(\theta, D) = \sigma(\theta + d\theta, D + dD) \quad \text{or} \quad \left(\frac{\partial \sigma}{\partial \theta}\right)_D d\theta + \left(\frac{\partial \sigma}{\partial D}\right)_\theta dD = 0.$$

Using the Maxwell relation $(\partial \sigma / \partial D)_\theta = -(\partial E / \partial \theta)_D$ we obtain

$$d\theta = \theta \left(\frac{\partial E}{\partial \theta}\right)_D \frac{dD}{\varrho_0 C_D(\theta)}. \tag{5.2.34}$$

Similarly,

$$d\theta = -\theta \left(\frac{\partial D}{\partial \theta}\right)_E \frac{dE}{\varrho_0 C_E(\theta)}. \tag{5.2.35}$$

These equations relate the changes, $d\theta$, dD in one case and $d\theta$, dE in the other, which occur in an infinitesimal isentropic transformation.

5.3 Electroelastic Coupling

5.3.1 Introduction

In this section we discuss the effect of the coupling between the electrical and mechanical properties on the behaviour of the system introduced in § 5.1.2. We make no attempt to provide a general discussion but merely consider two particular phenomenological models to illustrate the problems involved and the method of tackling them.

† To date the electrocaloric effect has been used to measure spontaneous polarization only in the case of Rochelle salt (Wiseman and Kuebler, 1963). These authors observe some deviation from linearity near the D^2-axis which they attribute to domain effects.

The two models are described by the following equations of state:

(A)
$$T = c^D(S-\gamma^0)+\tilde{q}D^2, \tag{5.3.1}$$
$$E = \beta^S D+2qSD+\xi^S D^3+\zeta^S D^5, \tag{5.3.2}$$

and

(B)
$$T = c^D S-hD, \tag{5.3.3}$$
$$E = -hS+\beta^S D+\xi^S D^3+\zeta^S D^5. \tag{5.3.4}$$

T represents some stress tensor component and S the conjugate strain. Because we are particularly interested in the effect of the change of the mechanical and electric constraints on the coefficients appearing in the representation for the equations of state, the superscripts S, T, D and E are displayed explicitly. In model (A) there is a non-vanishing normal thermal expansion γ^0 and the coupling between S and D is electrostrictive. We recognize the fact that the electrostrictive coefficients appearing in the equations for E and for T may not be the same by adding a tilde to one, §4.2. In model (B) there is no normal thermal expansion and there is a linear piezoelectric coupling between S and D. We shall assume in this latter model that T is a shear stress, vanishing when the stress is hydrostatic. The results for model (A) are easily extended to describe the case of barium titanate and TGS, and those for model B apply immediately to Rochelle salt [$T \to T_4$, $S \to S_4$, $E \to E_1$, $D \to D_1$, eqns. (1.3.1) and (1.3.2)] and KDP [$T \to T_6$, $S \to S_6$, $E \to E_3$, $D \to D_3$, eqns. (1.2.1) and (1.2.2)].

Solving for S and expressing E as a function of T and D, we obtain for these models equations of state of the following form:

(A)†
$$S = \gamma^0+s^D T-\tilde{Q}D^2, \tag{5.3.5}$$
$$E = \beta^T D+2QTD+\xi^T D^3+\zeta^T D^5, \tag{5.3.6}$$

† It is instructive to apply the type of analysis described in § 3.1.4 to these equations of state. For example, eqn. (5.3.6) represents the dielectric equation of state $E(T, D, \theta)$ along a path $T =$ constant in the $T-S$ plane. In a first approximation the dielectric equation of state along a path $S =$ constant is

$$E(S, D, \theta) = E(S, D, \theta)\Big|_{S=S(T)} + \left(\frac{\partial E}{\partial S}\right)_{D,\,\theta}\Big|_{S=S(T)} [S-S(T)]$$

[cf. eqn. (3.1.38b)]. Using eqns. (5.3.5) and (5.3.6), the right-hand side of this equation

where

$$s^D = 1/c^D, \qquad \tilde{Q} = \tilde{q}/c^D, \qquad Q = q/c^D, \tag{5.3.7}$$

$$\beta^T = \beta^S + 2q\gamma^0, \quad \xi^T = \xi^S - \frac{2q\tilde{q}}{c^D}, \quad \zeta^T = \zeta^S, \tag{5.3.8}$$

and

(B)

$$S = s^D T - gD, \tag{5.3.9}$$

$$E = \beta^T D - gT + \xi^T D^3 + \zeta^T D^5, \tag{5.3.10†}$$

where

$$s^D = 1/c^D, \qquad g = -h/c^D, \tag{5.3.11}$$

$$\beta^T = \beta^S - h^2/c^D, \quad \xi^T = \xi^S, \quad \zeta^T = \zeta^S. \tag{5.3.12}$$

The coefficients in eqns. (5.3.1) to (5.3.4) and in the inverted forms eqns. (5.3.5), (5.3.6), (5.3.9) and (5.3.10) are at most functions of θ.

5.3.2 Spontaneous Strain

When $T = 0$ and $E = 0$ the equation determining D reduces to the form

$$0 = \beta^T D + \xi^T D^3 + \zeta^T D^5$$

for both models. Suppose $\beta^T = A^T(\theta - \theta_0^T)$ [cf. eqn. (5.1.3)], where $1/A^T$ and θ_0^T are respectively the Curie–Weiss constant and Curie–Weiss temperature of the free crystal. This equation has globally stable,

---

takes the form

$$\beta^T D + 2QTD + \xi^T D^3 + \zeta^T D^5 + 2QD\frac{1}{s^D}[S - (\gamma^0 + s^D T - \tilde{Q}D^2)]$$

$$= \left(\beta^T - \frac{2\gamma^0 Q}{s^D}\right) D + \frac{2QSD}{s^D} + \left(\xi^T + \frac{2Q\tilde{Q}}{s^D}\right) D^3 + \zeta^T D^5 = \beta^S D + 2qSD + \xi^S D^3 + \zeta^S D^5$$

in agreement with eqn. (5.3.2). In the last step we have used eqns. (5.3.7) and (5.3.8). It is clear from this example that if the elastic and dielectric equations of state are represented by polynomials then the approximation procedure described at the end of § 3.1.4 is fully equivalent to the inversion of the equations of state, such as that carried out in going from eqns. (5.3.1) and (5.3.2) to eqns. (5.3.5) and (5.3.6). The *exact* identities of § 3.1.4 are in fact most useful when the range of field values being considered would require many terms for an accurate polynomial representation (see, for example, Gill and Grindlay, 1966).

† When $T \neq 0$ model (B) exhibits a non-ferroelectric phase change. This is easily verified. From eqn. (5.3.10), $E(D, T)$ is not an odd function of D, and thus the spontaneous polarization values are not distributed symmetrically about zero as required for a ferroelectric phase. Non-ferroelectric phase changes lie outside the scope of this book.

non-zero roots $D = \pm P^0(\theta)$ for a range of θ values below some critical temperature θ_c (see § 5.1). $P^0(\theta)$ is the spontaneous polarization. Substituting $D = +P^0(\theta)$, $T = 0$ into eqns. (5.3.5) and (5.3.9) we get the following expressions for the thermal expansion $\gamma = S(E = 0, T = 0, \theta)$ in the ferroelectric phase:

$$\text{(A)} \qquad \gamma = \gamma^0(\theta) + S^0(\theta), \quad \text{where} \quad S^0(\theta) = -\tilde{Q}(\theta)[P^0(\theta)]^2, \qquad (5.3.13)$$

$$\text{(B)} \qquad \gamma = S^0(\theta), \quad \text{where} \quad S^0(\theta) = -g(\theta)P^0(\theta). \qquad (5.3.14)$$

$S^0(\theta)$ is the spontaneous strain. Thus because of the electroelastic coupling in these models the spontaneous polarization provides a contribution† to the thermal expansion in addition to the normal thermal expansion γ^0. In practice these expressions [eqns. (5.1.13) and (5.1.14)] provide a means of obtaining the values of $\tilde{Q}(\theta)$ or $g(\theta)$ (whichever is appropriate) from the experimental values for γ, γ^0 and P^0 [see discussion in § (4.3.4)].

Using the results of § 5.1 we may make the following predictions about the behaviour of $S^0(\theta)$ in the vicinity of θ_c^T. For the case $\xi^T > 0$ the form of $P^0(\theta)$ near θ_c^T is given by eqn. (5.1.12).

Thus

$$\text{(A)} \qquad S^0(\theta) = -\frac{\tilde{Q}(\theta_c)}{\xi^T(\theta_c)} A^T(\theta_c - \theta) + 0[(\theta_c - \theta)^2] \quad (\theta \leqslant \theta_c),$$

$$S^0(\theta) \equiv 0 \quad (\theta \geqslant \theta_c),$$

and

$$\text{(B)} \qquad S^0(\theta) = -g(\theta_c)\sqrt{\frac{A^T(\theta_c - \theta)}{\xi^T(\theta_c)}}\{1 + 0(\theta_c - \theta)\} \quad (\theta \leqslant \theta_c),$$

$$S^0(\theta) \equiv 0 \quad (\theta \geqslant \theta_c).$$

Provided $\gamma^0(\theta)$ is a smooth function of θ (this appears to be true for the ferroelectrics) then by inspection we see that in both models $\gamma(\theta)$ is continuous through the Curie point while $d\gamma/d\theta$ is discontinuous. In (A) this discontinuity is finite and in (B) it is infinite (Fig. 5.3.1).

† In ferroelectric crystals this contribution leads to a lowering of the symmetry of the physical shape of the crystal as it is cooled through the Curie point.

Similarly, for $\xi^T < 0$, using the results given in eqn. (5.1.24), we get

(A) $$S^0(\theta) = \frac{3}{4}\tilde{Q}(\theta_c)\frac{\xi^T(\theta_c)}{\zeta^T(\theta_c)} + 0(\theta_c - \theta) \quad (\theta \leqslant \theta_c),$$

$$S^0(\theta) \equiv 0 \quad (\theta \geqslant \theta_c).$$

(B) $$S^0(\theta) = -g(\theta_c)\sqrt{-\frac{3}{4}\frac{\xi^T(\theta_c)}{\zeta^T(\theta_c)}}\,[1 + 0(\theta_c - \theta)] \quad (\theta \leqslant \theta_c),$$

$$S^0(\theta) \equiv 0 \quad (\theta \geqslant \theta_c).$$

In this case for (A) and (B) both $\gamma(\theta)$ and $d\gamma/d\theta$ suffer finite discontinuities at the Curie point (Fig. 5.3.1).

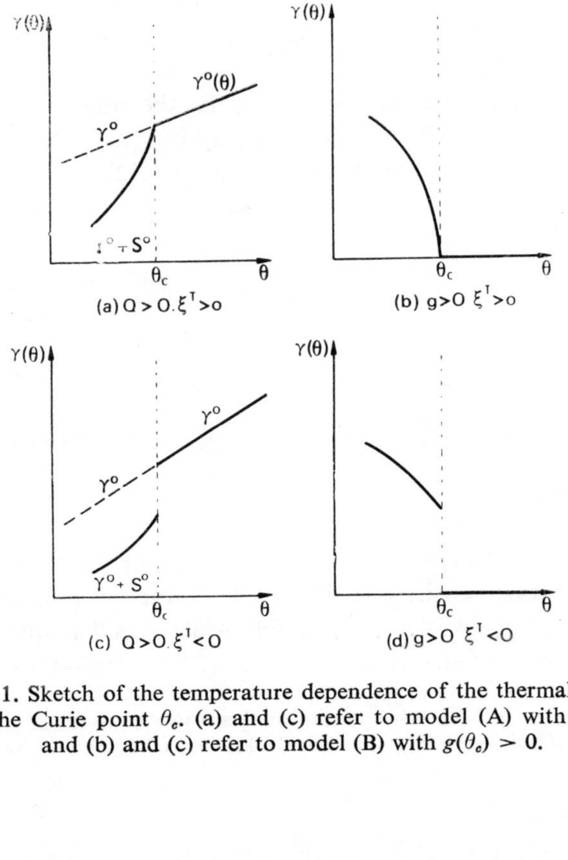

FIG. 5.3.1. Sketch of the temperature dependence of the thermal expansion γ near the Curie point θ_c. (a) and (c) refer to model (A) with $\tilde{Q}(\theta_c) > 0$, and (b) and (c) refer to model (B) with $g(\theta_c) > 0$.

When E and T are small, $(S-\gamma)$ and $(D-P_0)$ are small and the linear approximation to the equations of state for model (A) takes the form

$$S-\gamma = s^D T-(2\tilde{Q}P^0)(D-P^0), \tag{5.3.15}$$

$$E = \beta^T D+(2QP^0)T. \tag{5.3.16}$$

Hence below the Curie point the crystal displays linear piezoelectricity in the sense that $S-\gamma$ is a linear function of $(D-P^0)$ and E is a linear function of T. This linear piezoelectric effect is induced by the spontaneous polarization, and the induced linear piezoelectric coefficients are $2\tilde{Q}P^0$ and $2QP^0$ respectively. By measuring these coefficients and $P^0(\theta)$ one may determine $\tilde{Q}(\theta)$ and $Q(\theta)$.

5.3.3 The Free Crystal and the Clamped Crystal (A)

In model (A) if the crystal is clamped to keep $S \equiv 0$, the dielectric equation of state is given by (5.3.2) with $S = 0$, i.e.

$$E = \beta^S D+\xi^S D^3+\zeta^S D^5, \tag{5.3.17}$$

whereas for a free crystal ($T \equiv 0$), from eqn. (5.3.6),

$$E = \beta^T D+\xi^T D^3+\zeta^T D^5. \tag{5.3.18}$$

Equations (5.3.17) and (5.3.18) represent the dielectric equation of state along the paths $S = 0$ and $T = 0$ respectively in the T, S plane (Fig. 5.3.2). The relationship of the set of coefficients β^S, ξ^S, ζ^S to β^T, ξ^T, ζ^T is given in eqn. (5.3.8). Setting $\gamma^0(\theta) = \alpha(\theta-\theta_i)$, where α is the coefficient of the normal thermal expansion and θ_i the temperature of the initial state and neglecting the temperature variation of q, we may write β^S in the form

$$\beta^S = A^S(\theta-\theta_0^S), \tag{5.3.19}$$

where, from eqn. (5.3.8),

$$A^S = A^T-2\alpha q(\theta_c), \tag{5.3.20}$$

$$\theta_0^S = \theta_0^T+(\theta_0^T-\theta_i)\Big/\left[\frac{A^T}{2\alpha q(\theta_c)}-1\right]. \tag{5.3.21}$$

For example, in the case of barium titanate $\alpha \sim 10^{-5}$, $q \sim 10^{10}$ and

$A \sim 6\times10^5$, in the MKS system; hence $A^T/2\alpha q \sim 6$. Thus the difference between θ_0^S the Curie–Weiss temperature of the clamped crystal and θ_0^T the Curie–Weiss temperature of the free crystal can be numerically significant especially if $\theta_0^T-\theta_i$ is large, i.e. if the initial state has a temperature well away from θ_0^T. From eqn. (5.3.21) $\theta_0^S = \theta_0^T$ when $\theta_i = \theta_0^T$. Referring to eqns. (5.3.8), (5.3.20) and (5.3.21) we see that in general the Curie–Weiss constant, the Curie–Weiss temperature

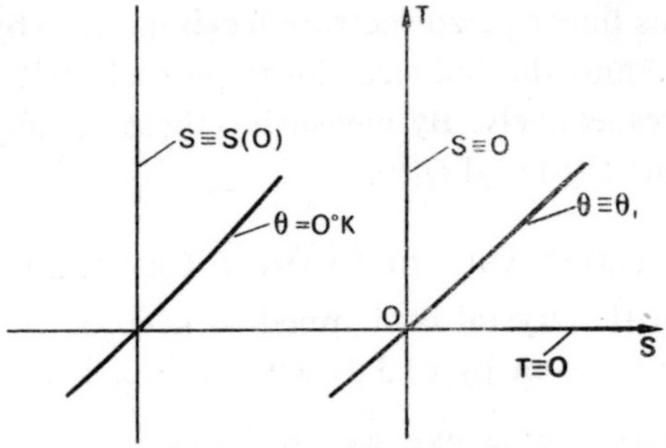

FIG. 5.3.2. A stress–strain diagram displaying the various paths discussed in the text. The diagram is drawn on the assumption that the strain at absolute zero is negative.

and the ξ-coefficient of the clamped crystal differ from those of the free crystal while the ζ-coefficient is the same for both constraints, at least in the approximation being considered here. Therefore a change in the mechanical constraint (from clamped at $S = 0$ to free or vice versa) can alter the order of the phase transition (if ξ^S and ξ^T have different signs) or destroy the phase transition (if the new *Curie temperature* is negative).

Suppose now that the crystal is clamped such that $S = S(0)$, the spontaneous strain at 0°K. From eqn. (5.3.2)

$$E = [\beta^S+2q(\theta)S(0)]D+\xi^S D^3+\zeta^S D^5.$$

Therefore, from eqn. (5.3.8), with an obvious extension of notation,

$$\beta^{S(0)}(\theta) = \beta^S(\theta)+2q(\theta)S(0) = \beta^T(\theta)+2q[S(0)-\gamma(\theta)], \quad (5.3.22)$$

$$\xi^{S(0)}(\theta) = \xi^S(\theta) = \xi^T(\theta)+\frac{2q(\theta)q(\theta)}{c^D(\theta)}, \quad (5.3.23)$$

$$\zeta^{S(0)}(\theta) = \zeta^S(\theta) = \zeta^T(\theta). \quad (5.3.24)$$

Equations (5.3.22) to (5.3.24) provide relationships† between the coefficients appearing in the dielectric equations of state appropriate to the three mechanical constraints, zero stress, zero strain and clamped with $S = S(0)$. The remarks made above for the case $S = 0$ apply also when $S = S(0)$, that is clamping the crystal, $S = S(0)$, may alter the order of or destroy the phase transition exhibited by the stress free crystal and conversely.

A *caveat* is in order at this point: the expressions listed above in eqns. (5.3.22) to (5.3.24) are valid only if the representation of the equations of state (5.3.1) and (5.3.2) are correct at the stress values required to clamp the crystal.

5.3.4 The Free Crystal and the Clamped Crystal (B)

In the model (B) the dielectric equation of state for the free crystal and the clamped crystal ($S = 0$) are

$$E = \beta^T D + \xi^T D^3 + \zeta^T D^5 \quad \text{and} \quad E = \beta^S D + \xi^S D^3 + \zeta^S D^5$$

respectively; the relations between β^T, ξ^T, ζ^T and β^S, ξ^S, ζ^S are given in eqn. (5.3.12). Setting $\beta^S = A^S(\theta - \theta_0^S)$ and $\beta^T = A^T(\theta - \theta_0^T)$ and replacing $[g^2(\theta)/s^D(\theta)]$ by the linear approximation $L + M(\theta - \theta_0^T)$, where

$$L = \frac{g^2(\theta_0^T)}{s^D(\theta_0^T)}, \quad M = \frac{d}{d\theta}\left[\frac{g^2(\theta)}{s^D(\theta)}\right]\Bigg|_{\theta=\theta_0^T},$$

we get

$$A^S = A^T - M, \tag{5.3.25}$$

and

$$\theta_0^S = \theta_0^T + \frac{L}{A^T - M}. \tag{5.3.26}$$

Thus clamping ($S = 0$) in this model may destroy the phase transition (if $\theta_c^S < 0$), but unlike the case of model (A) it will not change the order of the phase transition [since $\xi^S = \xi^T$, eqn. (5.3.12)]. For clamp-

† These results are particularly significant for a comparison between microscopic theory and experiment. Generally speaking the microscopic theory of the dielectric properties of crystals is easier to deal with if the model is kept clamped at the equilibrium volume and shape of the stress free crystal at 0°K, whereas the experimental determination of the dielectric properties of a crystal are measured at low stress values (usually zero) because the clamped crystal almost invariably shatters.

ing with $S = S(0)$, the dielectric equation of state reads

$$E+h(\theta)S(0) = \beta^S D+\xi^S D^3+\zeta^S D^5. \tag{5.3.27}$$

Because of the presence of the term $hS(0)$, the equation $E = 0$ has no roots of the form $D = \pm P^0$, and hence this equation describes a non-ferroelectric pyroelectric.

5.3.5 Hydrostatic Pressure

Let us suppose that a hydrostatic pressure P is applied to a crystal described by the equations of state for model (A). Setting $T = -P$, we can write the equation for E (5.3.6) in the form

$$E = (\beta^T - 2QP)D+\xi^T D^3+\zeta^T D^5. \tag{5.3.28}$$

Above the Curie point

$$\left(\frac{\partial E}{\partial D}\right)\bigg|_{E=0} = (\beta^T - 2QP).$$

Expressing the electrostrictive coefficient Q as a linear function of θ

$$Q = Q(\theta_0^T)+\left(\frac{dQ}{d\theta}\right)\bigg|_{\theta=\theta_0^T}(\theta-\theta_0^T),$$

we write the permeability in zero electric field in the form

$$\left(\frac{\partial E}{\partial D}\right)\bigg|_{E=0} = A(P)[\theta-\theta_0(P)], \tag{5.3.29}$$

where

$$A(P) = A^T+\left(\frac{dQ}{d\theta}\right)\bigg|_{\theta=\theta_0^T} P, \tag{5.3.30}$$

$$\theta_0(P) = \theta_0^T - \frac{Q(\theta_0^T)P}{A^T+\left(\frac{dQ}{d\theta}\right)\bigg|_{\theta=\theta_0^T} P} \cong \theta_0^T - \frac{Q(\theta_0^T)P}{A^T}. \tag{5.3.31}$$

Thus by measuring the variation of the Curie–Weiss temperature and Curie–Weiss constant with pressure, one may obtain the value of the electrostrictive coefficient Q (not $\tilde{Q}$!) and its temperature derivative at θ_0^T (Fig. 5.3.3). In model (B) we have assumed that the stress compo-

nent T vanishes in the case of hydrostatic pressure and so within the framework of the representation given by eqns. (5.3.9) and (5.3.10) hydrostatic pressure does not affect the zero field permeability. If we were to extend the framework of model (B) to include an electrostrictive term which couples the hydrostatic pressure with the displacement D, then the model would predict shifts in A^T, θ_0^T with pressure of the

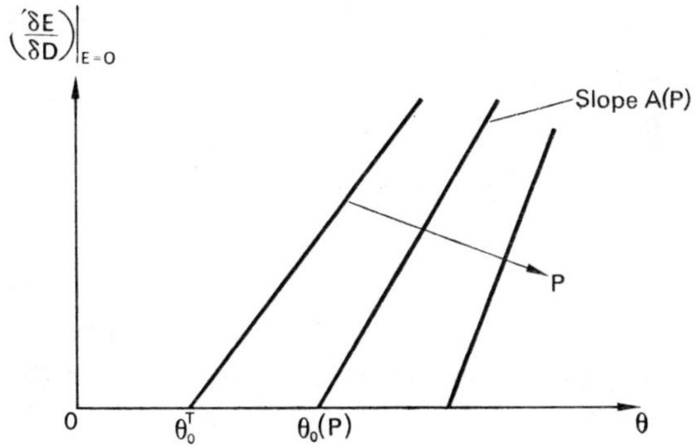

FIG. 5.3.3. The family of straight lines $(\partial E/\partial D)_{\theta, P}\,|_{E=0}$ versus θ for different pressures, drawn on the assumption that $Q > 0$ and $(dQ/d\theta) > 0$ when $\theta = \theta_0^T$. The intercept on the θ-axis is $\theta_0(P)$ [eqn. (5.3.31)] and the slope is $A(P)$ [eqn. (5.3.30)].

form given in eqns. (5.3.30) and (5.3.31). This remark suggests that one might hope to detect electrostriction in piezoelectric KDP or Rochelle salt, for example, by looking for a variation in the zero field permeability with pressure.

5.4 Phase Transition

In this section we discuss the effects of stress and electric fields on the ferroelectric phase transition exhibited by model (A) and the effect of electric fields in the case of model (B). For the purposes of this section we shall assume that all the linear coefficients in eqns. (5.3.1) to (5.3.4) except β^T and γ^0 are independent of temperature. The following discussion provides both a summary of many of the results obtained in earlier sections and examples of the type of phase transitions envisaged by Ehrenfest (§§ 3.2.3 and 3.2.4).

Part I. Model (A)

5.4.1 Thermodynamic Potentials

In the quasi-linear approximation† ($Q \equiv Q$) the Helmholtz function $H(\theta, T, D)$ (§ 3.1.2) generates the equations of state

$$S = -\left(\frac{\partial H}{\partial T}\right)_{D,\theta}, \quad E = +\left(\frac{\partial H}{\partial D}\right)_{T,\theta}, \quad \sigma = -\left(\frac{\partial H}{\partial \theta}\right)_{T,D}. \tag{5.4.1}$$

By inspection, eqns. (5.3.5) and (5.3.6) are generated by a Helmholtz function of the form

$$H(\theta, T, D) = H_0(\theta) - \gamma^0 T - \frac{s^D T^2}{2} + QTD^2 + \frac{\beta^T D^2}{2} + \frac{\xi^T D^4}{4} + \frac{\zeta^T D^6}{6}, \tag{5.4.2}$$

where $H_0(\theta)$ is the additive constant in the T and D integrations (cf. § 5.2.1). Assuming that s^D, Q, ξ^T and ζ^T are independent of θ and $\gamma^0 = \alpha(\theta - \theta_i)$, $\beta^T = A^T(\theta - \theta_0^T)$ we obtain an expression for the entropy density of the form

$$\sigma = \sigma_0 + \alpha T - A^T D^2/2 \tag{5.4.3}$$

with $\sigma_0 = -(dH_0/d\theta)$.

For convenience in later references we list the equations of state for S and E; from eqns. (5.3.5) and (5.3.6)

$$S = \gamma^0 + s^D T - QD^2, \tag{5.4.4}$$

$$E = (\beta^T + 2QT)D + \xi^T D^3 + \zeta^T D^5. \tag{5.4.5}$$

Hence

$$\left(\frac{\partial \sigma}{\partial \theta}\right)_{T,E} = \left(\frac{d\sigma_0}{d\theta}\right) - A^T D \left(\frac{\partial D}{\partial \theta}\right)_{T,E}, \tag{5.4.6}$$

$$\left(\frac{\partial \sigma}{\partial T}\right)_{\theta,E} = \left(\frac{\partial S}{\partial \theta}\right)_{T,E} = \alpha - A^T D \left(\frac{\partial D}{\partial T}\right)_{\theta,E} = \alpha - 2QD\left(\frac{\partial D}{\partial \theta}\right)_{T,E}, \tag{5.4.7}$$

$$\left(\frac{\partial S}{\partial T}\right)_{E,\theta} = s^D - 2QD\left(\frac{\partial D}{\partial T}\right)_{E,\theta} = s^D - \frac{4Q^2}{A^T} D \left(\frac{\partial D}{\partial \theta}\right)_{E,T}, \tag{5.4.8}$$

$$\left(\frac{\partial S}{\partial E}\right)_{T,\theta} = \left(\frac{\partial D}{\partial T}\right)_{E,\theta} = -2QD\left(\frac{\partial D}{\partial E}\right)_{T,\theta}, \tag{5.4.9}$$

† The thermodynamic results derived in Chapter 3 and used here are based on this approximation.

$$\left(\frac{\partial \sigma}{\partial E}\right)_{\theta,\, T} = \left(\frac{\partial D}{\partial \theta}\right)_{E,\, T} = -A^T D\left(\frac{\partial D}{\partial E}\right)_{\theta,\, T}, \tag{5.4.10}$$

$$\left(\frac{\partial D}{\partial E}\right)_{T,\, \theta} = \frac{1}{(\beta^T+2QT)+3\xi^T D^2+5\zeta^T D^4}. \tag{5.4.11}$$

By definition the Gibbs functions $G = H - ED$; hence

$$G(\theta, T, E) = H_0(\theta) - \gamma^0 T - \frac{s^D T^2}{2} + QTD^2 + \frac{\beta^T D^2}{2} + \frac{\xi^T D^4}{4} + \frac{\zeta^T D^6}{6} - ED, \tag{5.4.12}$$

where $D(\theta, T, E)$ is a solution of eqn. (5.4.5).

5.4.2 $\xi^T < 0$

Consider a phase space describing the three variables θ, T, E and consider a plane T = const. in this space. Referring to § 5.1.4 we deduce that in the case of the shorted slab the equation of state (5.4.5) with $\xi^T < 0$ predicts a line of ferroelectric phase transitions in this plane. The line cuts the plane $E = 0$ in the point $[T, \theta_c(T)]$, where

$$\theta_c(T) = \theta_0^T - \frac{2Q}{A^T} T + \frac{3}{16} \frac{(\xi^T)^2}{A^T \zeta^T} \tag{5.4.13}$$

is the Curie point in zero electric field, eqn. (5.1.26), and terminates at a "critical temperature"

$$\theta_{cr}(T) = \theta_0^T - \frac{2QT}{A^T} + \frac{9}{20} \frac{(\xi^T)^2}{A^T \zeta^T}. \tag{5.4.14}$$

The value of E at this "critical point" is

$$E_{cr} = \frac{6}{25} \sqrt{\frac{3}{10} \frac{|\xi^T|^5}{(\zeta^T)^3}} \tag{5.4.15}$$

[see eqn. (5.1.4) and § 5.1.10]. E_{cr} is independent of T. As T varies the line generates a phase boundary in (θ, T, E) space. On the high temperature side of the boundary the states are paraelectric and on the low temperature side ferroelectric (Fig. 5.4.1).

The function $\theta_c(T, E)$ describing the phase boundary is determined by the condition for phase equilibrium. In the case of the plated slab

with an applied voltage the condition is $G_f(\theta, T, E) = G_p(\theta, T, E)$, where the subscripts f and p refer to the *functions* describing the ferroelectric and paraelectric phases respectively. For the model (A) the phase equilibrium condition has the form, eqn. (5.4.12),

$$\left.\begin{aligned}(\beta^T+2QT)\left(\frac{D_f^2}{2}\right)+\frac{\xi^T D_f^4}{4}+\frac{\zeta^T D_f^6}{6}-ED_f\\ =(\beta^T+2QT)\left(\frac{D_p^2}{2}\right)+\frac{\xi^T D_p^4}{4}+\frac{\zeta^T D_p^6}{6}-ED_p,\end{aligned}\right\}\qquad(5.4.16)$$

where D_p, D_f are the ferroelectric and paraelectric roots [i.e. $D_f(E \to 0) \not\equiv 0$, $D_p(E \to 0) \equiv 0$] respectively of eqn. (5.4.5). As far as we are

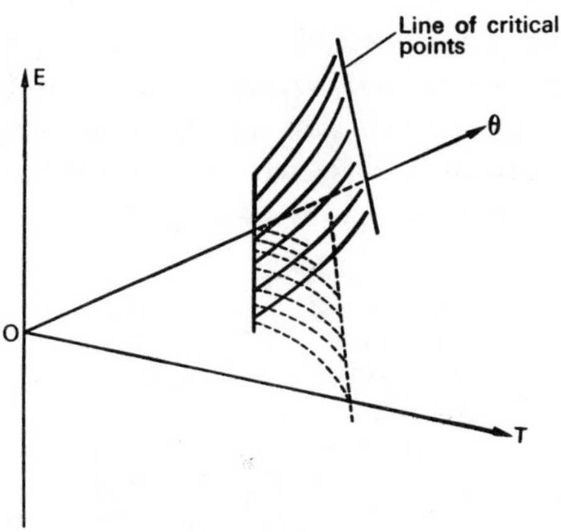

FIG. 5.4.1. The E, T phase space and the phase boundary for model (A) with $\xi^T < 0$.

aware the solution $\theta = \theta_c(T, E)$ cannot be written in closed form. We can, however, draw the following semi-quantitative conclusions about the points on the phase boundary. Referring to § 5.1.4, and more specifically to Fig. 5.1.9, it is clear that at each point on the phase boundary (except for the critical points) D undergoes a finite non-zero discontinuity. Because of the electromechanical and electrothermal coupling, this jump in D contributes finite non-zero discontinuities to σ and S [eqns. (5.4.3) and (5.4.4)]. Therefore this phase boundary describes

a set of phase transitions which are first order in the Ehrenfest classification. Moreover, we may apply the Clausius–Clapeyron equations, (3.2.12) and (3.2.13), to these transitions; thus

$$\left(\frac{\partial\theta_c}{\partial T}\right)_E = -\frac{(S_f - S_p)}{(\sigma_f - \sigma_p)}, \quad \left(\frac{\partial\theta_c}{\partial E}\right)_T = -\frac{(D_f - D_p)}{(\sigma_f - \sigma_p)}.$$

From eqns. (5.4.3) and (5.4.4)

$$\sigma_f - \sigma_p = -\frac{A^T}{2}(D_f^2 - D_p^2), \quad S_f - S_p = -Q(D_f^2 - D_p^2),$$

and hence

$$\left(\frac{\partial\theta_c}{\partial T}\right)_E = -\frac{2Q}{A^T}, \tag{5.4.17}$$

$$\left(\frac{\partial\theta_c}{\partial E}\right)_T = \frac{2}{A^T(D_f + D_p)}. \tag{5.4.18}$$

In the special case $E = 0$, we conclude that (a) eqn. (5.4.17) agrees with the result obtained from eqn. (5.4.13), and (b)

$$\left(\frac{\partial\theta_c}{\partial E}\right)_T\bigg|_{E=0} = \frac{2}{A^T P^0(T, \theta_c)}$$

or

$$\left(\frac{\partial\theta_c}{\partial E}\right)_T\bigg|_{E=0} = \frac{4}{A^T}\sqrt{\frac{\zeta^T}{3|\xi^T|}}, \quad \text{from eqn. (5.1.24).} \tag{5.4.19}$$

5.4.3. $\xi^T > 0$

When $\xi^T > 0$, the dielectric equation of state

$$E = [A^T(\theta - \theta_0^T) + 2QT]D + \xi^T D^3 + \zeta^T D^5$$

predicts a ferroelectric phase transition if and only if $E = 0$ (§ 5.1.3). Thus in this case the phase boundary in the T, E, θ space is a line in the plane $E = 0$ (Fig. 5.4.2). The Curie temperature $\theta_c(T)$ is determined by the vanishing of the coefficient of the term linear in D (§ 5.1.3). The equation of the line is

$$\theta = \theta_c(T) = \theta_0^T - \frac{2QT}{A^T}. \tag{5.4.20}$$

The slope of this line in the $E = 0$ plane is

$$\left(\frac{\partial \theta_c}{\partial T}\right) = -\frac{2Q}{A^T}\,. \tag{5.4.21}$$

Close to this phase boundary the spontaneous polarization is small (Fig. 5.1.14a). Suppose that P^0 is sufficiently small that $|\zeta^T(P^0)^5| \ll |\xi^T(P^0)^3|$, then

$$(P^0)^2 \cong -[A^T(\theta-\theta_0^T)+2QT]/\xi^T = \frac{A^T}{\xi^T}[\theta_c(T)-\theta]. \tag{5.4.22}$$

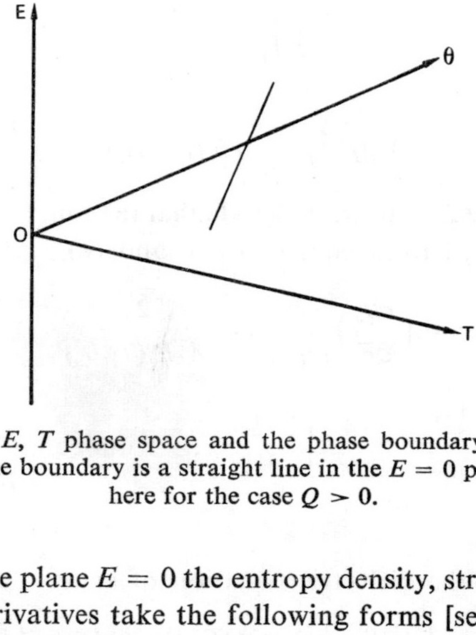

FIG. 5.4.2. The E, T phase space and the phase boundary for model (A) with $\xi^T > 0$. The boundary is a straight line in the $E = 0$ plane. It is drawn here for the case $Q > 0$.

For a point in the plane $E = 0$ the entropy density, strain, displacement and their six derivatives take the following forms [see eqns. (5.4.3) to (5.4.11)]:

$$\sigma_p = \sigma_0+\alpha T; \quad \sigma_f = \sigma_0+\alpha T-\frac{A^T(P^0)^2}{2} \cong \sigma_0+\alpha T-\frac{(A^T)^2}{2\xi^T}[\theta_c(T)-\theta]. \tag{5.4.23}$$

$$S_p = \gamma^0+s^DT; \quad S_f = \gamma^0+s^DT-Q(P^0)^2 \cong \gamma^0+s^DT-\frac{QA}{\xi^T}[\theta_c(T)-\theta]. \tag{5.4.24}$$

$$D_p \equiv 0; \quad D_f = \pm P^0 \cong \pm \sqrt{\frac{A^T}{\xi^T}[\theta_0(T)-\theta]}. \tag{5.4.25}$$

$$\left(\frac{\partial \sigma_p}{\partial \theta}\right)_{T,E} = \left(\frac{d\sigma_0}{d\theta}\right); \quad \left(\frac{d\sigma_f}{d\theta}\right)_{T,E} = \left(\frac{d\sigma_0}{d\theta}\right) - \frac{A^T}{2}\left[\frac{\partial}{\partial \theta}(P^0)^2\right]_{T,E}$$
$$\cong \left(\frac{d\sigma_0}{d\theta}\right) + \frac{(A^T)^2}{2\xi^T}. \tag{5.4.26}$$

$$\left(\frac{\partial \sigma_p}{\partial T}\right)_{\theta,E} = \alpha; \quad \left(\frac{\partial \sigma_f}{\partial T}\right)_{\theta,E} = \alpha - \frac{A^T}{2}\left[\frac{\partial}{\partial T}(P^0)^2\right]_{\theta,E} \cong \alpha + \frac{A^T Q}{\xi^T}. \tag{5.4.27}$$

$$\left(\frac{\partial S_p}{\partial T}\right)_{\theta,E} = s^D; \quad \left(\frac{\partial S_f}{\partial T}\right)_{\theta,E} = s^D - \frac{2Q^2}{A^T}\left[\frac{\partial}{\partial \theta}(P^0)^2\right]_{T,E} \cong s^D + \frac{2Q^2}{\xi^T}. \tag{5.4.28}$$

$$\left(\frac{\partial S_p}{\partial E}\right)_{\theta,E} = 0; \quad \left(\frac{\partial S_f}{\partial E}\right)_{\theta,T} = -2Q\frac{P^0}{\varkappa(P^0,\theta)} \cong -\frac{Q}{\sqrt{A^T\xi^T[\theta_c(T)-\theta]}}. \tag{5.4.29}$$

$$\left(\frac{\partial \sigma_p}{\partial E}\right)_{\theta,T} = 0; \quad \left(\frac{\partial \sigma_f}{\partial E}\right)_{\theta,T} = -A^T\frac{P^0}{\varkappa(P^0,\theta)} \cong -\frac{1}{2}\sqrt{\frac{A^T}{\xi^T[\theta_c(T)-\theta]}}. \tag{5.4.30}$$

$$\left(\frac{\partial D_p}{\partial E}\right)_{T,\theta} = \frac{1}{A[\theta-\theta_c(T)]}; \quad \left(\frac{\partial D_f}{\partial E}\right)_{\theta,T} = \frac{1}{\varkappa(P^0,\theta)} \cong \frac{1}{2A^T[\theta_c(T)-\theta]}. \tag{5.4.31}$$

As before, the subscripts p and f refer to the functions describing the behaviour of the system in the paraelectric and ferroelectric phases respectively. $\varkappa(P^0, \theta)$ is the zero field permeability in the ferroelectric phase [eqn. (5.1.38)]. The approximate expressions on the right are valid if the point under consideration is close to the phase boundary. It is evident from these results that on the boundary $\theta = \theta_c(T)$, the entropy density, strain and displacement of the system are continuous functions of temperature whereas the six compliances are not. $(\partial\sigma/\partial\theta)_{T,E}$, $(\partial\sigma/\partial T)_{E,\theta}$ and $(\partial S/\partial T)_{E,\theta}$ suffer finite discontinuities; $(\partial S/\partial E)_{T,\theta}$ $(\partial\sigma/\partial E)_{T,\theta}$ and $(\partial D/\partial E)_{T,\theta}$ suffer infinite discontinuities. The transition is clearly second order in the Ehrenfest classification scheme defined

at the end of § 3.2.5. Because certain of the second-order derivatives of G suffer infinite discontinuities at the phase boundary, the analysis described in § 3.2.4 and the resulting Ehrenfest relations eqns. (3.2.22) and (3.2.23) are not immediately applicable to the present model. However, the analysis is appropriate in a narrower sense, as we now show. If $E = 0$ then, by eqn. (5.4.12),

$$G_p(\theta, T, 0) = H_0(\theta) - \gamma^0 T - \frac{s^D}{2} T^2 \tag{5.4.32}$$

and

$$G_f(\theta, T, 0) = H_0(\theta) - \gamma^0 T - \frac{s^D T^2}{2} + QT(P^0)^2 + \frac{\beta^T (P^0)^2}{2} + \frac{\xi^T (P^0)^4}{4} + \frac{\zeta^T (P^0)^6}{6}. \tag{5.4.33}$$

Because $P^0(\theta, T)$ is imaginary when $\theta > \theta_c$ [eqn. (5.4.22)], the function $G_f(\theta, T, 0)$ is defined only in the ferroelectric region and on the phase boundary in the plane $E = 0$. In contrast G_p is defined at all points in the plane $E = 0$. Consider the difference

$$G_f(\theta, T, 0) - G_p(\theta, T, 0) = (\beta^T + 2QT) \frac{(P^0)^2}{2} + \frac{\xi^T (P^0)^4}{4} + \frac{\zeta^T (P^0)^6}{6}.$$

Close to the phase boundary we may neglect the term of order six in P^0; hence, on substituting for (P^0) from eqn. (5.4.22), we get

$$G_f(\theta, T, 0) - G_p(\theta, T, 0) = -\frac{(AT)^2}{4\xi^T} [\theta - \theta_c(T)]^2. \tag{5.4.34}$$

$(G_f - G_p)$ is a well-behaved function of θ and T on the phase boundary and to the ferroelectric side of this boundary. Differentiating with respect to θ and T, we get

$$\sigma_f - \sigma_p = -\left[\left(\frac{\partial G_f}{\partial \theta}\right) - \left(\frac{\partial G_p}{\partial \theta}\right)\right] = \frac{(A^T)^2}{2\xi^T} [\theta - \theta_c(T)], \tag{5.4.35}$$

$$S_f - S_p = -\left[\left(\frac{\partial G_f}{\partial T}\right) - \left(\frac{\partial G_p}{\partial T}\right)\right] = -\frac{(A^T)^2}{2\xi^T} [\theta - \theta_c(T)] \left(\frac{d\theta_c}{dT}\right), \tag{5.4.36}$$

$$\left[\left(\frac{\partial\sigma_f}{\partial\theta}\right)-\left(\frac{\partial\sigma_p}{\partial\theta}\right)\right]=\frac{(A^T)^2}{2\xi^T},\quad \left(\frac{\partial\sigma_f}{\partial T}-\frac{\partial\sigma_p}{\partial T}\right)=-\frac{(A^T)^2}{2\xi^T}\left(\frac{d\theta_c}{dT}\right),$$

$$\left(\frac{\partial S_f}{\partial T}-\frac{\partial S_p}{\partial T}\right)=\frac{(A^T)^2}{2\xi^T}\left(\frac{d\theta_c}{dT}\right)^2, \tag{5.4.37}$$

where $d\theta_c/dT = -2Q/A^T$ [eqn. (5.4.21)]. These results could have been obtained directly from eqns. (5.4.23), (5.4.24), (5.4.26), (5.4.27) and (5.4.28). By inspection,

$$\left(\frac{\partial\sigma_f}{\partial T}-\frac{\partial\sigma_p}{\partial T}\right)^2=\left(\frac{A^TQ}{\xi^T}\right)^2=\left(\frac{\partial\sigma_f}{\partial\theta}-\frac{\partial\sigma_p}{\partial\theta}\right)\left(\frac{\partial S_f}{\partial T}-\frac{\partial S_p}{\partial T}\right)>0$$

and

$$\left(\frac{\partial\theta_c}{\partial T}\right)=-\left[\left(\frac{\partial\sigma_f}{\partial T}-\frac{\partial\sigma_p}{\partial T}\right)\Big/\left(\frac{\partial\sigma_f}{\partial\theta}-\frac{\partial\sigma_p}{\partial\theta}\right)\right]=-2Q/A^T. \tag{5.4.38}$$

Thus these second derivatives of G exhibit finite discontinuities at the transition and satisfy the Ehrenfest relations (§ 3.2.4). It is clear from these results that as far as the variables T and θ are concerned, the phase transitions in the plane $E = 0$ are second order in the sense first introduced by Ehrenfest, i.e. along the phase boundary $\sigma_f - \sigma_p \equiv 0$, $S_f - S_p \equiv 0$.

5.4.4 Adiabatic Elastic and Piezoelectric Compliances

In this subsection expressions for the adiabatic elastic compliance $(\partial S/\partial T)_{E,\sigma}$ and the adiabatic piezoelectric coefficient $(\partial S/\partial E)_{T,\sigma}$ are obtained for the case of model (A). The adiabatic permeability is discussed in § 5.1.9.

From eqns. (3.1.19) and (3.1.20)

$$\left(\frac{\partial S}{\partial T}\right)_{E,\sigma}=\left(\frac{\partial S}{\partial T}\right)_{E,\theta}-\frac{\theta}{\varrho_0 C_{E,T}}\left(\frac{\partial S}{\partial\theta}\right)^2_{E,T}, \tag{5.4.39}$$

$$\left(\frac{\partial S}{\partial E}\right)_{T,\sigma}=\left(\frac{\partial S}{\partial E}\right)_{T,\theta}-\frac{\theta}{\varrho_0 C_{E,T}}\left(\frac{\partial S}{\partial\theta}\right)_{E,T}\left(\frac{\partial D}{\partial\theta}\right)_{E,T}. \tag{5.4.40}$$

Substituting for $(\partial S/\partial T)_{E,\,\theta}$, $(\partial S/\partial \theta)_{E,\,T}$ and $(\partial D/\partial \theta)_{E,\,T}$ from eqns. (5.4.7), (5.4.8) and (5.4.10) respectively, we get

$$\left(\frac{\partial S}{\partial T}\right)_{E,\,\sigma} = \left(s^D - \frac{\theta\alpha^2}{\varrho_0 C_{E,\,T}}\right) + A^T \frac{Q^2 D^2}{\varkappa}\left(1 - \frac{\theta A^T \alpha}{\varrho_0 C_{E,\,T} Q}\right) - \frac{4\theta}{\varrho_0 C_{E,\,T}}\left(\frac{QA^T D^2}{\varkappa^2}\right)^2, \tag{5.4.41}$$

$$\left(\frac{\partial S}{\partial E}\right)_{T,\,\sigma} = 2\frac{QD}{\varkappa}\left(-1 + \frac{\theta\alpha A^T}{2\varrho_0 C_{E,\,T} Q}\right) + \frac{2\theta}{\varrho_0 C_{E,\,T}}\left(\frac{A^T}{\varkappa}\right)^2 QD^3. \tag{5.4.42}$$

5.4.5 Neighbourhood Temperature

It is instructive to use some of the results obtained in previous sub-sections to obtain explicitly in the case of model (A) the terms appearing in two of the neighbourhood temperature identities of § 3.4.4. We restrict the discussion to the case $\xi^T > 0$ with $E = 0$.

By definition the neighbourhood temperature $\tau = \theta - \theta_c$. From eqn. (5.4.20), $\theta_c = \theta_0^T - 2QT/A^T$; hence

$$\left(\frac{\partial \theta}{\partial T}\right)_\tau = \left(\frac{d\theta_c}{dT}\right) = \frac{-2Q}{A^T} \quad \text{and} \quad \left(\frac{\partial \theta}{\partial E}\right)_\tau \equiv 0. \tag{5.4.43}$$

The neighbourhood identity relating $(\partial S/\partial \theta)_{T,\,E}$ and $(\partial \sigma/\partial \theta)_{T,\,E}$ is

$$\left(\frac{\partial S}{\partial \theta}\right)_{T,\,E} = \left(\frac{\partial \sigma}{\partial T}\right)_{\tau,\,E} - \left(\frac{\partial \sigma}{\partial \theta}\right)_{T,\,E}\left(\frac{d\theta_c}{dT}\right) \tag{5.4.44}$$

[see eqn. (3.4.12)].

Using eqns. (5.4.3), (5.4.6) and (5.4.7) we find that

$$\left(\frac{\partial \sigma_f}{\partial \theta}\right)_{T,\,E} = \left(\frac{d\sigma_0}{d\theta}\right) - A^T P^0 \left(\frac{\partial P^0}{\partial \theta}\right)_{T,\,E}, \tag{5.4.45}$$

$$\left(\frac{\partial S_f}{\partial \theta}\right)_{T,\,E} = \alpha - 2QP^0 \left(\frac{\partial P^0}{\partial \theta}\right)_{T,\,E}, \tag{5.4.46}$$

and

$$\left(\frac{\partial \sigma}{\partial T_f}\right)_{\tau,\,E} = \left(\frac{d\sigma_0}{d\theta}\right)\left(\frac{\partial \theta_c}{\partial T}\right)_{\tau,\,E} + \alpha - A^T P^0 \left(\frac{\partial P^0}{\partial T}\right)_{\tau,\,E}. \tag{5.4.47}$$

Now the equation of state $E = E(D, T, \theta)$ for model (A), eqn. (5.3.6),

can be written in form $E = A^T\tau D + \xi^T D^3 + \zeta^T D^5$ and hence

$$\left(\frac{\partial P^0}{\partial T}\right)_{\tau,\, E} = 0. \tag{5.4.48}$$

Therefore

$$\left(\frac{\partial \sigma_f}{\partial T}\right)_{\tau,\, E} = \left(\frac{d\sigma^0}{d\theta}\right)\left(\frac{\partial \theta_c}{\partial T}\right)_{\tau,\, E} + \alpha. \tag{5.4.49}$$

From an inspection of eqns. (5.4.43), (5.4.45), (5.4.46) and (5.4.49) we see that the identity given in eqn. (5.4.44) is indeed satisfied. We note that the strong temperature dependence occurs in $(\partial S/\partial \theta)_{T,\, E}$ and $(\partial \sigma/\partial \theta)_{T,\, E}$ through the quantity $P^0(\partial P^0/\partial \theta)$ [see Fig. 5.2.1a and eqn. (5.2.7)], whereas the temperature dependence of $(\partial \sigma/\partial T)_{T,\, E}$ [eqn. (5.4.49)] is not coupled (at least thermodynamically) to that of the spontaneous polarization. Similarly, for the terms in the identity

$$\left(\frac{\partial D}{\partial T}\right)_{\theta,\, E} = \left(\frac{\partial D}{\partial T}\right)_{\tau,\, E} - \left(\frac{\partial D}{\partial \theta}\right)_{T,\, E}\left(\frac{d\theta_c}{dT}\right) \quad \text{[eqn. (3.4.21)]}, \tag{5.4.50}$$

we find

$$\left(\frac{\partial D_f}{\partial T}\right)_{\theta,\, E} = -2QP^0\left(\frac{\partial P^0}{\partial E}\right)_{T,\, \theta}, \quad \left(\frac{\partial D_f}{\partial \theta}\right)_{T,\, E} = -A^T P^0\left(\frac{\partial P^0}{\partial E}\right)_{\theta,\, T},$$

$$\left(\frac{\partial D_f}{\partial T}\right)_{\tau,\, E} = 0.$$

Part II. Model (B)

5.4.6 Model (B)

For model (B) the Helmholtz function is

$$H(\theta,\, T,\, D) = H_0(\theta) - \frac{s^D T^2}{2} + gTD + \frac{\beta^T D^2}{2} + \frac{\xi^T D^4}{4} + \frac{\zeta^T D^6}{6} \tag{5.4.51}$$

[see eqns. (5.3.9) and (5.3.10)].

Assuming that s^D, g, ξ^T and ζ^T are constants independent of θ and that $\beta^T = A^T(\theta - \theta_0^T)$, we obtain the following expression for the entropy density:

$$\sigma = -\left(\frac{\partial H}{\partial \theta}\right)_{T,\, D} = \sigma_0 - \frac{A^T D^2}{2}, \tag{5.4.52}$$

where $\sigma_0 = -(dH_0/d\theta)$. From eqns. (5.3.9), (5.3.10) and (5.4.52),

$$\left(\frac{\partial\sigma}{\partial\theta}\right)_{T,\,E} = \left(\frac{d\sigma_0}{d\theta}\right) - A^T D\left(\frac{\partial D}{\partial\theta}\right)_{T,\,E} = \left(\frac{d\sigma_0}{d\theta}\right) + \frac{(A^T D)^2}{\varkappa(D,\,\theta)}, \quad (5.4.53)$$

$$\left(\frac{\partial\sigma}{\partial E}\right)_{T,\,\theta} = \left(\frac{\partial D}{\partial\theta}\right)_{E,\,T} = -\frac{A^T D}{\varkappa(D,\,\theta)}, \quad (5.4.54)$$

$$\left(\frac{\partial\sigma}{\partial T}\right)_{E,\,\theta} = \left(\frac{\partial S}{\partial\theta}\right)_{E,\,T} = -g\left(\frac{\partial D}{\partial\theta}\right)_{T,\,E} = g\frac{A^T D}{\varkappa(D,\,\theta)}, \quad (5.4.55)$$

$$\left(\frac{\partial S}{\partial T}\right)_{E,\,\theta} = s^D - g\left(\frac{\partial D}{\partial T}\right)_{E,\,\theta} = s^D + \frac{g^2}{\varkappa(D,\,\theta)}, \quad (5.4.56)$$

$$\left(\frac{\partial S}{\partial E}\right)_{T,\,\theta} = \left(\frac{\partial D}{\partial T}\right)_{E,\,\theta} = -\frac{g}{\varkappa(D,\,\theta)}, \quad (5.4.57)$$

$$\left(\frac{\partial D}{\partial E}\right) = \frac{1}{\varkappa(D,\,\theta)}, \quad (5.4.58)$$

where the dielectric permeability

$$\varkappa(D,\,\theta) = A^T(\theta-\theta_0^T)+3\xi^T D^2+5\zeta^T D^4. \quad (5.4.59)$$

When $T \neq 0$, model (B) exhibits a non-ferroelectric phase change (see footnote on p. 220); hence we shall restrict the remainder of the discussion in this subsection to the case $T = 0$. The phase diagrams for model (B) in the form of a plated slab are sketched in Fig. 5.1.11. There are two cases: (i) if $\xi^T > 0$ there is a single ferroelectric phase transition which occurs when $E = 0$, and (ii) if $\xi^T < 0$ there is a line of ferroelectric phase transitions terminating in a critical point. Consider first the case in which $\xi^T > 0$. When $T = 0$ and $E = 0$ we obtain the following results from eqns. (5.3.9), (5.3.10) and (5.4.52) through (5.4.58) [cf. eqns. (5.4.23) et seq.]:

$$\sigma_p = \sigma_0; \quad \sigma_f = \sigma_0 - \frac{A^T}{2}(P^0)^2 \cong \sigma_0 - \frac{(A^T)}{2\xi^T}(\theta_0^T-\theta). \quad (5.4.60)$$

$$S_p = 0; \quad S_f = -gP^0 \cong \mp g\sqrt{\frac{A^T}{\xi^T}(\theta_0^T-\theta)}. \quad (5.4.61)$$

$$D_p = 0; \quad D_f \cong \pm P_0 = \pm\sqrt{\frac{A^T}{\xi^T}(\theta_0^T-\theta)}. \quad (5.4.62)$$

$$\left(\frac{\partial\sigma_p}{\partial\theta}\right)_{T,E}=\left(\frac{d\sigma_0}{d\theta}\right),\quad \left(\frac{\partial\sigma_f}{\partial\theta}\right)_{T,E}=\left(\frac{d\sigma_0}{d\theta}\right)+\frac{(A^T)^2(P^0)^2}{\varkappa(P^0,\theta)}\cong\frac{d\sigma_0}{d\theta}+\frac{(A^T)^2}{2\xi^T}. \tag{5.4.63}$$

$$\left(\frac{\partial\sigma_p}{\partial E}\right)_{\theta,T}=0;\quad \left(\frac{\partial\sigma_f}{\partial E}\right)_{\theta,T}=\mp\frac{A^TP^0}{\varkappa(P^0,\theta)}\cong\mp\frac{1}{2}\sqrt{\frac{A^T}{\xi^T}\frac{1}{(\theta_0^T-\theta)}}. \tag{5.4.64}$$

$$\left(\frac{\partial\sigma_p}{\partial T}\right)_{\theta,E}=0;\quad \left(\frac{\partial\sigma_f}{\partial T}\right)_{\theta,E}=g\frac{A^TP^0}{\varkappa(P^0,\theta)}\cong\mp\frac{g}{2}\sqrt{\frac{A^T}{\xi^T}\cdot\frac{1}{(\theta_0^T-\theta)}}. \tag{5.4.65}$$

$$\left(\frac{\partial S_p}{\partial T}\right)_{\theta,E}=+s^D+\frac{g^2}{A^T(\theta-\theta_0^T)};\quad \left(\frac{\partial S_f}{\partial T}\right)_{\theta,E}=s^D+\frac{g^2}{\varkappa(P^0,\theta)}$$

$$\cong s^D+\frac{g^2}{2A^T(\theta_0^T-\theta)}. \tag{5.4.66}$$

$$\left(\frac{\partial S_p}{\partial E}\right)_{T,\theta}=-\frac{g}{A^T(\theta-\theta_0^T)};\quad \left(\frac{\partial S_f}{\partial E}\right)_{T,\theta}=\frac{-g}{\varkappa(P^0,\theta)}\cong\frac{-g}{2A^T(\theta_0^T-\theta)}. \tag{5.4.67}$$

$$\left(\frac{\partial D_p}{\partial E}\right)_{T,\theta}=\frac{1}{A^T(\theta-\theta_0^T)};\quad \left(\frac{\partial D_f}{\partial E}\right)_{T,\theta}=\frac{1}{\varkappa(P^0,\theta)}\cong\frac{1}{2A^T(\theta_0^T-\theta)}. \tag{5.4.68}$$

$\varkappa(P^0, \theta)$ is the zero field permeability in the ferroelectric phase [eqn. (5.1.38)]. The approximate expressions to the right are valid when $\theta \precsim \theta_0^T$. It is evident from the results listed above that for $T = 0$, $E = 0$, the entropy density σ, the strain S and the displacement D of the absolutely stable states are continuous functions of θ through the transition temperature θ_0^T, whereas all six compliances $(\partial\sigma/\partial\theta)_{T,E}$, $(\partial\sigma/\partial E)_{\theta,T}$, etc., are discontinuous at θ_0^T. Of these six quantities only one, $(\partial\sigma/\partial\theta)_{T,\theta}$, suffers a finite discontinuity, the remaining five discontinuities are infinite. This transition is clearly second order in the Ehrenfest classification.

In the case $\xi^T < 0$ the displacement suffers a finite discontinuity across the phase boundary in the $T = 0$ plane (Fig. 5.1.9), and hence, from eqns. (5.3.9) and (5.4.52), S and σ also suffer finite discontinuities; thus this phase boundary describes a line of first-order phase transitions.

By eqn. (3.2.23) the slope of the phase boundary is

$$\left(\frac{\partial\theta_c}{\partial E}\right)_T\bigg|_{T=0} = -\left(\frac{D_f-D_p}{\sigma_f-\sigma_p}\right) = \frac{2}{A^T}\left(\frac{1}{D_f+D_p}\right)$$

from eqn. (5.4.52) in agreement with eqn. (5.4.18).

5.4.7 Adiabatic Elastic and Piezoelectric Compliances

In this subsection we determine expressions for the adiabatic elastic compliance $(\partial S/\partial T)_{E,\sigma}$ and the adiabatic piezoelectric coefficient $(\partial S/\partial E)_{T,\sigma}$.

Substituting for $(\partial D/\partial\theta)_{E,T}$, $(\partial S/\partial T)_{E,\theta}$ and $(\partial S/\partial\theta)_{E,T}$ from eqns. (5.4.54) to (5.4.56) in eqns. (5.4.39) and (5.4.40) we get

$$\left(\frac{\partial S}{\partial T}\right)_{E,\sigma} = s^D + \frac{g^2}{\varkappa(D,\theta)} - \frac{\theta}{\varrho_0 C_{E,T}}\left[\frac{gA^T D}{\varkappa(D,\theta)}\right]^2, \tag{5.4.69}$$

$$\left(\frac{\partial S}{\partial E}\right)_{T,\sigma} = -\frac{g}{\varkappa(D,\theta)} + \frac{\theta g}{\varrho_0 C_{E,T}}\left[\frac{A^T D}{\varkappa(D,\theta)}\right]^2, \tag{5.4.70}$$

where $\varkappa(D,\theta) = A^T(\theta-\theta_0^T)+3\xi^T D^2+5\zeta^T D^4$.

5.5 A Multi-axial Ferroelectric Model

In this section we discuss the model of a cubic ferroelectric with several ferroelectric axes. The associated analysis is considerably more complicated in comparison with the analysis of the uniaxial models described above, and we shall merely indicate how the calculation is carried out without entering into details.

5.5.1 Equations of State

For material symmetry $m3m$ the Helmholtz function for zero stress has the form [cf. eqn. (3.3.89)]

$$\left.\begin{aligned} H(\theta, D_1, D_2, D_3, T_\alpha \equiv 0) = H_0(\theta) + \tfrac{1}{2}\beta_{11}^T[D_1^2+D_2^2+D_3^2] \\ +\tfrac{1}{4}\xi_{11}^T(D_1^4+D_2^4+D_3^4)+\tfrac{3}{2}\xi_{12}^T(D_1^2D_2^2+D_2^2D_3^2+D_3^2D_1^2) \\ +\tfrac{1}{6}\zeta_{111}^T(D_1^6+D_2^6+D_3^6)+\tfrac{5}{2}\zeta_{112}^T[D_1^4(D_2^2+D_3^2) \\ +D_2^4(D_3^2+D_1^2)+D_3^4(D_1^2+D_2^2)]+15\zeta_{123}^T D_1^2D_2^2D_3^2. \end{aligned}\right\} \tag{5.5.1}$$

The equations of state for the components of electric field are

$$\left.\begin{aligned} E_1 = {} & \beta_{11}^T D_1 + \xi_{11}^T D_1^3 + 3\xi_{12}^T D_1(D_2^2 + D_3^2) + \zeta_{111}^T D_1^5 \\ & + 5\zeta_{11}^T [2D_1^3(D_2^2 + D_3^2) + D_1(D_2^4 + D_3^4)] \\ & + 30\zeta_{123}^T D_1 D_2^2 D_3^2, \end{aligned}\right\} \quad (5.5.2)$$

$$\left.\begin{aligned} E_2 = {} & \beta_{11}^T D_2 + \xi_{11}^T D_2^3 + 3\xi_{12}^T D_2(D_3^2 + D_1^2) + \zeta_{111}^T D_2^5 \\ & + 5\zeta_{112}^T [2D_2^3(D_3^2 + D_1^2) + D_2(D_3^4 + D_1^4)] \\ & + 30\zeta_{123}^T D_1^2 D_2 D_3^2, \end{aligned}\right\} \quad (5.5.3)$$

$$\left.\begin{aligned} E_3 = {} & \beta_{11}^T D_3 + \xi_{11}^T D_3^3 + 3\xi_{12}^T D_3(D_1^2 + D_2^2) + \zeta_{111}^T D_3^5 \\ & + 5\zeta_{112}^T [2D_3^3(D_1^2 + D_2^2) + D_3(D_1^4 + D_2^4)] \\ & + 30\zeta_{123}^T D_1^2 D_2^2 D_3. \end{aligned}\right\} \quad (5.5.4)$$

The permeability tensor components are

$$\left.\begin{aligned} \frac{\partial E_1}{\partial D_1} = \varkappa_{11} = {} & \beta_{11}^T + 3\xi_{11}^T D_1^2 + 3\xi_{12}^T (D_2^2 + D_3^2) + 5\zeta_{111}^T D_1^4 \\ & + 5\zeta_{112}^T [6D_1^2(D_2^2 + D_3^2) + (D_2^4 + D_3^4)] + 30\zeta_{112}^T D_2^2 D_3^2, \end{aligned}\right\} \quad (5.5.5)$$

$$\frac{\partial E_1}{\partial D_2} = \varkappa_{12} = 6\xi_{12}^T D_1 D_2 + 20\zeta_{112}^T (D_1^3 D_2 + D_2^3 D_1) + 60\zeta_{123}^T D_1 D_2 D_3^2. \quad (5.5.6)$$

The remaining components can be easily obtained by permuting the coordinate subscripts 1, 2, 3 in these expressions.

When $E_i = 0$ $(i = 1, 2, 3)$, eqns. (5.5.2) to (5.5.4) provide three non-linear equations for the three unknowns D_i $(i = 1, 2, 3)$. We note that if $\boldsymbol{D} = \boldsymbol{P}^0$ is a root of these equations then so also is $\boldsymbol{D} = -\boldsymbol{P}^0$ (i.e. the non-zero $\boldsymbol{D}$ roots or spontaneous polarization vectors occur in anti-parallel pairs of vectors of equal magnitude), in other words the non-zero roots describe ferroelectric states. Each anti-parallel pair of vectors defines a ferroelectric axis in the material. Because of the symmetry apparent in these three equations the $\boldsymbol{D}$ roots fall into one of the following four categories:

(i) All components of $\boldsymbol{D}$ zero—(0, 0, 0).

(ii) One component non-zero and two zero. There are six roots, namely

$$(\pm P_a,\ 0,\ 0)\ (0,\ \pm P_a,\ 0),\ (0,\ 0,\ \pm P_a),$$

with P_a a positive real root of

$$\beta_{11}^T + \xi_{11}^T P_a^2 + \zeta_{111}^T P_a^4 = 0. \tag{5.5.7}$$

(iii) One component zero and two non-zero. There are twelve roots, namely

$$(\pm P_b,\ \pm P_b,\ 0)\ (\pm P_b,\ 0,\ \pm P_b)\ (0,\ \pm P_b,\ \pm P_b),$$
$$(\pm P_b,\ \mp P_b,\ 0)\ (\pm P_b,\ 0,\ \mp P_b)\ (0,\ \pm P_b,\ \mp P_b),$$

with P_b a positive real root of

$$\beta_{11}^T + (\xi_{11}^T + 3\xi_{12}^T)P_b^2 + (\zeta_{111}^T + 15\zeta_{112}^T)P_b^4 = 0. \tag{5.5.8}$$

(iv) Three components non-zero. There are eight roots, namely

$$(\pm P_c,\ \pm P_c,\ \pm P_c)\ (\pm P_c,\ \pm P_c,\ \mp P_c),$$
$$(\pm P_c,\ \mp P_c,\ \mp P_c)\ (\pm P_c,\ \mp P_c,\ \pm P_c),$$

with P_c a positive real root of

$$\beta_{11}^T + (\xi_{11}^T + 6\xi_{12}^T)P_c^2 + [\zeta_{111}^T + 30(\zeta_{112}^T + \zeta_{123}^T)]P_c^4 = 0. \tag{5.5.9}$$

For these categories of solutions the Helmholtz function takes the ollowing forms:

(i) $H = H_0(\theta)$. (5.5.10)

(ii) $H = H_0(\theta) + \frac{1}{2}\beta_{11}^T P_a^2 + \frac{1}{4}\xi_{11}^T P_a^4 + \frac{1}{6}\zeta_{111}^T P_a^6$. (5.5.11)

(iii) $H = H_0(\theta) + \beta_{11}^T P_b^2 + \dfrac{1}{2}(\xi_{11}^T + 3\xi_{12}^T)P_b^4 + \left(\dfrac{\zeta_{111}^T}{3} + 5\zeta_{112}^T\right)P_b^6$. (5.5.12)

(iv) $H = H_0(\theta) + \frac{3}{2}\beta_{11}^T P_c^2 + (\frac{3}{4}\xi_{11}^T + \frac{9}{2}\xi_{12}^T)P_c^4$
$+ [\frac{1}{2}\zeta_{111}^T + 15(\zeta_{112}^T + \zeta_{123}^T)]P_c^6$. (5.5.13)

The associated permeability components are:

(i) $\varkappa_{11} = \varkappa_{11} = \varkappa_{11} = \beta_{11}^T, \quad \varkappa_{12} = \varkappa_{22} = \varkappa_{31} = \varkappa_{31} = 0,$ (5.5.14)

(ii) $\varkappa_{11} = \beta_{11}^T + 3\xi_{11}^T P_a^2 + 5\zeta_{111}^T P_a^4, \quad \varkappa_{12} = \varkappa_{23} = \varkappa_{31} = 0;$
$\varkappa_{22} = \varkappa_{33} = \beta_{11}^T + 3\xi_{12}^T P_a^2 + 5\zeta_{112}^T P_a^4,$ (5.5.15)

with $\boldsymbol{D} = (P_a, 0, 0)$.

$$\left.\begin{aligned}
&\text{(iii)}\ \varkappa_{11} = \varkappa_{22} = \beta_{11}^T + 3(\xi_{11}^T + \xi_{12}^T)P_b^2 + (5\zeta_{111}^T + 35\zeta_{112}^T)P_b^4,\\
&\varkappa_{12} = 6\xi_{12}^T P_b^2 + 40\zeta_{112}^T P_b^4,\\
&\varkappa_{13} = \varkappa_{23} = 0, \quad \varkappa_{33} = \beta_{11}^T + 3\xi_{12}^T P_b^2 + 5\zeta_{112}^T P_b^4,
\end{aligned}\right\} \tag{5.5.16}$$

with $\boldsymbol{D} = (P_b, P_b, 0)$.

$$\left.\begin{aligned} \text{(iv)}\ \varkappa_{11} = \varkappa_{22} = \varkappa_{33} &= \beta^T_{11} + (3\xi^T_{11} + 6\xi^T_{12})P^2_c \\ &\quad + (5\zeta^T_{111} + 70\zeta^T_{112} + 30\zeta^T_{123})P^4_c, \\ \varkappa_{12} = \varkappa_{23} = \varkappa_{31} &= 6\xi^T_{12}P^2_c + (40\zeta^T_{112} + 60\zeta^T_{123})P^4_c. \end{aligned}\right\} \quad (5.5.17)$$

In categories (ii) and (iii) the form of the permeability tensor is a function of the direction of the spontaneous polarization vector. The expressions given above refer to $\boldsymbol{D} = (P_a, 0, 0)$ and $\boldsymbol{D} = (P_b, P_b, 0)$ respectively. The other forms are easily generated from these results.

5.5.2 Stability Conditions

A state is locally stable if at each point in the material the leading minors of the matrix $(\partial E_i/\partial D_j)$ are positive (§ 3.2.2), i.e. the leading minors of $\varkappa_{ij}$ are positive. The conditions for local stability in each of the four categories are easily obtained from the corresponding expressions for $\varkappa_{ij}$ given above: hence

$$\text{(i)}\ \beta^T_{11} > 0. \quad (5.5.18)$$

$$\left.\begin{aligned} \text{(ii)}\ &\beta^T_{11} + 3\xi^T_{11}P^2_a + 5\zeta^T_{111}P^4_a > 0, \\ &\beta^T_{11} + 3\xi^T_{12}P^2_a + 5\zeta^T_{111}P^4_a > 0. \end{aligned}\right\} \quad (5.5.19)$$

$$\left.\begin{aligned} \text{(iii)}\ &\beta^T_{11} + 3(\xi^T_{11} + \xi^T_{12})P^2_b + (5\zeta^T_{111} + 35\zeta^T_{112})P^4_b > 0, \\ &\beta^T_{11} + 3\xi^T_{12}P^2_b + 10\zeta^T_{112}P^4_b > 0, \\ &|\beta^T + 3(\xi^T_{11} + \xi^T_{12})P^2_b + 5(\zeta^T_{111} + 7\zeta^T_{112})P^4_b| \\ &\qquad > |6\xi^T_{12}P^2_b + 40\zeta^T_{112}P^4_b|. \end{aligned}\right\} \quad (5.5.20)$$

$$\left.\begin{aligned} \text{(iv)}\ &\beta^T_{11} + (3\xi^T_{11} + 6\xi^T_{12})P^2_c + (5\zeta^T_{111} + 70\zeta^T_{112} + 30\zeta^T_{123})P^4_c > 0, \\ &\beta^T_{11} + (3\xi^T_{11} + 18\xi^T_{12})P^2_c + (5\zeta^T_{111} + 150\zeta^T_{112} + 150\zeta^T_{123})P^4_c > 0, \\ &|\beta^T_{11} + (3\xi^T_{11} + 6\xi^T_{12})P^2_c + (5\zeta^T_{111} + 70\zeta^T_{112} + 30\zeta^T_{123})P^4_c| \\ &\qquad > |6\xi^T_{12}P^2_c + (40\zeta^T_{112} + 60\zeta^T_{123})P^4_c|. \end{aligned}\right\} \quad (5.5.21)$$

These conditions are necessary and sufficient for local stability in the respective categories (i)–(iv).

Consider a homogeneous dielectric described by the equations of state listed in (5.5.2) to (5.5.4). If $\boldsymbol{E}$ vanishes everywhere in the dielectric the associated displacement field is uniform and the components D_i are solutions of the equations $E_j = 0$ $(j = 1, 2, 3)$; $\boldsymbol{D} = 0$ is always

a solution. Other solutions exist when one or more of the equations belonging to the categories (ii), (iii) and (iv) have two or more real roots [see eqns. (5.5.7) to (5.5.9) above]. Whether or not real roots occur in these categories depends of course on the numerical values of β_{11}^T, ξ_{11}^T, etc. A solution may describe a physical state provided the state is locally stable. The inequalities which ensure this condition are listed above. Thus, for example the state described by the solution $\boldsymbol{D} = 0$ is locally stable if and only if β_{11}^T is positive. If there are no locally stable states then there can be no uniform field solution $E \equiv 0$ to the problem. If two or more locally stable states occur, then the absolutely stable state is determined by the relative sizes of the space integrals of the appropriate thermodynamic potential (§ 3.2.2).

Devonshire (1949) applied the equation of state representation and the analysis described above to the case of barium titanate [for example in this crystal $\beta_{11}^T = A^T(\theta - \theta_0)$, § 1.4]. By choosing numerical values for the coefficients, he obtained a semi-quantitative agreement with the experimental observations on the three-phase transitions and the temperature dependence of the spontaneous polarization and permeability of barium titanate. In particular the four phases with symmetries *m*3*m*, 4*mm*, *mmm*, 3*m* (§ 1.4) are found to correspond respectively to the four categories of roots (i), (ii), (iii) and (iv) introduced above. By adding electrostrictive terms (Devonshire, 1951) he was also able to account for the temperature dependence of the spontaneous strains. We refer the reader to these papers for further details.

References

von Arx, A. and Bantle W. (1944) *Helv. Phys. Acta* **17**, 298.

Bantle, W. and Caflisch, C. (1943) *Helv. Phys. Acta* **16**, 235.

Barkla, H. M. and Finlayson, D. M. (1953) *Phil. Mag.*, Ser. 7, **44**, 109.

Becker, R. (1964) *Electromagnetic Fields and Interactions*, Vol. 1: *Electromagnetic Theory and Relativity*, ed. F. Sauter, translated by A. W. Knudsen (Blackie, London).

Berlincourt, D. and Jaffe, H. (1958) *Phys. Rev.* **111**, 143.

Birss, R. R. (1963) *Rep. Prog. Phys.* XXVI, 307, ed. A. C. Stickland (Institute of Physics and Physical Society, London).

Buckingham, M. J. and Fairbanks, W. M. (1961) *Prog. Low Temp. Physics* **3**, 80, ed. C. J. Gorter (North Holland Pub. Co.).

Burfoot, J. C. (1967) *Ferroelectrics* (Van Nostrand, New York).

Busch, G. (1938) *Helv. Phys. Acta* **11**, 269.

Cady, W. G. (1946) *Piezoelectricity* (McGraw-Hill, New York).

Craig, P. P. (1966) *Phys. Letters* **20**, 140.

Devonshire, A. F. (1949) *Phil. Mag.*, Ser. 7, **40**, 1040.

Devonshire, A. F. (1951) *Phil. Mag.*, Ser. 7, **42**, 1065.

Devonshire, A. F. (1954) *Adv. in Physics* **3**, 85.

Devonshire, A. F. (1964) *Rep. Prog. Phys.* XXVII, 1, ed. A. C. Stickland (Institute of Physics and Physical Society, London).

Drougard, M. E. and Young, D. R. (1954) *Phys. Rev.* **74**, 1561.

Drougard, M. E. and Huibgreste, E. J. (1957) *IBM J. Res. Devel.* **1**, 318.

Ehrenfest (1933) *Comm. Leiden*, Suppl., 756.

Epstein, P. S. (1957) *Textbook of Thermodynamics* (Wiley, New York).

Eringen, A. C. (1963) *Int. J. Eng. Sci.* **1**, 127.

Fano, R. M., Chu, J. J. and Adler, R. B. (1960) *Electromagnetic Fields, Forces and Energy* (Wiley, New York).

Fattuzzo, E. and Merz, W. J. (1967) *Ferroelectricity: Selected Topics in Solid State Physics*, Vol. VII, ed. E. P. Wohlfarth (Wiley, New York).

Fisher, M. E. (1966) *Conference on Phenomena in the Neighbourhood of Critical Points*, ed. M. S. Green and J. V. Sengers (National Bureau of Standards Miscellaneous Publication 273).

Forsbergh, Jr., P. W. (1956) *Handbuch der Physik*, Vol. XVII (Springer-Verlag, Berlin).

Fuller-Brown, Jr., W. (1956) *Handbuch der Physik*, XVII (Springer-Verlag, Berlin).

Fumi, E. G. (1952) *Acta Crysta.* **5**, 44.

Gill, E. K. and Grindlay, J. (1966) *Can. J. Phys.* **44**, 431.

Ginzburg, V. L. (1946) *Zhur. Eksp. i Teoret. Fiz.* **15**, 739.

Grindlay, J. (1965a) *Phys. Rev.* **139**, 1603.

Grindlay, J. (1965b) *Phys. Letters* **18**, 239

References

GRINDLAY, J. (1966a) *Phys. Rev.* **149**, 637.

GRINDLAY, J. (1966b) *Proc. Int. Meet. on Ferroelectricity*, Vol. 1, 433, ed. V. Dvorak (Czechoslovakian Institute of Physics, Prague).

GRINDLAY, J. (1967) *Phys. Rev.* **160**, 698.

GRINDLAY, J. and WONG H. C. (1968), *Can. J· Phys.* **46**, 2331.

GROT, R. A. and ERINGEN, A. C. (1966) *Int. J. Eng. Sci.* **4**, 611, 639.

HABLÜTZEL, J. (1939) *Helv. Phys. Acta* **12**, 489.

HAJICEK, P. (1967) *Phys. Letters* **25A**, 36.

VON HIPPEL, A. (1950) *Rev. Mod. Phys.* **22**, 221.

HOSHINO, S., MITSUI, T., JONA, F. and PEPINSKY, R. (1957) *Phys. Rev.* **107**, 1255.

HUNTINGTON, H. B. (1958) *Solid State Physics*, Vol. 7, 214, ed. F. Seitz and D. Turnbull (Academic Press Inc. New York).

JANOVEC, V. (1966) *J. Chem. Phys.* **45**, 1874.

JAYNES, E. T. (1953) *Ferroelectricity* (Princeton University Press, Princeton).

JONA, F. and SHIRANE, G. (1962) *Ferroelectric Crystals* (Pergamon Press, Oxford).

KAY, H. F. and VOUSDEN, P. (1949) *Phil. Mag.* **40**, 1019.

KELLOGG, O. D. (1926) *Foundations of Potential Theory* (Springer, Berlin).

LANDAU, L. D. and LIFSHITZ, E. M. (1960) *Electrodynamics of Continuous Media* (Pergamon Press, Oxford).

LAX, M. (1963) *Lattice Dynamics*, ed. R. F. Wallis (Pergamon Press, Oxford).

MASON, W. P. (1946) *Phys. Rev.* **69**, 173.

MASON, W. P. (1950) *Piezoelectric Crystals and their Application to Ultrasonics* (Van Nostrand, New York).

MASON, W. P. (1966) *Crystal Physics of Interaction Processes* (Academic Press, New York).

MEGAW, H. D. (1957) *Ferroelectricity in Crystals* (Methuen, London).

MERZ, W. J. (1949) *Phys. Rev.* **76**, 1221.

MERZ, W. J. (1953) *Phys. Rev.* **91**, 513.

MERZ, W. J. (1962) *Progress in Dielectrics* **4**, ed. J. Birks (Heywood & Co., London).

MUELLER, H. (1940) *Phys. Rev.* **57**, 829.

NYE, J. F. (1964) *Physical Properties of Crystals* (Oxford University Press, Oxford).

PEARSON, C. E. (1959) *Theoretical Elasticity* (Harvard University Press, Cambridge, Mass.).

PIPPARD, A. B. (1957) *The Elements of Classical Thermodynamics* (Cambridge University Press, London).

DE QUERVAIN, M. (1944) *Helv. Phys. Acta* **17**, 509.

REESE, W. and MAY, L. F. (1967) *Phys. Rev.* **162**, 510.

STEPHENSON, C. C. and HOOLEY, J. G. (1944) *Am. Chem. Soc.* **66**, 1397.

TOUPIN, R. A. (1956) *J. Rat. Mech. Anal.* **5**, 849.

TOUPIN, R. A. (1963) *Int. J. Eng. Sci.* **1**, 101.

TRIEBWASSER, S. (1960) *Phys. Rev.* **118**, 1001.

TRUESDELL, C. and TOUPIN, R. A. (1960) *Handbuch der Physik*, III/1 (Springer-Verlag, Berlin).

TRUESDELL, C. and NOLL, W. (1965) *Handbuch der Physik*, III/3 (Springer-Verlag, Berlin).

WEYL, H. (1939) *The Classical Groups; their Invariants and Representations* (Princeton University Press, Princeton).

WISEMAN, G. G. and KUEBLER, S. K. (1963) *Phys. Rev.* **131**, 2023.

Glossary of Symbols

A reference after an entry signifies that this usage occurs only in the chapter or section indicated.

a	plate area of condenser
a	invariant of the full orthogonal group (Chapter 4)
A	inverse Curie–Weiss constant
A	free space region in an arbitrary state
A_0	free space region in the initial state
$c_{\alpha\beta}^{x,y}$	linear elastic stiffness at constant x and y
C	Curie–Weiss constant
$C_{x,y}$	specific heat at constant x and y
C	region occupied by the conductor in an arbitrary state
C_0	region occupied by the conductor in the initial state
d	reduced electric displacement field (Chapter 5)
$d_{i\alpha}^{x}$	linear piezoelectric coefficient at constant x
$\boldsymbol{D}, D_i$	electric displacement field
e	reduced electric field (Chapter 5)
$e_{i\alpha}^{x}$	linear piezoelectric coefficient at constant x
e_{ijk}	Levi–Civita density
E	coercive field
$\boldsymbol{E}, E_i$	electric field
$\boldsymbol{E}^s, E_i^s$	self-field
$\boldsymbol{E}^0, E_i^0$	thermal electric field (§ 3.3)
$\mathcal{E}$	applied electric field
$\mathcal{E}'$	electric field (§ 2.4.3)
f	reduced free energy (Chapter 5)
$\boldsymbol{f}, f_i$	body force density (Chapter 2)
F	free energy
g	reduced Gibbs function (Chapter 5)
$g_{i\alpha}^{x}$	linear piezoelectric coefficient at constant x
G	Gibbs function
$\mathcal{G}$	integral of Gibbs function
$h_{i\alpha}^{x}$	linear piezoelectric coefficient at constant x
H	Helmholtz energy
H_{KLM}^{α}	coefficient of polynomial representation of the energy density
$\mathcal{H}$	integral of the Helmholtz energy
J	thermodynamic potential (Chapter 3)
J_{MN}	coefficient of polynomial representation of the Maxwell stress
L_i	depolarization factor

Glossary of symbols

dm	mass element
$\boldsymbol{n}, n_i$	unit vector parallel to outward normal
p^0	reduced spontaneous polarization (Chapter 5)
P	pressure
$\boldsymbol{P}, P_i$	polarization
$\boldsymbol{P}^0, P_i^0$	spontaneous polarization
$q_{ij\alpha}$	electrostrictive coefficient
Q	charge on conductor
Q	transition heat (§ 5.2.4)
δQ	heat (Chapter 3)
$Q_{ij\alpha}$	electrostrictive coefficient
$\boldsymbol{R}_{ij}$	element of rotation matrix
$s_{\alpha\beta}^{x,y}$	linear elastic compliance at constant x and y
S_α, S_{ij}	infinitesimal strain
S_α^0, S_{ij}^0	spontaneous strain
t	time (Chapter 2)
t	reduced temperature (Chapter 5)
$\boldsymbol{t}^0, t_i^0$	surface traction
T_α, T_{ij}	stress tensor
T_α^0, T_{ij}^0	thermal stress
$\boldsymbol{u}, u_i$	displacement vector
$u_{ij}, u_{i;K}$	displacement gradient
$\mathcal{U}$	energy
v	arbitrary volume of dielectric
V	e.m.f.
V	volume or region occupied by the dielectric in an arbitrary state
V_0	volume or region occupied by the dielectric in the initial state
$dW, d\mathcal{W}$	work
$\boldsymbol{x}, x_i$	position vector in arbitrary state
$x_{i;K}$	deformation gradient
$\boldsymbol{X}, X_K$	position vector in initial state
$d\boldsymbol{x}, d\boldsymbol{X}$	line elements
z	plate separation in condenser
α	coefficient of thermal expansion
β_{ij}^x	linear permeability at constant x
γ_α	thermal expansion
$\boldsymbol{\Gamma}$	torque
δ_{ij}	Kronecker delta
ε_{ij}^x	linear permittivity at constant x
$\zeta_{ijklm}, \zeta_{\alpha\beta\gamma}$	non-linear coefficient in dielectric equation of state
η_{KJ}	Lagrangian strain tensor
η_{ijk}	non-linear coefficient in the dielectric equation of state
θ	temperature
θ_0	Curie–Weiss temperature
θ_c	Curie temperature
θ_i	initial temperature
$\boldsymbol{\varkappa}$	permeability

Glossary of symbols

$\varkappa$	reduced permeability
μ_{ijklm}	non-linear coefficient in the dielectric equation of state
ξ_{ijkl}, $\xi_{\alpha\beta}$	non-linear coefficient in the dielectric equation of state
Π_i	invariant of the full orthogonal group (Chapter 4)
ϱ	mass density in an arbitrary state
ϱ_0	mass density in the initial state
σ, σ'	entropy density
Σ	transition entropy
τ	neighbourhood temperature
ϕ	electrostatic potential
ϕ_0	electrostatic potential of the conductor
ψ	energy density in dielectric
Ψ	energy density (Chapter 4)
ω	surface charge density on the conductor

Index

Index

Index

OTHER TITLES IN THE SERIES IN NATURAL PHILOSOPHY